KINEMATICS AND DYNAMICS OF DIFFUSE ASTROPHYSICAL MEDIA

Edited by

J. E. DYSON

The University, Manchester,
England

AND

E. B. CARLING

The University, Manchester,
England

Reprinted from *Astrophysics and Space Science*
Volume 216, Nos. 1–2, 1994

SPRINGER SCIENCE+BUSINESS MEDIA, B.V.

A C.I.P. Catalogue record for this book is available from the Library of Congress.

ISBN 978-94-010-4399-1   ISBN 978-94-011-0926-0 (eBook)
DOI 10.1007/978-94-011-0926-0

Printed on acid-free paper

TABLE OF CONTENTS

TABLE OF CONTENTS

Preface

The area of diffuse astrophysical media is enormous and ranges over circumstellar to extragalactic scales. The physical conditions can vary from cool dusty gases to collections of relativistic particles. Flows in such media are set up by energy and momentum injection from winds, jets and explosions. The study of these phenomena involves physics, chemistry and, inevitably, hydrodynamics. One of the most important aspects of this study is the ever increasing overlap between theory and observation. Indeed, it can be argued that the only way to really understand these complex flows which can never be duplicated under terrestrial conditions, is to encourage this overlap, and this was one major aim of this Conference.

Because of the long theoretical and observational association of the Manchester Group with this general area, Manchester seemed an appropriate venue for this Conference. But in fact this long association and the actual year of the Conference are connected. In 1951 Franz Kahn joined the Astronomy Department at Manchester University and immediately the study of diffuse media, particularly the hydrodynamic aspects, commenced and has flourished ever since. Franz became Head of the Astronomy Department in 1981 following the retirement of Professor Z. Kopal, who founded the Department and was instrumental in attracting Franz to it. In 1993, Franz retired from this position and a most serendipidous coincidence was his election to the Royal Society announced shortly before the Conference. The second major aim of this Conference was therefore to mark this very special year in Franz's life and to demonstrate that the subject area in which he has made so many distinguished contributions is alive and well. The contributions to the Conference amply demonstrated that this is indeed the case.

We would like to thank the participants in the meeting, many of whom have known and worked with Franz over the years, both for their attendance and the way in which most of them met the submission deadline for manuscripts. Conference organization is exacting and Melanie Thomas' efforts are warmly appreciated.

Because we wish to emphasise the similarities of phenomena over different scales, we have chosen not to divide the proceedings into separate sections, but have ordered the papers in roughly the direction of increasing scale size. We also felt that no introduction given by us would match Henk van de Hulst's contribution which gives a most apposite historical and personal overview of cosmical gas dynamics, which underpins this meeting.

John Dyson   Ellen Carling

Astrophysics and Space Science **216**: vii, 1994.

List of Participants

C. F. Akujor	Onsala Space Observatory, Sweden
D. J. Axon	Nuffield Radio Astronomy Laboratories, University of Manchester, U.K.
M. E. Bailey	School of Computing and Mathematical Sciences, Liverpool John Moores University, U.K.
A. C. Baker	Institute of Astronomy, Cambridge, U.K.
B. Balick	Osservatorio Arcetri, Firenze, Italy and Astronomy Department, University of Washington, Seattle, Washington, U.S.A.
A. R. Bell	Department of Physics, Imperial College, London, U.K.
S. Biro	Department of Astronomy, University of Manchester, U.K.
R. Blomme	Royal Observatory of Belgium, Brussels, Belgium
M. F. Bode	School of Chemical and Physical Sciences, Liverpool John Moores University, U.K.
M. Bowman	Nuffield Radio Astronomy Laboratories, University of Manchester, U.K.
D. Breitschwerdt	Max-Planck-Institut für Kernphysik, Heidelberg, Germany
M. Bryce	Department of Astronomy, University of Manchester, U.K.
H. Castañada	Instituto de Astrofisica de Canarias, La Laguna, Tenerife, Spain
S. J. Chapman	Department of Physics & Astronomy, University of Wales College of Cardiff, U.K.
L. Chernin	Centre for Astrophysics, Cambridge, Massachusetts, U.S.A.
Y.-H. Chu	Department of Astronomy, University of Illinois, Urbana, Illinois, U.S.A.
M. Collins	Institute of Mathematics, University of Oxford, U.K.
L. Cuesta Crespo	Instituto de Astrofisica de Canarias, La Laguna, Tenerife, Spain
S. Curiel	Center for Astrophysics, Cambridge, Massachusetts, U.S.A.
A. A. da Costa	Instituto Superior Tecnico, Lisbon, Portugal
R. J. Davis	Nuffield Radio Astronomy Laboratories, University of Manchester, U.K.
J. E. Drew	Department of Astrophysics, University of Oxford, U.K.
J. E. Dyson	Department of Astronomy, University of Manchester, U.K.
J. Eislöffel	Dublin Institute for Advanced Studies, School of Cosmic Physics, Dublin, Ireland

Astrophysics and Space Science **216**: ix-xii, 1994.

S. A. E. G. Falle	Department of Mathematics, University of Leeds, U.K.
P. R. A. Farquar	Department of Mathematics, UMIST, Manchester, U.K.
D. A. Green	MRAO, Cambridge, U.K.
H. J. Habing	Leiden Observatory, The Netherlands
R. Hanuschik	Astronomisches Institut, Ruhr Universität Bochum, Germany
T. W. Hartquist	Max-Planck-Institut für Extraterrestrial Physics, Garching bei München, Germany
M. P. Hobson	Cavendish Laboratory, Cambridge, U.K.
D. A. Howe	Department of Mathematics, UMIST, Manchester, U.K.
W. Hummel	Astronomisches Institut, Ruhr Universität Bochum, Germany
D. Hutsemékers	Institut d'Astrophysique, Université de Liège, Belgium
D. E. Innes	Max-Planck-Institut für Aeronomie, Katlenburg-Lindau, Germany
R. A. James	Department of Astronomy, University of Manchester, U.K.
F. D. Kahn	Department of Astronomy, University of Manchester, U.K.
S. Komissarov	Department of Astronomy, University of Manchester, U.K.
M. J. Kukula	Nuffield Radio Astronomy Laboratories, University of Manchester, U.K.
S. A. Lamb	NORDITA and Niels Bohr Institute, Copenhagen, Denmark
V. Laspias	Department of Astronomy, University of Manchester, U.K.
A. Lazarian	DAMPT, Cambridge, U.K.
A. Lim	Department of Mathematics, UMIST, Manchester, U.K.
H. M. Lloyd	School of Chemical and Physical Sciences, Liverpool John Moores University, U.K.
R. Lopez	Departament d'Astronomia, Universitat de Barcelona, Spain
S. Lucek	Department of Physics, Imperial College, London, U.K.
L. Lucy	ST-ECF, Garching bei München, Germany
E. Lüdke	Nuffield Radio Astronomy Laboratories, University of Manchester, U.K.
M. T. Malone	Department of Astronomy, University of Manchester, U.K.
R. M. Massey	Department of Astronomy, University of Manchester, U.K.
M. McNaughton	Inveralligin, Wester Ross, U.K.
J. Meaburn	Department of Astronomy, University of Manchester, U.K.
T. J. Millar	Department of Mathematics, UMIST, Manchester, U.K.

A. Moffat	Département de Physique, Université de Montréal, Canada
D. Moss	Department of Mathematics, University of Manchester, U.K.
M. Noguchi	Astronomical Institute, Tohoku University, Sendai, Japan
T. J. O'Brien	School of Computing and Mathematical Sciences, Liverpool John Moores University, U.K.
C. R. O'Dell	Space Physics & Astronomy, Rice University, Houston, Texas, U.S.A.
R. Padman	Cavendish Laboratory, Cambridge, U.K.
J. W. Palmer	Department of Astronomy, University of Manchester, U.K.
J. J. Perry	Institute of Astronomy, Cambridge, U.K.
A. Poll	Institut für Astrophysik und Extraterrestriche Forschung, Bonn, Germany
J. Porter	Astrophysics, Nuclear and Astrophysics Laboratory, University of Oxford, U.K.
J. J. Quenby	Astrophysics Group, Department of Physics, Imperial College, London, U.K.
A. C. Raga	Department of Mathematics, UMIST, Manchester, U.K.
H. Rauer	Max-Planck-Institut für Aeronomie, Katlenburg-Lindau, Germany
J. Rawlings	Department of Mathematics, UMIST, Manchester, U.K.
T. P. Ray	Dublin Institute for Advanced Studies, Ireland
J. C. Raymond	Center for Astrophysics, Cambridge, Massachusetts, U.S.A.
A. Riera	Departament d'Astronomia, Universitat de Barcelona, Spain
A. B. Romeo	Onsala Space Observatory, Chalmers University of Technology, Onsala, Sweden
I. Sakelliou	Neos Kosmos, Athens, Greece
W. L. W. Sargent	Astronomy Department, Caltech, Pasadena, California, U.S.A.
S. M. Scarrott	Department of Physics, University of Durham, U.K.
L. J. Smith	Department of Physics & Astronomy, University College, London, U.K.
N. Solomos	Hellenic Naval Academy, Department of Physics, Piraeus, Greece
G.-X. Song	Shanghai Observatory, Academia Sinica, Shanghai, China
S. Taylor	Department of Mathematics, UMIST, Manchester, U.K.
R. T. Templeman	Department of Astronomy, University of Manchester, U.K.
J. M. Torrelles	Instituto de Astrofisica de Andalucia, Granada, Spain
J. A. Turner	Department of Physics & Astronomy, University

	College of Wales College of Cardiff, U.K.
H. C. van de Hulst	Sterrewacht, Leiden, The Netherlands
H. van Woerden	Kapteyn Institute, Groningen, The Netherlands
R. Wagenblast	Department of Mathematics, UMIST, Manchester, U.K.
A. Whitworth	Department of Physics & Astronomy, University of Wales College of Cardiff, U.K.
A. Wilkinson	Department of Astronomy, University of Manchester, U.K.
K. Willacy	Department of Mathematics, UMIST, Manchester, U.K.
D. A. Williams	Department of Mathematics, UMIST, Manchester, U.K.
R. Williams	Institute of Astronomy, Cambridge, U.K.
J. Wiseman	Center for Astrophysics, Cambridge, Massachusetts, U.S.A.
L. Woltjer	Observatoire de Haute Provence, St. Michel, France
K. Wood	Department of Physics & Astronomy, University of Glasgow, U.K.

Cosmical Gas Dynamics, Why Was It So Difficult?

H. C. van de Hulst
Sterrewacht, P.O.Box 9513, 2300 RA Leiden, The Netherlands.

1   Introduction

When the invitation for this meeting reached me, I did not hesitate. Of course, I wished to be present when it was Franz Kahn's turn for the formal move to say goodbye. Like many of us, he will probably sneak in again through the back door and continue working.

The choice of a subject was more of a problem. I am not an expert in gas flows on a galactic scale, and I missed (largely or wholly) two relevant symposia: "*The interstellar disk-halo connection in galaxies*" (Leiden, 1990), (Bloemen, 1991) and "*Back to the Galaxy*" (Maryland, 1992) (Holt and Verter, 1993). Yet I chose this subject, because for a very limited number of years in my career, I had ambitious plans for research in this direction.

My interest started early 1949, when Jan Oort gently put me to work for the Paris symposium on Cosmical Gas Dynamics, going to be held that summer. It ended 5–10 years later in steps. First, the discovery of the 21 cm line drew me into practical radio astronomy. Secondly, when Lo Woltjer became a professor at Leiden, I had learned enough to know how difficult these problems of gas dynamics are, and was happy to leave that part to him. Thirdly, COSPAR drew me from astrophysics into politics. Excuse me for mentioning these personal details.

And this, incidentally, explains the "was" in my title. The problems may be difficult even now, but I cannot tell, because I did not try.

During the eight years mentioned, I participated intensively, as editor or as organizer, in three symposia organized jointly by the International Astronomical Union IAU (read: Jan H.Oort) and the International Union for Theoretical and Applied Mechanics IUTAM (read Jan M.Burgers). The venues, titles, and references are:

I   Paris (1949), *"Problems of Cosmical Aerodynamics"*.
J.M.Burgers and H.C. van de Hulst, editors;
Central Air Documents office, Dayton (Ohio), 1951.

II   Cambridge UK (1953), *"Gas Dynamics of Cosmic Clouds"*,
IAU Symposium No.2, J.M.Burgers and H.C. van de Hulst, editors,
North-Holland Publishing, Amsterdam, 1955.

III Cambridge Mass (1957), *"Third Symposium on Cosmical Gas Dynamics"*

Astrophysics and Space Science **216**: 1–12, 1994.
© 1994 *Kluwer Academic Publishers.*

J.M.Burgers and R.N.Thomas, editors;
Rev. Modern Phys, **30**, 905–1108, 1958.

The numbers I, II, III, will be used in the quotations below.

When Oort asked me to help him and Burgers with the preparation of Symposium I, I had barely finished my post-doc years. Franz Kahn, who gave **three** papers in Symposium II and **two-and-a-half** in Symposium III, must also have been a fresh Ph.D.

It was a great privilege for me to mix with the giants of the time, and, as an editor, to try to make sense of deep but very confused discussions. And—by the way—the private tutoring from both Burgers and Oort was invaluable.

Generally, I am not given to history writing. But in this particular instance, where most of the persons have changed, but many of the problems are still the same, I thought it useful and amusing to comment on the way the problems were seen at that time. History **can** be helpful, for it bares the roots of our present concepts, and thereby gives us a firmer ground from where to assess the ambitions, irritations, and misunderstandings that plague us today.

I shall work mostly by literal quotations from these three volumes. You probably understand that these are embellished versions of what actually has been said. In one of the discussions a participant accused another one of throwing mud in his eyes. The most amusing line from the original transcript of the taped record was:

Mayall: *Do you refer to secrets?* Oort: *Yes.*

Magnetic tapes were quite new then, and an American Air Force Office had offered to make and transcribe these records. "Secrets" is what they heard. Unfortunately, in the edited text (I, p.185) it had to be corrected to *Seyferts*.

2 The General Plan

Let me read to you the words by which Burgers prefaced the first symposium.

Burgers (I,i): *The plan to organize a Symposium on borderline problems between astrophysics and gas dynamics originated in 1948, as a result of several informal discussions on the subject treated by Professor J. H. Oort in the George Darwin Lecture for 1947 before the Royal Astronomical Society in London, entitled "Some Phenomena connected with Interstellar Matter." The peculiar conditions presented by gaseous matter in cosmic space appeared to open promising new fields for study. While these conditions necessitate a far-reaching application of molecular and atomic physics, they reveal at the same time many features which are of great interest in connection with present-day developments in hydro- and aerodynamics. In particular, the problems of turbulence and those of expansion phenomena and of shock waves immediately come to the foreground.* Please note a slight error; the actual year of Oort's lecture was 1946 (Oort, 1946).

Sir G. I. Taylor, in introducing the second symposium, adopted a slightly more sceptical tone.

<u>Taylor</u> (II,1): *The fact that hydrodynamicists have solved a limited number of problems of continuous media gave rise to the idea in the mind of Burgers that they may have something to contribute to cosmological theory and so, being a man who never lets grass grow under his feet when he sees a chance of promoting useful scientific cooperation, he arranged the first symposium in Paris. Since that meeting, and no doubt largely because of it, there have been developments in theory.*

The mutual understanding did not always come smoothly. The easy part of it related to factual information and led to many straight questions and answers. It is a pleasure to reread these discussions and to see how competently such questions were handled. I just quote some examples from (II,112) and (II,242).

<u>Savedoff</u>: *How does one obtain the spectra of luminous edges?*

<u>Minkowski</u>: *With high-aperture spectrographs and preferably not too big telescopes because the objects are fairly large.*

<u>Van de Hulst</u>: *In one spectrum of the Cygnus loop taken at McDonald I found a very strong change of the ratio of Hγ to 4363 along the length of the slit. How should this be explained?*

<u>Minkowski</u>: *This may be due to a local variation in electron temperature. Perhaps there is a hot spot.*

<u>Liepmann</u>: *Is it certain that the physical condition of the Cygnus wisps is stationary?*

<u>Minkowski</u>: *Yes, the relaxation times are of the order of a few years, while the loop has been observed for more than 60 years.*

<u>Hoyle</u>: *Another problem that should be settled by observations is the Chandrasekhar-Fermi theory of the magnetic field along a spiral arm. If the suggestion that the motion of the clouds is only strong enough to cause a simple corrugation of a homogeneous field is correct, this should show up in the statistics of cloud motions. Or are they sufficiently random to tell that this is not the case? Is there a hope to find out ?*

<u>Oort</u>: *There is a definite possibility by looking perpendicular to the arm and along it. So far, no evidence of a difference is available.*

<u>Von Karman</u> (II,180; after having received the answer that a suggestion he made about the flatness of galaxies was ruled out by the observations): *My imagination is not handicapped by any knowledge of the facts.*

This all referred to the 'easy' part. When it came to the 'deep' questions, there was repeatedly a tendency to jump to conclusions. This, in turn incited other aerodynamicists to serious warnings that the astronomers should not do that. In symposium III the irritation after two days of papers and discussions had grown so large that the organizing committee decided to cancel all papers that had been scheduled for the third day and, instead, inserted an interim summary discussion that took the full morning. Batchelor introduced this discussion as follows (III,994).

<u>Batchelor</u> (III,994): *Last night, Burgers, van de Hulst, Thomas and myself, feeling a little dissatisfied with the way the sessions had gone up to that time,*

decided that it would be useful to have some stock-taking, in order to see what has been accomplished so far.

It was then my turn to present a list of questions which had emerged from the brainstorming of the night before (questions 1–6), to which Batchelor added a further one (question 7). They were carefully distinguished into astronomical and physical questions, and to each category a list of 'permissible' answers was spelled out. For the present review it suffices to quote only the questions (III,994–995).

1. *How precisely can we estimate the conversion from energy produced internally by hot stars into kinetic energy of turbulent motion (or cloud motion) in the interstellar gas?*

2. *Does a magnetic field effectively inhibit dissipation of energy from turbulent motion (or cloud motion) into thermal motion?*

3. *Differential galactic rotation: Given a distribution of mass in the galactic system that defines a gravitational potential, and given in this gravitational potential field a disk-like distribution of gas, in which the gas moves in laminar flow with circular orbits, is this motion unstable and will it develop turbulence?*

4. *Is a galactic halo, which is half ionized, half neutral, as proposed by Pickelner, compatible with the observations of the 21-cm line?*

5. *Do the observations give convincing proof that the magnetic lines of force run along spiral arms?*

6. *Equipartition: does it establish itself, and how rapidly?*

7. *How literally may we regard the gas clouds in the spiral arms as discrete?*

Permit me to add the wish that organizing committees would envisage taking such liberties more often . A "workshop" usually is an euphemism for just another collection of pre-arranged papers. Drastic moves should remain possible and in a true workshop the printed program and schedule should not be considered to be sacrosanct.

The following sections deal with the **contents** of what was discussed in these symposia. I have arranged the topics under a few separate headings, which may be considered to be the main themes.

3 MHD Waves, a Problem that could be Solved

Well before the first symposium, Oort had suggested that I should study in particular the Alfvén waves or magneto-hydrodynamic waves. This concept was about 10 years old and unchallenged, but it had hardly penetrated into general astrophysical theory. Oort was worried that Alfvén, who would be one of the participants, might not get much of a response in the discussion.

I duly followed Oort's advice and studied the MHD waves, not stopping at the usual place where the velocity is determined as an eigenvalue from the dispersion relation, but actually continuing—as an exercise—to find the amplitudes and phases of the field and velocity components, and the precise exchanges between the different forms of energy.

The actual history took a quite different turn from what Oort had feared. Early summer 1949 interstellar polarization was discovered by Hiltner and Hall. From the first day of the first Symposium, all speakers included magnetic fields in their theories and speculations. Evidently, many of the statements made were what now is called handwaving. This was fully recognized during the symposium. It became even clearer when, after the symposium, I was given the job, together with Burgers, to edit the hours of confused discussion from the taped record.

It then occurred to me that the least I could do was to extend my preparatory 'exercise', which I had to write up for publication anyhow, by including the effect of compressibility. That should lead to a restricted problem, the exact solution of which should be feasible. So I worked for a few months that same fall, mostly in switching the orders of the unknowns and of the equations around, until I obtained a pleasantly symmetric form. I still do not know **why** that is possible. Finally (I,50), the determinant equations, from which the eigenvalues can be solved, took the form reproduced in Fig.1.

Clearly, if you put the imposed, constant magnetic field zero, the coupling is gone and the matrices separate, as shown by the dashed rectangles. The solution then consists of unrelated wave forms. But the magnetic field terms couple them all. It gave me a big kick for the first time to have a set of equations at hand that in one distinct limit described light and in another distinct limit described sound.

4 Shocks and other fronts

The small-amplitude waves in a homogeneous medium with an imposed homogeneous magnetic field, which I just discussed, were only an exercise. That much was clear. The real world would be more violent, leading to non-linear phenomena. Shock waves were the classical example, and the astronomers had the benefit of hearing excellent tutorial papers both on the theory and on the laboratory experiments with shock waves.

Let me read part of the history lesson, which Von Neumann gave us (I,75).

Taylor: *Riemann also inferred, essentially by physical insight, what happens when the continuous solution ceases to exist. He made it very plausible that a discontinuity of a certain type, a "shock wave" develops.*

This was subsequently independently rediscovered and further developed, by Hugoniot. It is also true that in the entire literature up to 1910, i.e., up to the time of the work of Rayleigh and G.I.Taylor, there was a considerable confusion and disagreement between the authors on exactly what the shock looks like.

Several years later, a new generation of astrophysicists had caught on to this subject. Symposium III contains papers by Kahn (III,1058 and 1069) and Goldsworthy (III,1062) on shock waves and ionization fronts. What I remember most clearly, however, is how I sweated over the elaborate papers of Goldsworthy and Axford in the Philosophical Transactions (Goldsworthy, 1960; Axford, 1960). Why were they so difficult?

I now feel that part of the reason was that I still was too attached to the basic simplicity of the linear wave theory. I just did not have enough patience, or enough guidance, to get to grips with the more general problems. An added reason may have been that I was used to making checks and double-checks by looking at the same problem in different coordinate frames. This had the consequence, that words like "before" or "after" the shock often left me in a state of confusion. I am told that this is not an uncommon experience for a beginning student.

5 Numerical solutions

Nowadays any student knows that a problem which is too difficult to be solved analytically has a good chance of being amenable to a numerical solution, given enough computer time. On looking back at the 8 years which my review covers, it is remarkable that there were so few numerical solutions or simulations. In these three symposia they played no role at all. This has to do, among other things, with the enormous memory space that a true fluid dynamics problem in more than one dimension takes, if solved on a computer. It just could not be done yet! Other speakers at this symposium will inform you about what can be done now.

Personally, I never became actively engaged in numerical gas dynamics, although I made certain preparations and remember several visits and conversations which made me aware of its large potential. Let me, in this connection, read you one further quotation. The development in this field was rather accurately foreseen by A.N.Lowan. The date happens to be also 1949, but the source (Lowan, 1949) does not have any relation to these symposia.

A.N.Lowan: *When the Computation Laboratory has been thoroughly mechanized (perhaps "electronized" would be a truer description), it may be expected that the chief emphasis will be not on the preparation of new basic tables but rather on the solution of physical problems whose treatment by present-day methods is an unsurmountable task. Even then, however, it may be confidently predicted that there will still be a need for mathematically trained human operators equipped with desk calculators for the purpose of carrying on work of an exploratory nature and tasks which are either too small for high-speed calculators or which require the type of human intelligence, discrimination, and initiative which may not yet have been incorporated in the design of electronic computers.*

6 Turbulence

The word "turbulence" had been loosely used in the astrophysical literature— and is still being used that way—as a word for any irregular gas flow that causes non-thermal Doppler effects. We had to learn from the gas dynamicists that in their vocabulary the word "turbulence" had a far more precise meaning. Yet— so we learned—even the simplest case of homogeneous, isotropic turbulence in an incompressible fluid did not have a complete theory. Only the Kolmogoroff spectrum was generally accepted.

Starting from this incomplete theory, the discussion in Symposium I branched out primarily into two directions: turbulence with magnetic fields and turbulence with compressibility.

When, at the second symposium, it was time to take stock, the status of these two topics was summarized as follows.

Batchelor (II,117): *Firstly I should like to remind people that the determination of the asymptotic level of magnetic energy in a medium of high conductivity which is in statistically steady homogeneous turbulent motion, is still an unsolved problem— or, at any rate, is still a disputed problem—and it would be useful to hear of any recent developments.*

Lighthill (II,121): *If any applications of turbulence theory to astrophysics are to be made, it will be necessary to consider what effect compressibility may have on the postulated motions. This paper describes what is known, or can reasonably be conjectured, about the influence of compressibility on the turbulent motion of fluids, but makes no attempts to apply the results to any astrophysical problem.*

A detailed quantitative theory of the process, with good experimental backing, exists only when the root mean square Mach number of the turbulence is small compared with 1. For larger values of the root mean square Mach number one can make only tentative conjectures, ...

According to these conjectures, the influence of compressibility becomes dominant for root mean square Mach numbers comparable with 1, or greater.

Misgivings about 'supersonic turbulence'.

My further comments will be confined to the second topic. The notion of 'compressible turbulence' in the discussions soon got the name of 'supersonic turbulence'. Some of the bolder thinkers interpreted it as becoming the same thing as 'independently moving discrete clouds'. These speculations in turn led to cautionary remarks and serious warnings by the more conservative participants. I cannot give you a well-organized review but simply wish to read a few quotations to illustrate this confusion.

Von Karman (I,213): *I don't understand why you say that if the flow is supersonic, the disturbance goes faster than sound. I think that this has no physical meaning. First, it has no meaning that the flow is supersonic. What does that mean? One can as easily say that the motion of the Earth is supersonic, because its velocity is 30 km/sec. The real statement should be this: We have supersonic motion of a solid body through a medium when the velocity of the body relative to the medium is greater than the velocity of sound, the latter being defined as the velocity of propagation of infinitely small disturbances. (The propagation of a finite disturbance will have a velocity which is larger than that of infinitely small disturbances.)*

But if we are talking of turbulence, I don't think there is any difference between "supersonic" and "subsonic" turbulence.

Gold (II,238-239):   *The occurrence of high Mach numbers can be more easily understood when we consider that such forces as gravitation, and not only pressure gradients, may be responsible for the motion. Lighthill's idea that turbulence at high Mach numbers may be interpreted in terms of an assembly of shock waves seems to me an extremely fortunate one for explaining a variety of astronomical features. It has been pointed out by Kantrowitz that shock waves appear to make order from chaos; they can lead to the appearance of smooth shapes and contours which would fit the observations much better than the chaos of ordinary turbulence.*

Liepmann (II,241):   *I wish to protect—for the time being—the shock waves from misuse. We should not suddenly expect everything from shock waves. For example, shock waves cannot be considered as a simple alternative to shear waves or turbulence. It does not seem possible to conceive an ensemble of shock waves of finite strength without a coupled shear field. The interaction between two strong shocks always creates a vortex layer and hence a random ensemble of shock waves will be coupled with a turbulent field of random vorticity.*

Nevertheless, toward the end of the second symposium, Batchelor saw some light.

Batchelor (II,242):   *The "statistical assembly of shock waves" that has been mentioned in earlier sessions is not a very clear concept for me. We are trying to imagine the properties of a turbulent motion in which the Mach number is very high. A way of doing this which may throw some light on the situation is to think of the high Mach number as being produced by a very small velocity of sound. In the limit of zero velocity of sound, or infinite Mach number, hydrodynamic pressure loses its meaning and the various particles of the gas move independently of each other. Thus it becomes necessary to think in terms of particle dynamics rather than in terms of hydrodynamics. From this point of view the picture of discrete clouds moving independently in a rarified background medium and occasionally making collisions may not seem to be so strange.*

But his assessment in the "inserted summarizing discussion" during the third symposium, four years later, was less positive.

Batchelor (III,997):   *I, for one, have never been clear about the explanation of the discreteness of the clouds. I do not see how any theory of turbulent motions of high intensity can lead to the view that there would be separate clouds of high density, proceeding more or less independently of each other.*

7   Conclusions

After this somewhat haphazard review I must try to come to conclusions. In tribute to Burgers, I like to start off with a quotation from his closing words of the first symposium.

Burgers (I,237):   *It is a curious thing that we can think and reason about the depths of the universe, in particular when one puts against it the material from which its structure and its laws are deduced: minute specks on photographic plates and human fantasy playing around with mathematical formulas.*

It has been said sometimes that the order we see in the universe is something we put in it ourselves. Certainly our thinking is a continuous striving for bringing order in what we experience.

That our minds may reach to such images and to conclusions as those about which we have heard in these lectures, is a gift for which we cannot be too grateful.

This quotation shows Burgers as a philosopher. For those interested I may point out that Burgers went more deeply into such philosophical questions in his book "Ervaring en Conceptie" (Burgers, 1956).

My conclusions will focus on the differences between the situation then and the situation now. I have tried to formulate these differences in the form of six key words or brief statements, to each of which I shall add my private comments.

1. <u>Proliferation of data</u>. Since 1950 the observational data have proliferated enormously, in quantity, in sky coverage, in wavelength coverage, and in precision. The theoretical understanding has also grown, but by no means in the same proportion.

One can hardly begin to list all the observational improvements: molecular clouds, hot X-ray emitting gas, amazing details in 21-cm maps, the entire infrared sky, gamma rays, etcetera.

In illustration of this point I showed 7 slides, which cannot be reproduced in this printed version. Three slides were true-colour pictures of wide areas in the sky: the molecular cloud near ρ Ophiuchi, the dark rim in Ara, and the Vela supernova remnant. Further, one result from the Hubble Space Telescope was shown, the 'spike' in the Cygnus veil (Raymond, 1994). Finally, three slides represented other wavelength regions, a local radio polarization map showing unexpected streaks (Wieringa, 1991), an IRAS 100 μm map processed to make the structure of the 'froth' visible (Waller and Boulanger, 1993), and the gamma-ray sky copied by a baker in coloured candy from the COS-B results in 1980, to which the remark was made that the very recent EGRET data on CGRO give a similar picture with far greater precision.

2. <u>Reduced confidence in Grand Theories</u>. At the time of the first symposium there seemed to be no reason to drop the notion of a general, pervasive, and cool, interstellar medium. The HII-regions, observed as faint Hα emission regions and predicted by theory as localized pockets of hot gas around groups of O- and B-stars, formed the only exception. Since observations outside the optical domain were still minimal, there was no serious suspicion, by theory or from observations, of an even hotter halo.

All of this changed during the fifties, and even more so when space astronomy started. It is understandable, however, that before that time the astronomers (or at least some of them) had pretty high-strung hopes that certain exact solutions worked out by the aerodynamicists, notably turbulence and shock waves, might be directly applicable to the interstellar gas.

The aerodynamicists responded to these hopes by being present as very active participants of these meetings. Although the healthy voice of scepsis can be heard in virtually every discussion in these books, the basic feeling was that we were jointly working towards an overall understanding.

3. <u>Do-it-yourself theorists.</u> The present astronomers may be compared with present home owners. They know that many new techniques have been developed and they usually (but not always) **do** take care to orient themselves on what is available. But then, instead of calling the mason and the carpenter in, they tend to act by the do-it-yourself principle. I feel that this change of attitude is understandable in view of the much more varied picture that the observations show today. Yet I have reason to wonder if the collection of independent handymen (which we are as astronomers) could not profit again by a much closer contact with the professional aerodynamicists, notably in the field of Computational Fluid Dynamics (CFD).

4. <u>Implantation of numerical computations</u>. I use deliberately the medical metaphor of implanting an organ into a living body. The living astrophysical body consists of observations, interpretations, hunches, scenarios and theories. Nobody doubts that this body can greatly benefit from the solution of well-defined numerical problems on a large computer. However, the danger is—like in surgery— that this solution may remain a corpus alienum, that eventually will be rejected before it has grown to be a part of the body and has truly started to perform the function for which it had been intended.

I signal this difficulty, but at the same time I cannot give you an authoritative account of the actual state of affairs regarding this potential hazard.

5. <u>Attachment to claims.</u> My impression regarding this point is hard to support by facts. I have the feeling that prestige-oriented attitudes, including personal claim-staking and not giving up on 'one's own' model, are stronger in present-day astronomy than they used to be. If this is the true situation, it is **not** a healthy one. The opposite situation of everyone adhering uncritically to a theory once it has come into fashion (the bandwagon effect) is equally undesirable.

6. <u>A true "work"shop is hard to organize.</u> We fight a continual battle against **sclerosis** in scientific meetings. The name 'symposium' was invented to make it sound less formal than a 'conference'. Subsequently, the name 'workshop' was invented to make it sound less formal than a symposium. The spontaneous action to paste certain new results on the wall was institutionalized into posters. And the programme with its order and allotted times, conceived as a guideline, degenerated into something **never** to deviate from. I shall be happy if my review of how things **were** has revived in you some appetite for an occasionally more spontaneous and less predictable style of scientific exchange.

References

I, II, III. See section 1 of this paper.

Bloemen, H. (ed.): 1991, *The Interstellar Disk-Halo Connection in Galaxies*, Kluwer Acad. Publ., Dordrecht.

Holt, S and Verter, F. (eds.): 1993, *Back to the Galaxy*, Woodbury N.Y. (American Institute of Physics).

Oort, J. H.: 1946, *Some Phenomena Connected with Interstellar matter*, Mon. Not. Roy. Astr. Soc. **106**, 159-179.

Goldsworthy, F. A.: 1960, *Ionization Fronts in Interstellar Gas and the Expansion of HII Regions*, Phil. Trans. Roy. Soc. A **253**, 277-300.

Axford, W. I.: 1960, *Ionization Fronts in Interstellar Gas: the Structure of Ionization Fronts*, Phil. Trans. Roy. Soc. A, **253**, 301-333.

Lowan, A. N.: 1949, *The Computation Laboratory of the National Bureau of Standards*, Scripta Mathematica **15**, *33-63*.

Burgers, J. M.: 1956, *Ervaring en Conceptie*, Van Loghum Slaterus, Arnhem.

Raymond, J.: 1994, paper in this volume.

Wieringa, M.: 1991, *327 MHz Studies of the High-Redshift Universe and the Galactic Foreground*, Thesis Leiden.

Waller, W. H. and Boulanger, F.: 1993, *Worms or Froth? Fine-scale Structure in the Far-Infrared Milky Way*, in *Back to the Galaxy*, eds. S. Holt and F. Verter, Woodbury N. Y. (American Institute of Physics).

Set I

	(hy)	(Ex)	(Ix)	(Vy)	(Vz)	(p)	(Uz)
(1)	$\frac{i\omega}{c}$,	$-ik$,					
(2)	ik ,	$-\frac{\epsilon i\omega}{c}$,	$-\frac{4\pi}{c}$,				
(3)		-1 ,	$\frac{1}{\sigma}$,	$-\frac{Hz}{c}$,	$\frac{Hy}{c}$,		
(4)			$\frac{Hz}{c}$,	$i\omega\rho + k^2\mu$,			
(5)			$-\frac{Hy}{c}$,		$i\omega\rho + \frac{4}{3}k^2\mu$,	$-ik$,	
(6)					1 ,		$-i\omega$
(7)						1 ,	$-ikC$

$=0$

Set II

	(hx)	(Ey)	(Iy)	(Vx)	(Iz)	(Ez)
(1)	$\frac{i\omega}{c}$,	ik ,				
(2)	$-ik$,	$-\frac{\epsilon i\omega}{c}$,	$-\frac{4\pi}{c}$,			
(3)		-1 ,	$\frac{1}{\sigma}$,	$\frac{Hz}{c}$,		
(4)			$-\frac{Hz}{c}$,	$i\omega\rho + k^2\mu$,	$\frac{Hy}{c}$,	
(5)				$-\frac{Hy}{c}$,	$\frac{1}{\sigma}$,	-1
(6)					$-\frac{4\pi}{c}$,	$-\frac{\epsilon i\omega}{c}$

$=0$

Fig. 1. The two determinant equations from which the various modes of magneto-hydrodynamic waves in a compressible fluid may be calculated (I,50).

Shaping Planetary Nebulae

Bruce Balick*
Osservatorio Arcetri, Largo Enrico Fermi 5, I 51025 Firenze Italy and University of Washington, Seattle, WA 98195 U.S.A.

Abstract. Hydrodynamical calculations are becoming increasingly successful at understanding the shapes and kinematics of planetary nebulae (PNs). The most successful models are two-dimensional interacting stellar wind models for which the PN nucleus is assumed to originally expel much or most of its mass in an equatorial waistband. The physics of the ensuing evolution seems to be explained nicely by a combination of hydrodynamics coupled with time-dependent stellar ionization and energy loss through nebular radiation. Recent radiation gas dynamic calculations are shown to yield excellent agreement with data.

1 Introduction

PNs have been used as testing grounds for physical and astrophysical theories, most recently as proving grounds for models of nebular ionization, heavy-element production, and the return of heavy elements and kinetic energy to the ISM. Their relatively simple geometries, high surface brightnesses, and low extinctions make PNs ideal for such purposes.

The striking symmetric shapes of PNs, along with the discovery of continuing mass loss of their nuclei, have opened the question of how the shapes of PNs are imprinted. In this talk I shall review the results of work primarily by my collaborators, Vincent Icke, Adam Frank and Garrelt Mellema, and I on the shaping of PNs. Finally I'll describe some of the gaps between the observations and our (or, to be more precise, my) understanding of them.

It is with some trepidation that I give this talk here in Manchester, in front of Franz Kahn whose successful and productive career we have come to celebrate. Bringing hydro to Manchester is really about as useful—and vain—as taking coals to Newcastle. I venture to guess that by the time the ink dries on the pages of this conference's proceedings Franz and his collaborators will have bridged several of the questions that my talk leaves behind in their characteristically innovative style.

2 Modelling the Hydrodynamic Evolution of PNs

Here's a brief overview of the history of the field. As you might expect, the modern form of the enterprise was driven by observations; *to wit.*, IUE observations of the nuclei of PNs, deep PN images, and spatially resolved kinematics (the first derivative of the shape). These clearly showed that the old picture of a burst of

---

* On leave from Astronomy Dept. FM-20, University of Washingron, Seattle, WA 98195, U.S.A.

14   Bruce Balick

ejection followed by ballistic (or simple adiabatic expansion) was far too simplistic
a concept for the formation and evolution of PNs.

Before continuing it is useful to define some terminology based on PN image
data. The essential features are the central optically dark "cavity" centered on
the PN's nucleus, a thin "bright rim" which forms the cavity's boundary, and a
relatively smooth nebular "shell" attached to the rim's outer edge. Outside the
shell's sharp outer edge may be a limb-brightened "halo". NGC 3242 and 6826
nicely illustrate most of these features.

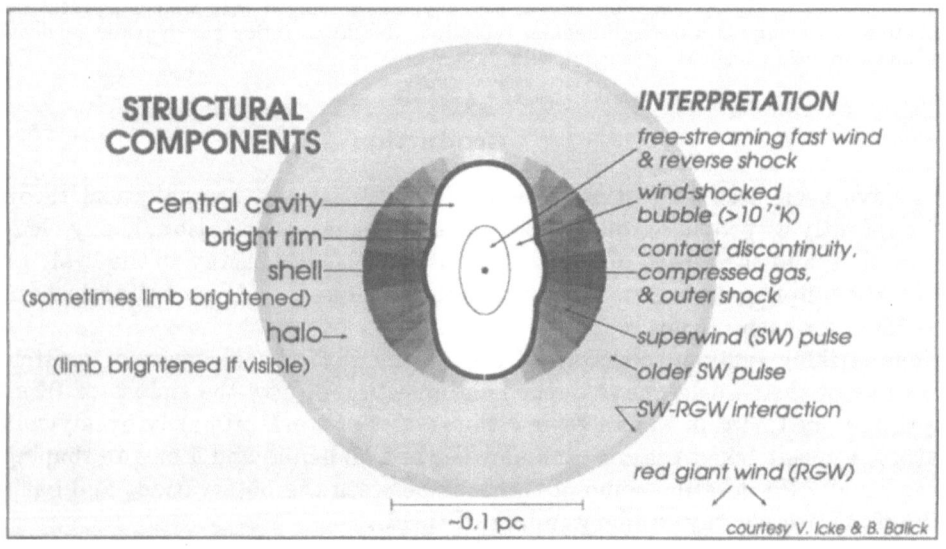

Fig. 1.   Functional definition of major morphological terms (left) and their hydro interpretation
(right).

IUE observations of mass loss from PN nuclei had already led to the "interacting-
winds scenario" (ISW) first proposed in a form applicable to PNs by Kwok and
collaborators and developed by them and many other theorists (cf. Kwok and Volk
1985). This pioneering work was based on principles developed in the late 1970s
by Mancunians John Dyson, Franz Kahn, and their collaborators for application
to the bubble nebulae of W-R stars.

The basic Dyson–Kahn idea of the ISW scenario is one-dimensional: a low-
mass-density "fast" wind (characterized by a speed of 10^3 km s^{-1} and a mass loss
rate $\dot{M} \approx 10^{-7} M_\odot$) from the PN nucleus. The fast wind rams into a denser,
cooler (10^4 K) and, for all practical purposes, stationary "slow wind" ejected 10^4
yr earlier by the nucleus in its AGB phase of evolution. The interaction is highly

supersonic, so an outer, outward-moving shock, quickly forms. Behind it gas is compressed forming a bright rim. Inside the outer shock is a high-pressure hot bubble whose internal energy is almost all thermal ($T \approx 10^{7-8}$ K), whose cooling time is longer than any dynamical time of the system, and whose relatively high pressure maintains the outer shock. Between the free-flowing fast wind from the nucleus and the hot bubble is an inner shock at which the kinetic energy of the wind is efficiently converted into thermal energy that maintains the pressure of the hot bubble. The very short sound-crossing time of the hot bubble insures that the bubble pushes outward fairly uniformly in all directions.

Based solely on image data, Balick (1987) found that few ($\approx 15\%$) of PNs are round and not readily described by any 1-D scenario. He speculated that PN shapes might be understood as tilted axi-symmetric figures of revolution in which the ram pressure of a dense equatorial wasteband was constraining the expansion nebular gas along this plane more so than along the perpendicular polar axis. He argued that the critical parameters of the evolution were the equatorial-to-pole (e-p) density contrast and the fast wind pressure (really, its luminosity $1/2\dot{M}v^2$).

This idea was not new at the time—Kahn and West (1987), had already published some analytical hydrodynamic models which formed peanut-shaped nebulae. However, the consistency of the constrained-flow concept with at least 75% of all PNs showed that the 1-D process was too naïve to be widely applicable.

Subsequently, various spatially-resolved observations of PN kinematics, (which are too numerous to mention individually) found that axisymmetric flows were quite standard for PNs, and that detailed 2-D hydrodynamic models might prove to be important and relevant.

Better and better 2-D models were quick in development, some Eulerian, some soft-particle hydro, and others Lagrangian in their formulation. Second-order Eulerian models are much better at following the evolution of the shocks and snowplows—features which are very prominent in the CCD images—so I shall focus only on them. A few models included possible magnetic effects (e.g. Pascoli, 1992).

Until recently most model computations, both analytical and numerical, assumed adiabatic expansion and cooling. For real nebulae, behind any shocks formed in regions already ionized by stellar UV photons, heat is very efficiently converted to (mostly) forbidden-line radiation that escapes, and isothermal shocks are more realistic than adiabatic ones. Realistic models had to await an accurate treatment of radiative cooling, called "radiation gas dynamics", or RGD. These fruits, largely the labours of Mellema and Frank, occupy the next few sections.

Just to close one more historical loop at this Manchester conference, Garrelt is expected to be joining the UMIST group later this year as a postdoc.

3 Methods

Adam and Garrelt developed entirely different numerical approaches to the numerical hydro computations. (Their efforts may sound redundant, but all sorts of
possible sneaky errors associated with numerical computations, such as numerical
diffusion through the finite grid cells, can cause quite plausible—and erroneous—
models to be produced.) The codes are fully second order. In addition, any
detailed predictions (e.g. the onset of instabilities) made by one model become far
more plausible when verified independently.

To be a little more concrete, Adam's hydro is based on a flux-correction algorithm developed by Icke (1991) with Adam's involvement. Garrelt uses a Roe-
solver approach which seems especially stable for handling shocks under "extreme"
conditions, such as the density jump of 10^5 and large upstream Mach numbers
(≤ 10) predicted for shocks within the context of PNs. Both of them jointly developed and then independently tested the radiative cooling algorithms. Getting the
radiative physics into the numerical code is tricky because the cooling times and
the natural hydro time scales are extremely different. Being conservative turns
Cray computers into smoking piles of silicon oxides, so one is forced to take risks
and to be clever. Both of them also added stellar UV photoionization and heating
to the hydro models since ionization and shocks associated with ionization fronts
need to be explicitly computed along with the effects of winds and hot bubbles.

Finally, Garrelt has implemented calculations with stars whose winds evolve
according to various stellar computations, whereas Adam turns on and maintains
a constant fast wind once the nucleus reaches a reasonable temperature. Adam's
thesis stresses morphological agreement with PNs, whereas Garrelt also focuses
more on a comparison of the observed internal motions of the visible nebula with
models.

Their methodologies and a comparison of their results has just been submitted
for publication (Mellema and Frank, 1993). The bottom line is that the two sets
of numerical models agree to 10% in most cases, and 20% in the worst case, for
identical initial and boundary conditions. (This represents a major milestone—
even if both models give wrong results!)

Let's skip to the results. Some of these have been reported in Frank et al.
(1993).

4 1-D RGD Models

It is instructive to begin with 1-D RGD models since these show many of the salient
results with far finer spatial resolution than is practical for 2-D computations.
Three impulses sweep through the nebula: the compression wave associated with
(1) the growth and expansion of the wind-heated interior hot bubble, and (2), (3)
the advancement of H° and He^+ ionization fronts (I.F.s). These three events occur
at different relative and absolute times for stars of various masses, and some waves

catch up to others causing interesting but temporary superposition effects. The details are complex, and depend on evolutionary models for the central stars, all of which are a bit uncertain themselves. We shall simply look at the highlights.

Of the three pressure waves, that of the hot bubble imparts the greatest amount of outward momentum to the warm (10^4 K), optically visible gas. This is expected, of course. Nonetheless, the pressure waves driven by the advancing ionization effects are significant also. In all three cases, each time a wave of compression works its way to the edge of the shell, a rarefaction wave begins moving back towards the star in its wake.

Each of the waves produces interesting effects. The initial radial profile of density of the shell of slow wind is assumed to fall as r^{-2}. Since the temperature of the ionized gas is constant, there is a built-in, strong pressure gradient which tries to relax at sonic speeds. However, much more important than this relaxation are the flows and ebbs driven through the shell by snowplows and the rarefaction waves associated with I.F.s and the expanding hot bubble. Gas in the shell is moved outwards, in some cases supersonically.

From the observer's point of view, although the (assumed) initial emission measure (EM) falls as r^{-3}, eventually the EM becomes *much* more gradual: as $A \bullet r^{+1}(A < 0)$! Frank, Balick and Riley (1990) first pointed out that such behavior is a very common feature of round and elliptical PNs. (The 2-D models yield much the same sort of result, at least at when the shell's e-p density contrast is fairly mild.) Furthermore, the velocity of the gas in the slow-wind shell tends to look something like a Hubble Law after a while—a common feature of many round and elliptical PNs seen by observers for many years.

A second impact of rarefaction waves concerns the manner in which the hot interior bubble expands. According to the numerical results, the pressure of the bubble exceeds that of the shell which confines it at all times. Before the first rarefaction wave returns the bubble is highly confined, and a bright rim of compressed gas forms. After the rarefaction wave returns and the forward density drops, the snowplow pushed forward by the hot bubble rushes ahead. The density of compressed gas ahead of the contact discontinuity decreases. The bright rim fades relative to the shell until the bubble-driven shock accretes more low-specific-momentum, slow-wind gas ahead of it.

In other words, there are times relatively early in the life of a PN (when it is still compact) when a long-slit echelle placed through the centre of a round PN would see the rim expanding faster than the shell which surrounds it. Later, the rim accelerates and moves radially at about the same speed as the now-expanding shell around it. Both types of motions have been observed in this manner by Chu and Jacoby (1988).

The one-dimensional models also show that the PN quickly becomes density bounded. The exception is for very massive progenitors whose nebular masses are large. These very rare PNs can remain ionization bounded throughout their lifetimes.

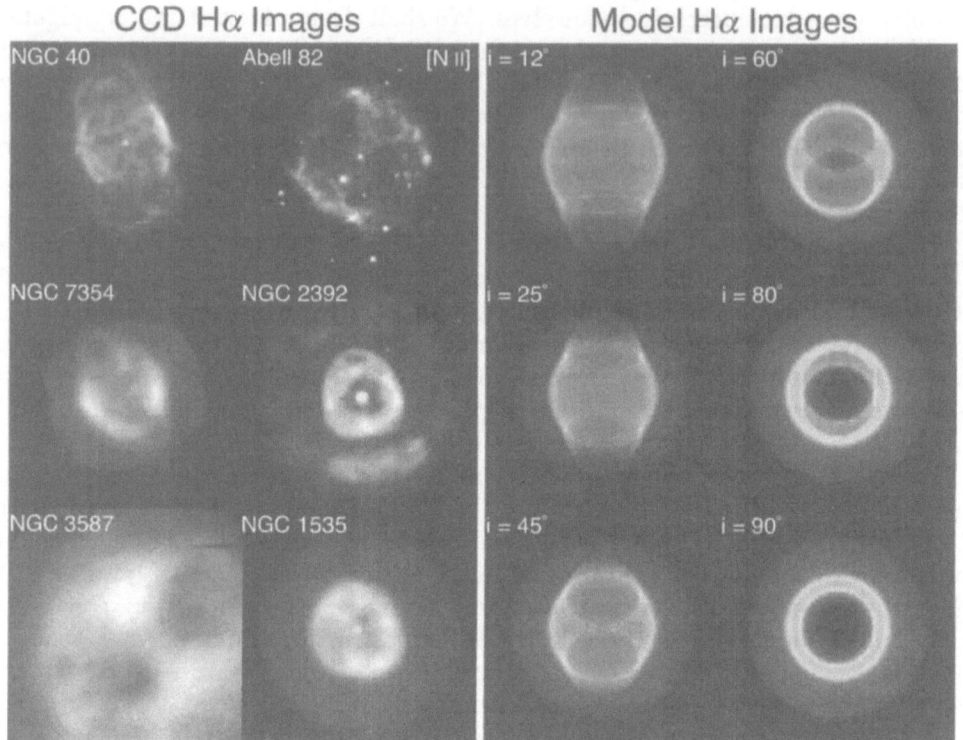

Fig. 2.   CCD and Model Hα images of various PNs (adapted from Frank *et al*, 1993).

The hydro models are providing a natural explanation for PN expansion which, if combined with other measured properties of the PN (like sizes), would help to provide the model builders with some information that might help them to understand empirically how the central star evolves! Wouldn't that be a nice "paleontological" discovery?

5   2-D RGD Models

At early times 2-D hydro models of elliptical PNs follow the routes expected of their 1-D counterparts. How long depends on the time required for the bubble to break through the weakest point in the constraining shell—i.e. the poles—which, in turn, is governed by the e-p density contrast and the bubble pressure (that is, the initial conditions). Once the shell springs a leak then all hell breaks loose.

Like the 1-D computations, the 2-D computations include the effects of stellar ionization, cooling by nebular lines, and of course the hydrodynamics, for each set of initial conditions. Most of the models are run on 50×50 grids for the sake of expediency, and key runs are verified on much finer ($\leq 250 \times 250$) grids. The output is a time series of density, velocity, energy density, and temperature distributions. In addition the 2-D line emissivities are computed for the brightest emission lines, one at every velocity. A set of synthetic CCD images and long-slit echelle profiles are made for direct comparison to observations.

Figure 2 shows an example of the model computations. Look first at the left panel of six actual Hα CCD images. What might these very-different- looking PNs have in common?

The answer, and one which nicely illustrates the success of numerical hydro calculations, is on the right side of the same figure. This shows one model result, an Hα emissivity distribution, spun about its symmetry axis into three dimensions, tilted in six different inclinations, and projected onto the sky to form synthetic Hα images.

Let us reconsider first the early development of elliptical PNs—before the shell is punctured. Generally speaking, the symmetry of the hot bubble (initially assumed spherical) begins to react to the axisymmetric distribution of pressure ahead of it. For shells whose temperatures are constant (10^4 K), the ram pressure follows the initial density distribution. Hence the bubble expands fastest along the symmetry axis where densities are lowest.

Eventually the hot bubble works its way to the polar edge of the shell where the pressure rapidly drops. Without a shell to constrain it, the hot bubble blows out. A bipolar flow of hot gas quickly forms on the symmetry axis, and (aside from the lobe's edges) behaves much like any adiabatically cooling flow constrained in its equatorial direction.

Invisible X-shocks and Mach disks form inside the hot, expanding lobes. Vortices of various sizes whirl away within. The rising hot gas ($\sim 10^7$ K) forms a mushroom cloud as little external pressure remains to constrain its lateral motions—a few bubble sound-crossing times. The details of the ensuing growth are sensitive to the structure assumed for the confining wasteband. Still, a few robust features of interest characterize the bipolar outflow.

The first of these is that the outer envelope of the lobe bears an amazingly strong resemblance to the predictions of analytical adiabatic models (based on the generalization of a method by Kompaneets for explosions) published in Icke, Preston and Balick (1989). Maybe one shouldn't be surprised that numerical and analytical models agree so well, but for those who have toiled at the computer for years this agreement is a dream come true! Insofar as observers are concerned, the simple analytical models can be used to relate the present shape of any adiabatically expanding bipolar nebula to its age, inclination, external pressure of confinement, and/or initial conditions of the confining torus.

In their theses, both Frank and Mellema find that a dense, warm (i.e not hot) blob tends to develop along the symmetry axis near the head of the expanding lobe. The blob tends to remain near the lobe's head as it grows. The origin of this blob is not yet clear, and may be in some doubt. But it is a fairly general feature of all of the 2-D computations for relatively large e-p contrast ratios, and it doesn't disappear when the number of grid points is increased from 50×50 to 250×250. Observers such as Balick *et al.*, (1993) have been finding pairs of high-velocity, small features along the major axes of several PNs, called FLIERs, which might be related to these ansae. It is too soon to tell. (More about FLIERs later.)

Another result, seen first in the adiabatic models (Icke *et al.*, 1992), is an autocollimation mechanism for bipolar lobes. The idea is this: when the shell is punctured the escaping hot gas scrapes against the puncture point in the shell. The shell consists of dense, relatively cool gas. It tends to become entrained along the base and outer surface of the rising column of gas forming a sheath of cooler, stiffer gas which acts somewhat like a chimney. Again, this is a general feature of the numerical models at all spatial resolutions, and inclusion of radiative energy losses do not significantly change the picture.

The collimation is reinforced by the mushroom cloud which forms at the head of lobes. The head of the mushroom cloud swirls around the outside of the rising column of gas and soon circles back around the outside of the chimney. In a manner of speaking the swirling vortex along the outside of the chimney acts a little like rollers on a conveyer belt, helping to maintain the collimation. How long this autocollimation process is effective depends on the pressure of gas assumed to exist in the medium far outside the torus—for which only wild guesses exist. Intriguingly, the process might be quite relevant along parts of collimated jets such as those found in Cyg A.

The successes of the 2-D hydro models are very impressive, as Garrelt's thesis will make clear in detail. There isn't space in this article to go into the many interesting details. However, the success of the models is primarily in the macroscopic realm. There are many details, which differ from nebula to nebula in the real world, and which make each them very individualistic. Many of these are not fully explained.

6   Haloes

Many PNs have large haloes (e.g. Balick *et al.*, 1990). When seen, they are universally edge-brightened and much rounder than the bright PN deep inside. The haloes are usually and reasonably attributed to early bursts of slow wind which have been coasting outward for about 10^4 yr During this time a bright leading edge will develop naturally from hydro processes (Frank, Balick and Riley 1990) as ram pressure of some external medium decreases the specific momentum of the leading edge of gas. Gas from deeper within the old ejecta catches up and slowly piles on to the slightly decelerating leading edge of the halo. Some back filling occurs. All points expand at about the same velocity which, in time, tends to circularize the halo. The central star ionizes the halo and makes it luminous.

What is wrong with this picture? Nothing that some *ad hoc* patches cannot fix. New H_2 images of NGC 6720 show that molecules surround both the halo and the shell deep inside of it. How can the H_2 around the shell survive in view of the intense uv photons streaming through it into the halo? Maybe NGC 6720 is really a bipolar seen pole-on. In this case the H_2 would be a sheath around the bipolar which, when projected onto the sky, might appear as a core plus halo, each surrounded by molecular gas. Support for this sort of picture is being presented elsewhere (Bryce, Balick, and Meaburn, this conference).

7 FLIERs

A variety of PNs, some elliptical and others bipolar, show a pair of small, bright knots of low-ionization gas along their major axes. In most cases the knots have highly supersonic velocities and, so, are called "Fast Low-Ionization Emission Regions", or FLIERs. Curiously, the kinematic ages of many FLIERs are less than that of the shell in which they are seen, suggesting that FLIERs are recent ejections or collimations of gas.

Space does not permit a detailed discussion of FLIERs here. Balick *et al.*, (1993) are finishing a long paper that reviews their many properties. In short, the emission-line spectrum closely resembles those of low- ionization H-H objects, including strong [OI], [SII], [NII], and [OII] and detectable [NI] and [MgI] lines. The chemical abundances appear to vary little from elsewhere in the nebula— except for nitrogen in which enhance ments by factors of about 5 are observed. The N enrichment suggests that FLIERs were recently flung directly from the star (somehow) and, so, are not formed hydrodynamically in the nebula. If so then FLIERs lie outside the scope of this review.

Perhaps. For now it is safe to urge that much more work be done on these very curious features.

8 Conclusions

Numerical hydrodynamics produces models consistent with the observed shapes and kine matics of PNs. In astronomy consistency is a close to proof as one gets, so we can rest quite satisfied with the choices of physical processes and initial conditions that have been made!

The logical consistency between observations and models can be improved further. To show that a series of PN images taken from a large catalogue can be matched to models is really no major feat. What is needed to tighten the consistency loop are agreements between the kinematic observations and model predictions for that set of PNs whose pictures match the synthesized images. This shows that the models can match both the structure and the present evolution of (hopefully) a broad class of PNs.

Adam and Garrelt have successfully incorporated the models and the morphologies of about 75% of the sample of PNs for which Balick published deep CCD images several years ago. This is very good progress, and some of the remaining 25% of the PNs can probably be understood within the existing picture. But wild celebrations are premature: not all PNs yet fit into the morphological scheme, let alone the kinematics. The most notorious outlaw is NGC 6543 whose intricate, symmetric morphologies and wonderfully complex but regular kinematics maintain a certain important level of humility among people who claim to know the path to a physical understanding of PN physics and evolution. IR images of NGC 7027 and 6720 are turning up major embarrassments as well.

Still, the physical models have reached the point where we can use the numerical models with some confidence, if not pride. PNs have been used to sharpen our physics and build our confidence, as it were—a role for which PNs have served faithfully and often in other contexts. It is time to plunge onward into the astrophysics of AGNs and YSOs where solid physical tools are more desperately needed.

The leadership needed is likely to emerge right here, from the top floor of this building, where Franz Kahn and his colleagues have long stood in the vanguard of understanding any medium whose atoms move with a short mean free path. It is with great admiration and respect that I wish him a long, healthy, and very active "retirement" with many more wonderful papers in the future.

References

Balick, B.: 1987, *Astron. J.* **94**, 671.
Balick, B., Gonzalez, G., Frank, A. and Jacoby, G.: 1992, *Astrophys. J.* **392**, 58.
Balick, B., Perinotto, M., Macchioni, A., Terzian, Y. and Hajian, A.: 1993. In preparation.
Chu, Y.-H. and Jacoby, G.H.: 1989, in S. Torres-Peimbert (ed.), IAU Symp. 131, *Planetary Nebulae*, Kluwer Academic Publishers, Dordrecht, p. 198.
Frank, A., Balick, B., Icke, V. and Mellema, G.: 1993, *Astrophys. J.* **404**, L25.
Frank, A., Balick, B. and Riley, J.: 1990, *Astron. J.* **100**, 1903.
Icke, V.: 1991, *Astron. Astrophys.* **251**, 369.
Icke, V., Mellema, G., Balick, B., Euderink, F. and Frank, A.: 1992, *Nature* **355**, 524.
Icke, V., Preston, H.L. and Balick, B.: 1989, *Astron. J.* **97**, 462.
Kahn, F. and West, K.A.: 1985, *Mon. Not. R. astr. Soc.* **212**, *837*.
Kwok, S. and Volk, K.: 1985, *Astron. Astrophys.* **153**, 79.
Mellema, G. and Frank, A.: 1993. Submitted to *Astron. Astrophys.*
Pascoli, G.: 1992, *Publ. Astron. Soc. Pacific* **104**, 350.

Investigating the Kinematics of the Faint, Giant Haloes of Planetary Nebulae

M. Bryce and J. Meaburn
Department of Astronomy, University of Manchester, Oxford Road, Manchester. M13 9PL, UK

B. Balick
*Astronomy Department, FM-20, University of Washington, Seattle. WA 98195, USA
and Arcetri Observatory, Largo Enrico Fermi 5, I-50125, Florence, Italy*

and

J.R. Walsh
*European Space Telescope Coordinating Facility, ESO, Karl-Schwarzchild-Strasse 2, D-8046
Garching bei München, Germany*

Abstract.
Some interesting results from a current program of high resolution spectroscopy of the faint, giant haloes of Planetary Nebulae are presented. Observations of optical emission line profiles show that these haloes have very little turbulent motion. The halo of NGC 6543 is thought to be a filamentary, thin, radially expanding, spherical shell and that of NGC 6826 is more probably a filled spherical shell. The halo of NGC 6720 contains inner and outer shells, the split profiles from the inner halo show that this part of the halo is bipolar in nature.

1 Introduction

The faint, giant haloes of Planetary Nebulae (PNe) are believed to be the remnants of red giant and multiple superwind mass loss events occuring as the progenitor stars evolve along the Asymptotic Giant Branch. Tracing the kinematic structure of the haloes is thus a probe of the mass loss paleontology.

We present results from an ongoing investigation into the haloes of PNe. High wavelength resolution, longslit spectral observations of emission lines from several haloes have been obtained with the Manchester echelle spectrometer (Meaburn *et al.*, 1984) at the INT. Analysis of Gaussian profiles fitted to the spatially resolved observed line profiles give kinematic and morphological information which can be used to calculate the expansion velocities and probe the dynamical structures present.

It is intriguing to note that the structures of the haloes as suggested by these kinematical data often differ from those suggested by direct, deep images.

2 Radial Expansion Velocities of NGC 6543 and NGC 6826

The haloes of both NGC 6543 and NGC 6826 are circular in appearance. That of NGC 6826 is strikingly spherically symmetric whereas that of NGC 6543 is very filamentary and contains a cluster of very bright knots to the west of the bright nebular core.

Astrophysics and Space Science **216**: 23–24, 1994.
© 1994 *Kluwer Academic Publishers.*

Longslit observations of the [OIII]5007Å emission line profiles were obtained from a number of $60\mu m$ wide ($\equiv 6kms^{-1}$) slit positions across the haloes of NGC 6543 and NGC 6826 Analysis of the observed line profiles show that they are remarkably narrow, indicating that there is very little turbulent motion ($\leq 7kms^{-1}$ for NGC 6543 (Meaburn *et al.*, 1991)) within these haloes. No clearly resolved line splitting was observed in any of the profiles obtained from these two haloes although there was definite evidence that the line profiles did contain two close components.

Two simple models were adopted to describe each halo; a thin, spherically symmetric, radially expanding shell and a spherically symmetric wind emitted at constant mass loss rate at constant velocity v_{exp}. It was found that the first model best described the halo of NGC 6543, giving a value of $v_{exp} \leq 4.6kms^{-1}$ (Bryce, Meaburn and Walsh, 1992a) whereas the second model was more appropriate to the halo of NGC 6826, giving $rv_{exp} \simeq 14kms^{-1}$ (Bryce, Meaburn and Walsh, 1992b).

3   The Ring Nebula - NGC 6720

The halo of the Ring nebula contains an inner, slightly elliptical, limb-brightened shell surrounded by a fainter, circlar outer region. There is a thin layer of H_2 emission surrounding the bright outer edge of the inner halo (Kastner *et al.*, Private communication).

The [NII]6584Å emission lines are the strongest of those observed from the inner halo and are up to three times brighter than those of H_α. Unlike the H_α profiles, the [NII] profiles show clear splitting of $\sim 30kms^{-1}$ (Bryce, Balick and Meaburn, 1993). The two kinematic components in [NII] are narrow and well separated right to the projected edge of the inner halo, as expected for an inclined bipolar. The theory that NGC 6720 is a bipolar nebula observed at high inclination angle was previously rejected (Balick *et al.*, 1992) on the grounds that the projections of the two two extended lobes should show two overlapping ellipses; in fact images of the inner halo cannot be resolved into two ellipses. Nonetheless, the present data are best interpreted as originating from an inclined bipolar where, as in almost all other bipolar lobes, [NII] emission arises along the surfaces of the lobes.

References

Balick, B., Gonzalez, G., Frank, A. and Jacoby, G.: 1992, *Astrophys. J.* **392**, 582.
Bryce, M., Balick, B. and Meaburn, J.: 1993, in Press.
Bryce, M., Meaburn, J., Walsh, J. R. and Clegg, R. E. S.: 1992a, *Mon. Not. Roy. Astr. Soc.* **254**, 477.
Bryce, M., Meaburn, J.and Walsh, J. R.: 1992b, *Mon. Not. Roy. Astr. Soc.* **259**, 634.
Kastner, Gatley, Merill and Weintraub, Priv. Comm.
Meaburn, J., Blundell, B., Carling, R., Gregory, D. F., Keir, D. and Wynne, C. G.: 1984, *Mon. Not. Roy. Astr. Soc.* **239**, 1.
Meaburn, J., Nicholson, R., Bryce, M., Dyson, J. E. and Walsh, J. R.: 1991, *Mon. Not. Roy. Astr. Soc.* **252**, 535.

Shock Modelling of Planetary Nebulae

L. Cuesta, J. P. Phillips and A. Mampaso
Instituto de Astrofísica de Canarias.
La Laguna. E-38200. Tenerife. Spain

June 11, 1993

Abstract.
The kinematics of Planetary Nebulae are analyzed in terms of the solutions to the equations of hydrodynamic equilibrium developed by J. Cantó. We apply our analysis to the Planetary Nebulae NGC 6905 and NGC 6537. A detailed spectroscopic study of these objects reveals the existence of high nuclear velocities, together with complex kinematic structures and unusual emission line intensities. Shock ionization clearly plays a key role in these nebulae. Remarkably good agreement is obtained when comparing the synthetic maps and spectra resulting from the shock solutions with the observational data.

Key words: Planetary Nebulae, Shocks

1 Theoretical Background

A large number of Bipolar Planetary Nebulae show certain peculiar characteristics, such as anomalous line ratios and enhanced H_2 molecular emission, and hence it is convenient to apply an analysis in terms of shock models. In this sense, two theoretical approaches are noteworthy. Firstly, Icke and collaborators developed a model to reproduce the aspherical shock bubble generated when an explosion occours in an inhomogeneous medium. Secondly, the solutions to the hydrodynamical equilibrium equation, developed by Cantó (1980), give the location of the shock front created when a supersonic stellar wind interacts with a dense cloud.

Let us assume a structure in which the nucleus of the planetary nebula ejects an isotropic wind towards a halo with density $\varrho(R, \alpha)$, where R and α are the polar coordinates, uniform temperature and, consequently, constant sound velocity c_s. Once hydrostatic equilibrium is reached, the cavity may be described by this equation (Cantó, 1980):

$$\varrho(R, \alpha) = \frac{\varrho_\star R_0^2}{R^2 + R'^2} \left[1 + \frac{R^2 + 2R'^2 - RR''}{\sqrt{R^2 + R'^2}} \frac{1}{R \sin \alpha} \int_{\alpha_0}^{\alpha} \frac{R' \sin \alpha}{\sqrt{R^2 + R'^2}} \mathrm{d}\alpha \right].$$

with $R_0 = \sqrt{\dot{M}_\star v_\star / 4\pi c_s^2 \varrho_\star}$, and where $\varrho_\star$ is the density that would have the cloud at the position of the star, $v_\star$ is the stellar wind velocity behind the shock and $\dot{M}_\star$ is the mass loss rate in the wind. R' and R'' are, respectively, the first and second derivative of R with respect to α. In α_0 we define the boundary conditions of the equation.

If θ is the inclination angle, the observed velocity at the shock front, is given by:

Astrophysics and Space Science **216**: 25–29, 1994.
© 1994 *Kluwer Academic Publishers.*

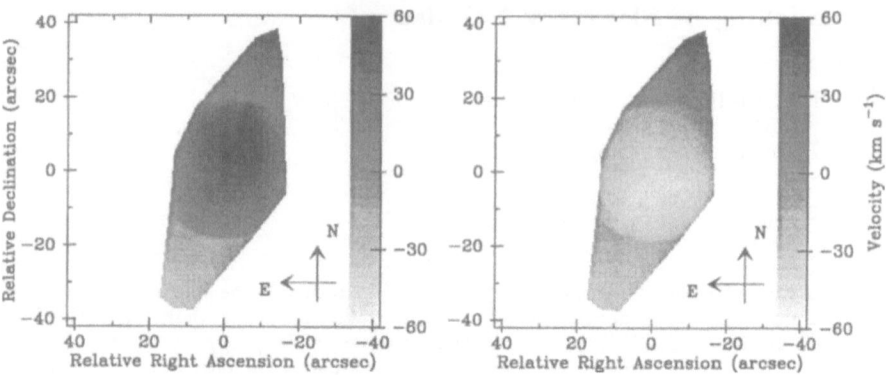

Fig. 1. Observed velocity maps for NGC 6905, one for the back rear side and the other for the front approaching side.

$$v_T = \frac{v_\star}{\cos\alpha} \left[\frac{R'\cos(\theta+\alpha) - R\sin(\theta+\alpha)}{\sqrt{R^2+R'^2}} \right] \int_{\alpha_0}^{\alpha} \frac{R'\sin\alpha}{\sqrt{R^2+R'^2}} d\alpha.$$

Let us now consider a distribution of density $\varrho(R,\alpha)$ with a disc of width $2W$ and whose initial density along the symmetry axis varies via a gaussian between a maximun value $\varrho_\star$ at the center, and a value ϱ_c at the edges. Outside the disc, the density of the halo is assumed to decrease radially, following a $R^{-\beta}$ law. The function that describes this variation is:

$$\varrho(R,\alpha) = \begin{cases} \varrho_\star\, e^{-\ln\left(\frac{\varrho_\star}{\varrho_c}\right)\frac{R^2\cos^2\alpha^2}{W^2}} & \text{if} \quad |R\cos\alpha| \leq W \\[2mm] \varrho_c\, e^{-\ln\left(\frac{\varrho_c R^\beta}{\varrho_m R_0^\beta}\right)\left(\frac{|R\cos\alpha|-W}{W_c}\right)^2} & \text{if} \quad W \leq |R\cos\alpha| \leq W+W_c \\[2mm] \varrho_m\left(\frac{R_0}{R}\right)^\beta & \text{if} \quad W+W_c \leq |R\cos\alpha| \end{cases}$$

where ϱ_m is the density of the halo at $R=R_0$. In order avoid discontinuities we have also introduced a small area of width W_c that connects the disc and the surroundings.

2 Applications

The Planetary Nebulae NGC 6905 and NGC 6537, were observed with the Isaac Newton Telescope at the Observatorio del Roque de los Muchachos (La Palma, Spain).

Both nebulae are bipolar, but clearly different. NGC 6905 is a high excitation planetary nebula, with a Wolf-Rayet OVI emission-line central star. Its shell shows two different kinematic zones. The inner region is rather regular, approximately

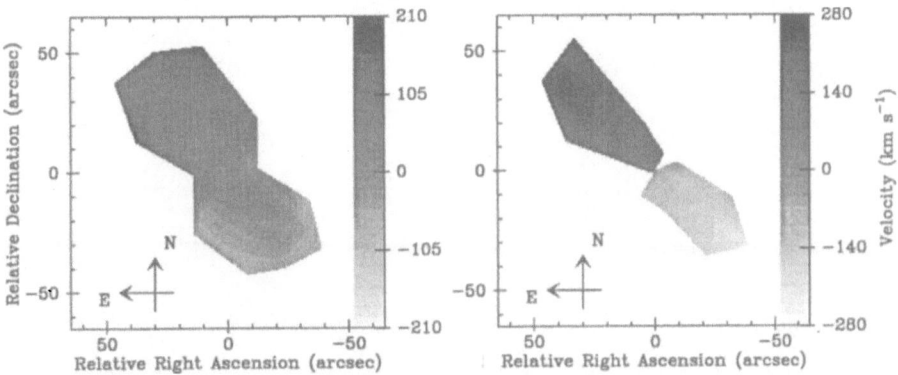

Fig. 2. Observed velocity maps for NGC 6537, one for the central component and the other for the lateral components.

elliptical, while the nebular extensions possess a highly complex velocity field and strong ansae-type formations. Figure 1 shows the velocity maps for this nebula, one for the back (receeding) side and the other for the front (approaching) side.

The other nebula, NGC 6537, possesses the typical hour-glass bipolar structure, with a rather compact and bright central nucleus, and two, much more extended, symmetrically arranged lobes. This is one of the highest excitation nebula yet found, showing line emission of [Si V] and [Si VI] in the inner regions. The central star is not visible, probably because of the high extinction towards the nucleus.

Our spectra for this nebula show a rather peculiar structure, with a central low velocity component and two other adjacent components with velocities of the order of $\Delta v = \pm 200$ km/s. The corresponding observed velocity maps are shown in Fig. 2, one for each component.

From our low resolution spectra it is apparent that the nuclear emission extends in both nebulae over a very large range of velocities. For NGC 6905, the strong ansae emission show up remarkably, especially in the low excitation transitions ([OI], [NII], [SII]). In the case of NGC 6537, there are a lot of emission lines, both nuclear and nebular, also showing anomalously strong emission in low excitation transitions.

All these arguments suggest that, in these nebulae, ionization is mainly due to shocks, and thus it is necessary to use a shock model in order to reproduce the kinematics of these sources.

The inner zone of NGC 6905 is adequately reproduced using a model in which an ellipsoid expands with radial velocity proportional to the distance from the centre. However, this simple model fails to explain the outer zones, and a shock model is developed here.

Assuming that the wind is constrained by the inner ionized shell, and escapes through openings in this shell to interact with the external medium, we obtain the synthetic velocity maps that are shown in Fig. 3. In Fig. 4 we illustrate the

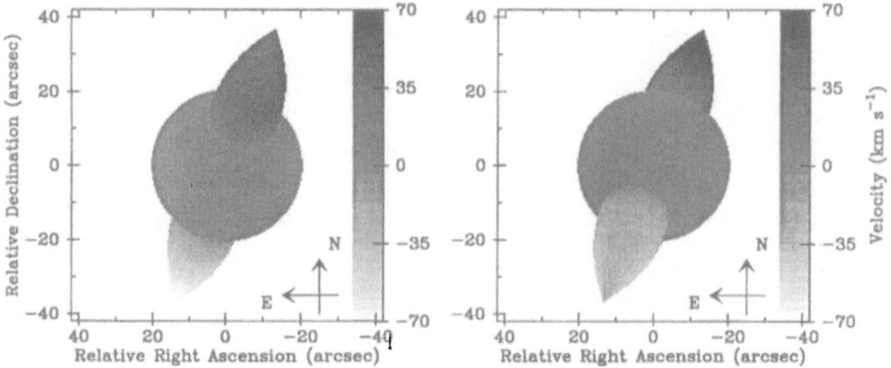

Fig. 3.   Theoretical velocity maps for the exterior shocked cavity of NGC 6905.

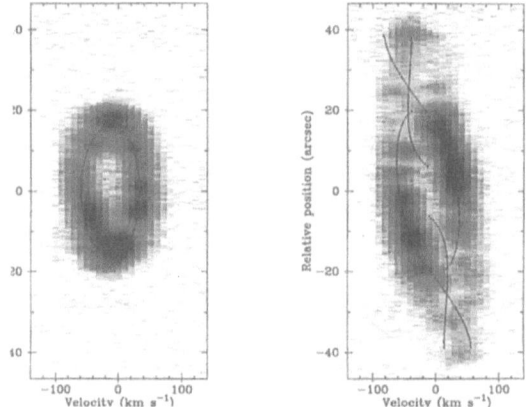

Fig. 4.   Synthetic spectra superimposed on the observed spectra in Hα for NGC 6905.

spectra generated by the model, superimposed on our observed spectra in Hα for angular positions 70°, and 340°. We have used the expanding ellipsoid model for the interior zone, and the appropiate shock model for the exterior regions.

In the case of NGC 6537, the shock region clearly extends over the whole nebula. We have to postulate the presence of a thin disc in order to produce the observed configuration. The resulting velocity maps are illustrated in Fig. 5, while Fig. 6 shows the synthetic spectra superimposed on the observed [NII] spectra for the position angles 31° and 71°.

To summarize, it is apparent that these models combining simple geometries plus expansion, with shock propagation into an inhomogeneous region, are able to reproduce very well the observed morphologies and kinematics of many Planetary Nebulae in which, as evidenced by their special kinematical characteristics and low excitation line ratios, the presence of a shock front plays a principal rôle in the nebular excitation.

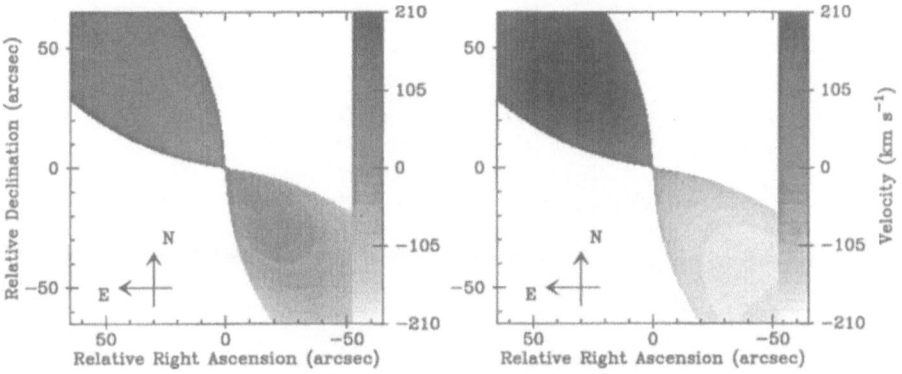

Fig. 5.   Theoretical velocity maps for NGC 6537.

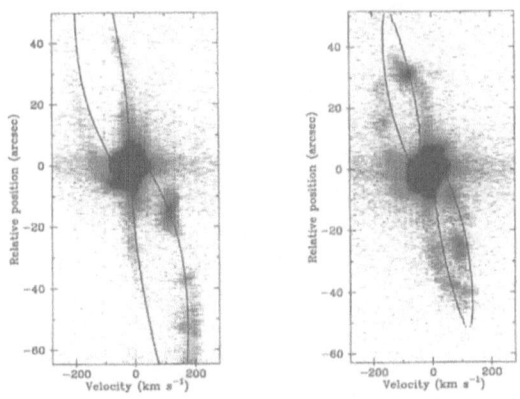

Fig. 6.   Synthetic spectra superimposed on the observed spectra in [NII] for NGC 6537.

References

Barral, J. F. and Cantó, J.: 1981, *Rev. Mex. Astron Astrofís.*, **5**, 101.

Cantó, J.: 1980, *Astron. Astrophys.*, **86**, 327.

Cuesta, L. and Phillips, J.: 1993, in I. A. U. Symp. 155, *Planetary Nebulae*, eds. R. Weinberger and A. Acker (Reidel, Dordrecht), in press.

Cuesta, L., Phillips, J. P. and Mampaso, A.: 1993, *Astron. Astrophys.*, **267**, 199.

Icke, V.: 1988, *Astron. Astrophys.*, **202**, 177.

Imaging Polarimetry of Proto-Planetary Nebulae

S. M. Scarrott and R. M. J. Scarrott
Physics Department, University of Durham, Durham, U.K.

and

R. D. Wolstencroft
Royal Observatory, Edinburgh, U.K.

Abstract. Polarization maps are presented for the three proto-planetary nebulae, Frosty Leo, MZ3 and M2-9. The structure of each object is discussed and geometries derived from these data are compared with those predicted by the currently fashionable two wind hydrodynamical models which account for many of the features of such transient nebulae.

1 Introduction

The post-main-sequence evolution which links AGB stars, Mira variables, IR/OH sources to planetary nebulae (PNs) involves a transient phase in which proto-planetary nebulae (PPNs) are formed. This phase is only short-lived, typically a few 1000 years, and the few known PPNs representing this stage of evolution have unusual characteristics. The common morphology is bipolar (sometimes called bowtie or butterfly) with a central star illuminating, and exciting, two diametrically opposed lobes of nebulosity. In the young PPNs (e.g. Frosty Leo) the central star is totally obscured from direct view by a dense circumstellar disc but light can escape along the poles of the disc to create the nebular lobes which are seen by reflected light. In the intermediately aged PPNs (e.g. IRAS 07131-0147, Scarrott *et al* 1990) the circumstellar material has been sufficiently dispersed to reveal the central star which again illuminates a bipolar reflection nebula. In an older PPN or young PN (e.g. MZ3 or M2-9 respectively), the central star ionizes the surrounding gaseous medium to create a nebula which is seen both in reflection and emission. In a PN the central star is stripped of circumstellar material and the nebular radiation is totally dominated by emission from a hot ionized gas. In this paper we present polarization maps for three PPNs seen at various stages of evolution and use the data to investigate their individual geometries.

2 The Frosty Leo

The Frosty Leo Nebula is an example of a very young PPN seen as a bipolar reflection nebula illuminated by a central star which is hidden from direct view at optical wavelengths by an obscuring circumstellar disc seen almost edge-on. Attention was first drawn to this system by the observation that the IRAS colours of its highly obscured central star, IRAS09371+1212, peaked at 60mm which led to the suggestion (Forveille *et al* 1987) that water ice was present in the circumstellar

Astrophysics and Space Science **216**: 31–42, 1994.
© 1994 *Kluwer Academic Publishers.*

material, possibly as a coating on preexisting silicate grains. The presence of water
ice was confirmed by the detection of the 3.1mm feature in the IR spectrum of
IRAS09371+1212 by Rouan *et al* (1988).

Figure 1 shows polarization and intensity contour maps for the Frosty Leo PPN.
The bright inner lobes are clearly discernible and Bare surrounded by fainter and
more amorphous nebulosity. Two ansae, bright localised knots of nebulosity, are
evident at $\sim \pm 20''$ offset along the major axis of the nebula. The general circular
pattern of polarization vectors confirms that the Frosty Leo is a reflection nebula
illuminated by a source located midway between the two inner lobes; this source
is not seen in our intensity image and is hidden from direct view even though light
can escape in other directions to create the visible nebulosity. The circular pattern
of polarization vectors breaks down in the very central area of the nebula where
it is replaced by a band of approximately parallel vectors and with null points
on on the minor axis of the nebula and equally spaced about the central star.
This feature is frequently seen in pre-main-sequence systems and is known as the
polarization disc and is attributed to the presence of a circumstellar disc around
the central star. The existence of such a feature in a post-main-sequence object can
also be taken as an indication of a circumstellar disc. Figure 2 shows a montage
of greyscale images of various parameters derived from our data. The polarization
image shows that generally high levels of polarization are seen throughout the
nebula and attain values of 55–60% in the two ansae. Levels of polarization of
$\sim 60\%$ indicate that the scattering grains are much smaller than those in the
general ISM and this confirms the idea that the atmospheres and outflows from
evolving stars represent the birthplace of dust grains and in the Frosty Leo we are
seeing the effects of nascent grains. The V-I colour map shows that the inner parts
of the nebula are highly reddened and this is further evidence for a concentration
of obscuring material in the form of a circumstellar disc.

3   The Proto-Planetary Nebula MZ3

This proto-planetary nebula is more evolved than the Frosty Leo and its central
star is directly visible as an ionized core which has created two lobes of nebulosity
seen predominantly by emission-line radiation. The general morphology of the
nebula and IR observations of the core again indicate the presence of a circumstellar
disc (Cohen *et al* 1978).

In fig. 3 we show polarization and intensity contour maps of MZ3. The central
frame shows the bipolar structure of the nebula with the central star/core clearly
visible. The left hand frame shows the original polarization map—it is quite different to the one for the Frosty Leo. The general level of polarization is much
lower and there are two distinct features in this map. In the region corresponding
to the bright lobes seen in the contour map the polarization levels are small (maximum $\sim 10\%$) and the pattern of orientations do not follow the expected circular
arrangement, furthermore, the central star is also polarized. Beyond the bright

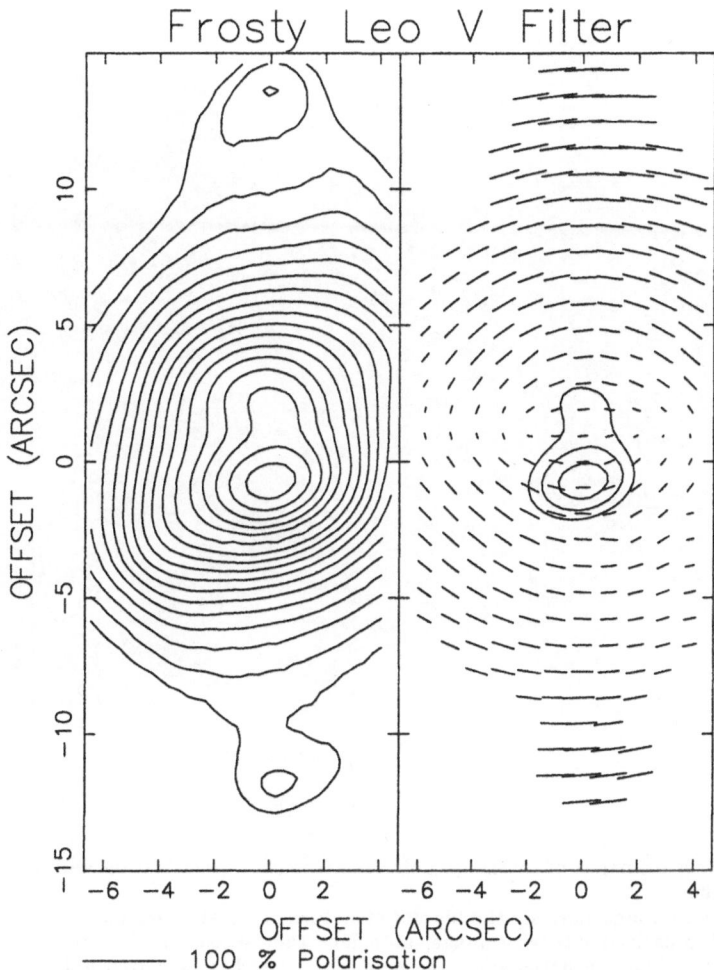

Fig. 1. V waveband polarization and intensity contour maps for the Frosty Leo Nebula.
Left: The intensity contour map. Note that the central star is not directly visible but is located between the two inner lobes of nebulosity. There are ansae located on the major axis of the nebula some ±12″ offset from the centre.
Right: The polarization map. Beyond ~ 5″ from the centre the pattern has the characteristic circular form of a reflection nebula illuminated from within. The high levels of polarization (up to 60% in the ansae) indicate that this nebula is seen solely by reflected light and there is no intrinsic emission from the nebular lobes. The non-circular pattern in the inner ±5″ is attributed to the effects of a circumstellar disc.

lobes the polarization levels are much higher (up to 30–4%) and form a circular pattern. What does this mean?

The low levels of polarization in the brighter lobes implies that most of the

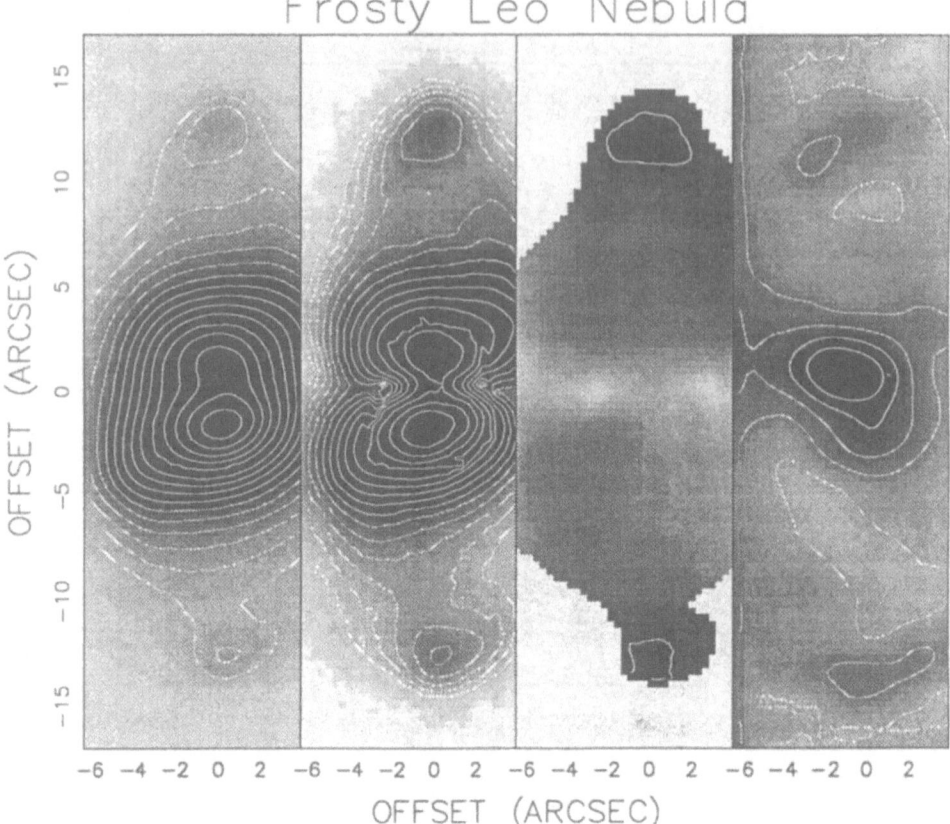

Fig. 2.   Greyscale images of intensity, polarized intensity, polarization and V-I colour for the
Frosty Leo Nebula.
Left: An intensity image showing the bipolar structure complete with ansae.
Centre left: A polarized intensity image, note how the "waist" of the bipolar, created by the
circumstellar disc, is accentuated when the nebula is viewed in polarized light.
Centre right: A polarization image where white and black correspond to 0 and 60% levels of
polarization respectively. Note that in the ansae the polarization levels are at a maximum ($\sim$ 60%)
which implies the presence of small nascent grains.
Right: A V-I colour map where dark corresponds to the reddened parts of the nebula. Note
how the inner lobes are much redder than the outer nebula; this results from obscuration by the
circumstellar disc.

radiation we receive from these areas is from emission by the hot nebular gas and
is unpolarized and only a small fraction of the radiation has been scattered. We
should still expect to see the circular pattern and its absence is due to the fact that
the light from the whole nebula and core has been extinguished in the interstellar
medium between MZ3 and ourselves. Since MZ3 lies in the galactic plane, this
extinction is dichroic and has polarized the the radiation (scattered and emission)

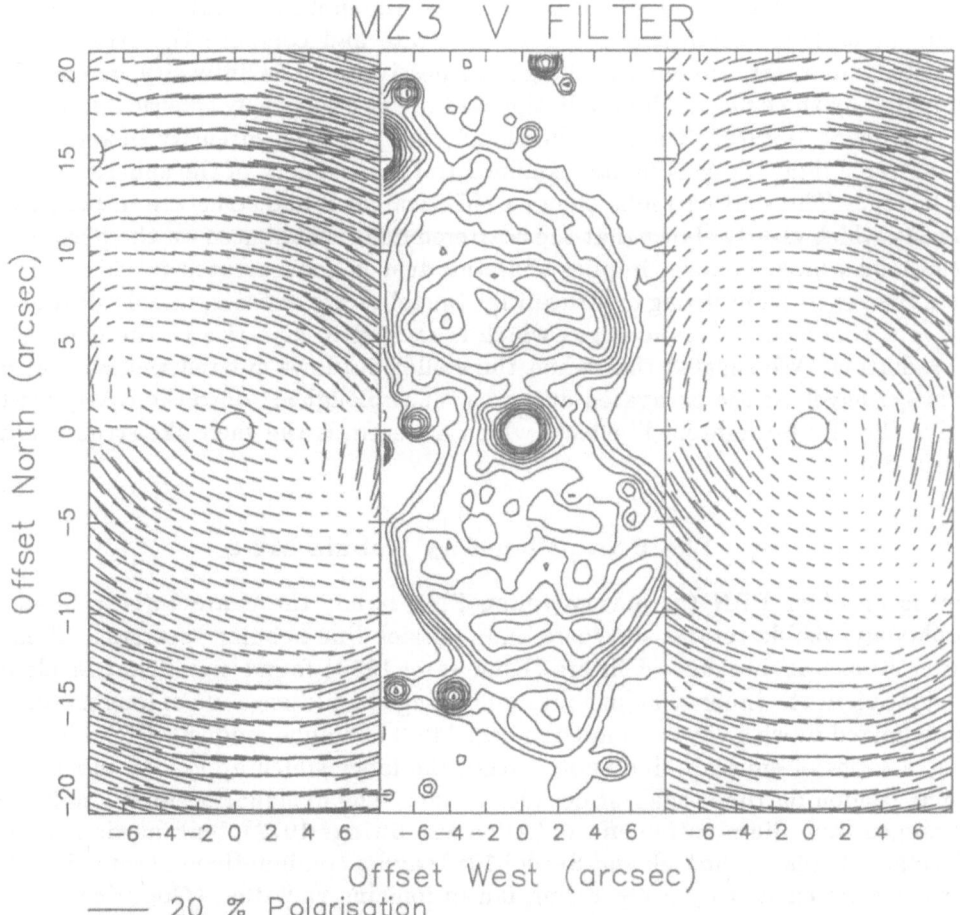

Fig. 3. V waveband polarization and intensity contour maps for MZ3. Left: The polarization map before correction for interstellar polarization. Centre: An intensity contour map. Note that the central star is directly visible and excites the two spherical lobes of nebulosity. Right: The polarization map after correction for the interstellar polarization. In all regions the pattern is now circular with the bright lobes showing even less polarization (most of the original polarization in these regions was due to the ISM) but beyond these regions the polarizations are typical of a simple reflection nebula. We have an emission-line bipolar nebula embedded in a much larger reflection nebula.

to a small, but significant, extent. We have corrected for this interstellar polarization and the right hand frame shows the final polarization map. In all regions the pattern is now circular with the bright lobes showing even less polarization (most of the original polarization in these regions was due to the ISM) but beyond these regions the polarizations are typical of a simple reflection nebula. MZ3 is an emission-line bipolar nebula embedded in a much larger reflection nebula.

In fig. 4 we show a montage of intensity and polarized intensity greyscale images at different brightness levels to illustrate and compare the structure of the nebula in total and polarized light respectively. The intensity images are dominated by the emission-line radiation and the two relatively spherical lobes are the prominent features. On the other hand, the polarized intensity images show the scattered light which is mainly confined to the periphery of the spherical lobes and beyond. The southern lobe shows a spherical shell-like structure in polarized intensity which clearly shows that the scattered light is enhanced on the periphery of the lobe which, in turn, implies that the dust is concentrated in this region. These images provide strong evidence that the lobes are cavities, mainly devoid of dust, but containing a tenuous gas which is excited to give rise to the emission-line radiation. We suggest that when the stellar outflows created the cavities in the larger envelope the excavated dust was "snowploughed" onto the cavity walls to give the "limb-brightened" effects which are seen as the shell-like structure in polarized light.

4   The Proto-Planetary Nebula M2-9

M2-9 is an older PPN (or a very young PN); it has the characteristic bipolar morphology, visible central star/core and emission-line nebular spectrum. Polarization studies in broadband filters (King *et al* 1980) found significant levels of polarization in the inner nebular lobes indicating the presence of copious amounts of dust mixed in with the line-emitting gas. M2-9 is unique amongst PPNs in that it has a series of emission-line knots within the lobes which appear to move in a regular fashion on time scales of decades or less. The explanation for these knots has ranged from illumination effects (Allen and Swings 1972), bulk motions of localised gas clouds (Kohoutek and Surdej 1980) and a combination of static density enhancements excited by a precessing fan of ionizing radiation (Goodrich 1991). Spectropolarimetric observations of localised regions in M2-9 show some evidence for scattered emission-line radiation (Schmidt and Cohen 1981) and both Carsenty and Solf (1983), and Icke, Preston and Balick (1989) later conjectured that some of the broad Hα emission emanating from the lobes was, in fact, scattered line-emission radiation produced in the central core. These conjectures prompted us to map M2-9 in the Ha emission line.

Figure 5 shows polarization and intensity contour maps of M2-9 in the Hα emission line. The centre frame illustrates the brightness structure of M2-9; it has the bright central core with two lobes of nebulosity which appear to contain a number of small knots. In terms of the brightness distribution and the location of the knots the whole system shows a basic reflection symmetry about the equatorial (E-W) plane with all the brighter nebulosity concentrated towards the eastern rim of each lobe.

The left hand frame shows the original polarization map and it suffers the same problem as the one for MZ3—a contribution of interstellar polarization which gives

MZ3 V Filter

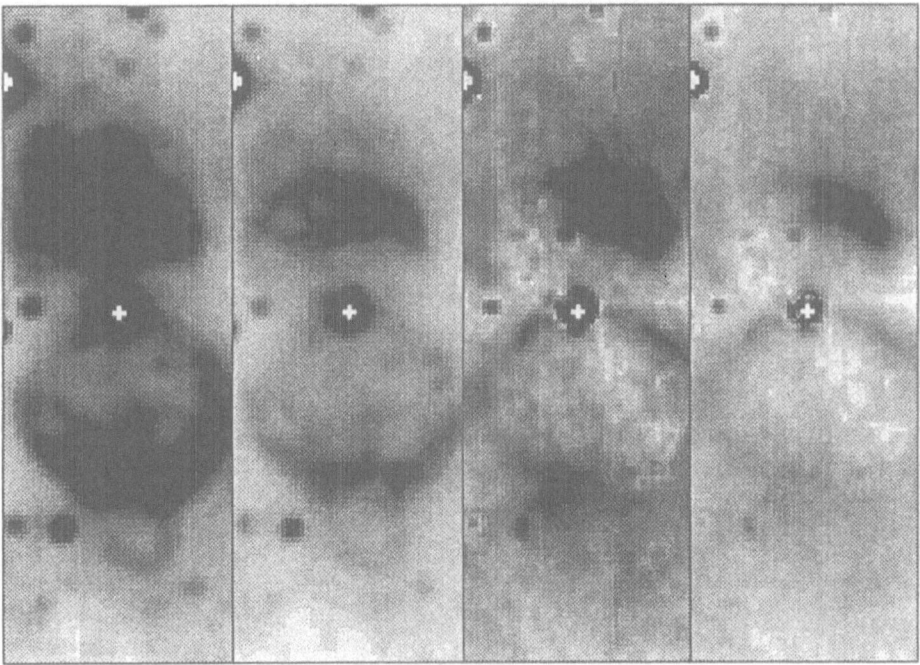

Fig. 4.   Greyscale images of intensity and polarized intensity for MZ3.
Left hand frames: Intensity images showing faint and bright regions respectively.
Right hand frames: Polarized intensity images showing faint and bright regions respectively. Note
how the southern lobe is shell-like when viewed in polarized light; this is due to the "snowplough-
ing" effect of the secondary wind excavating the large scale stellar envelope. The lobes appear to
be confined and have broken out beyond the envelope.

the impression that the core is extended (the elliptical inner polarization pattern)
and that the emission line knots are polarized. The right hand frame shows the
polarization results after correction for the interstellar contribution: the pattern is
now circular removing the need to invoke an extended source. The knots are also
unpolarized confirming they are pure emission effects. The levels of polarization
reach over 20% in places which suggests that even in the Hα emission-line parts
of M2-9 look like a simple reflection nebula. This is strong confirmation of the
prediction of copious scattering in the nebular lobes of emission-line radiation
originating in the core.

We have also made images of M2-9 in polarized intensity before and after correction for the ISP and these are shown in fig 6. In the uncorrected image (centre
frame) the knots on axis at $\sim \pm 14''$ are quite visible, yet on the corrected image
(right hand frame) these, and other small scale features, disappear to give a rather
smooth appearance to the image. This indicates that the knots are intrinsically

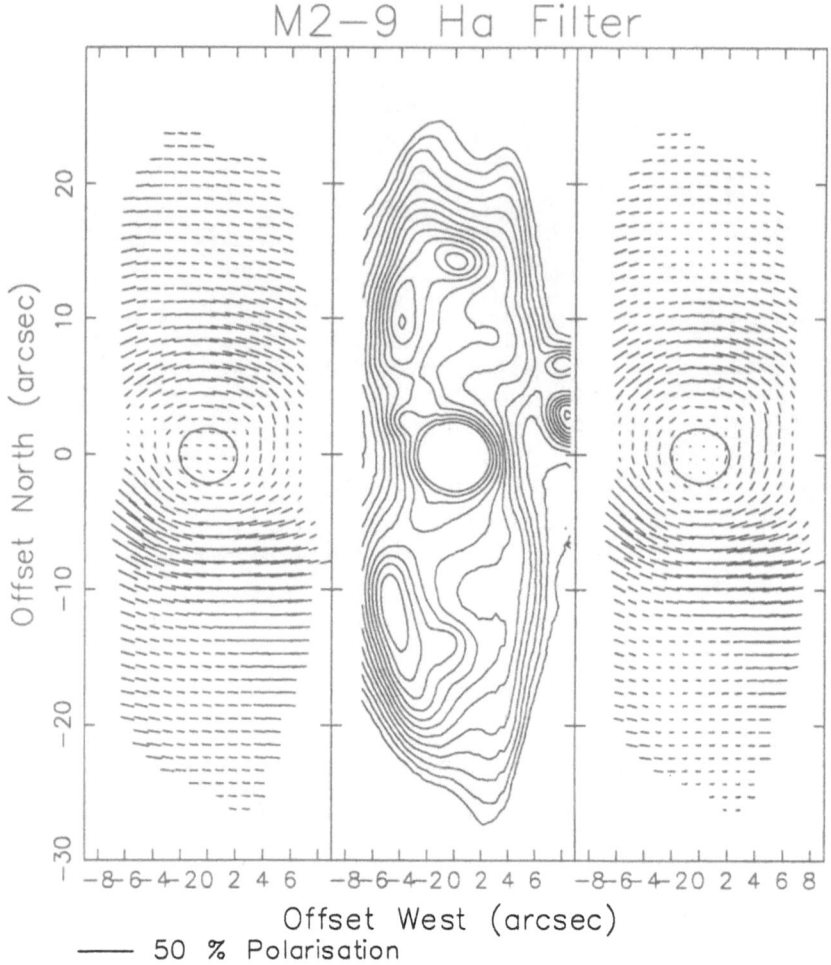

Fig. 5.   Polarization and intensity contour maps for M2-9 in the Hα emission-line.
Left: The polarization map before correction for interstellar polarization.
Centre: An intensity contour map.
Right: The polarization map after correction for the interstellar polarization. The central pattern
is quite circular which implies that the central core is not spatially and asymmetrically extended
on scale $\geq 1 - 2''$.

unpolarized but show through in polarized intensity images unless the effects of
the ISM are taken into account. The absence of any features in the corrected polarized intensity image corresponding to the ionized knots in the nebular lobes is
also significant since it implies that there is no concentration of dust at the position
of the knots. This, taken in conjunction with our observation that the knots have

not moved significantly relative to the core in the period spanning 1978.3 to 1986.7 and up to 1992.3, favours an origin for the knots in terms of sporadic illumination effects rather than any of the alternatives mentioned earlier (Scarrott, Scarrott and Wolstencroft 1992, discuss this in full detail).

5 Hydrodynamical Models of Proto-Planetary Nebulae

Current theoretical ideas invoked to explain the various observational features which chart the transitory progression from AGB stars to PNs involve channelled mass outflows in the form of stellar winds of various velocities and densities. The two-wind hydrodynamical models (e.g Balick, Preston and Icke 1987, Icke Preston and Balick 1989, and references quoted therein), suggest that the initial wind from the evolving red giant is slow and dense and not necessarily highly non-anisotropic but a later fast, tenuous and channelled wind "ploughs" into the existing extended stellar envelope, the remnants of the slow wind, to produce the observed bipolar morphologies. The various forms of PPNs result from briefly glimpsing these envelope-wind interactions at different stages of evolution.

6 Models Derived from Polarization Observations

Our results presented for three PPNs allows us to propose geometries for each of the objects and then compare these geometries with the predictions of the two-wind models. Our proposed geometries are illustrated in fig. 7 and in each case we give a cross-section through the major axis of the nebula.

The model for the Frosty Leo includes the ubiquitous disc which totally obscures the central star at optical wavelengths. The form of the lobes in polarized light suggests that they are illumination rather than density bounded, i.e. there is no evidence for a cavity surrounding the star. This indicates that the envelope, presumably formed in the early red giant mass loss phase, has not been significantly disrupted by any later fast wind, assuming that it has commenced.

The "snowploughed" dusty walls in the polarized intensity images of MZ3 provides strong evidence for bubble-like cavities surrounded by an extensive dust envelope in this object. The model illustrates our proposed geometry. The bubbles are still confined and have not broken through the outer envelope. This implies that the fast wind is in operation but has not totally disrupted the envelope.

Evidence for cavities with snowploughed walls is also found in M2-9 but in this case the cavities are not confined and have broken out through the large scale envelope so that they take on the open- ended champagne glass shape. We do not find any extensive outer dusty envelope which probably means that it is so dispersed that any reflection nebulosity is too faint to detect in our data.

These geometries based on the polarization data fit neatly into the two-wind models where we see the various stage of wind interaction; in the Frosty Leo the secondary fast wind has not started to be effective, in MZ3 the wind is in the

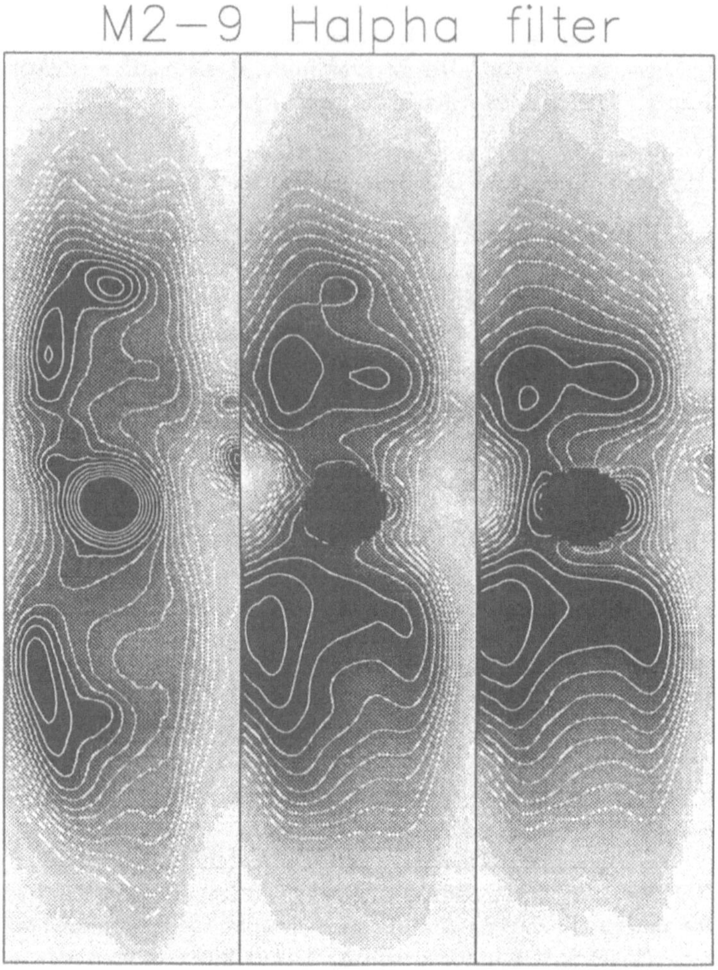

Fig. 6. Greyscale and contour maps of the bipolar nebula M2-9 in the light of the Hα emission line.
Left: Intensity image. Note the concentration of intensity on the eastern edges of the nebula and the remarkable reflection symmetry of the knotted structures in each of the nebular lobes about the equatorial (E-W) axis of the nebula.
Centre: Polarized intensity as measured and prior to correction for ISP. Note that the knotted structures in the intensity distribution are present in this image which might suggest they are polarized within M2-9 itself.
Right: As the centre frame but after correction has been made for the effects of the interstellar polarization. This image is basically free of the knotted structures indicating that they are seen by unpolarized intrinsic emission from the nebular lobes. This image illustrates the morphology of the reflection nebula associated with M2-9. The southern lobe shows a smooth champagne glass shape which we interpret as a cavity excavated in a larger envelope. The northern lobe is of similar form but is obscured by the circumstellar disc.

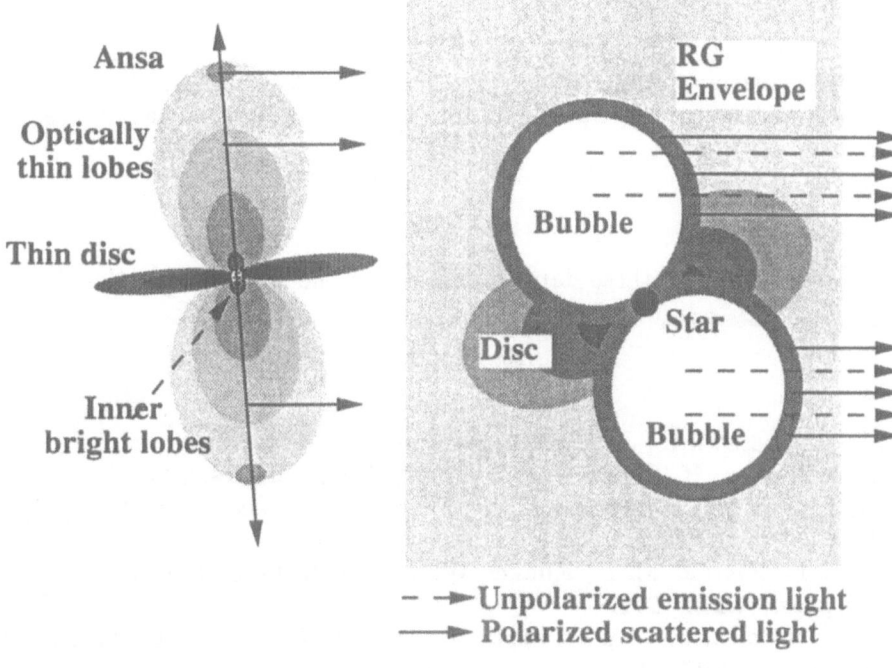

FROSTY LEO

Ansa

Optically
thin lobes

Thin disc

Inner
bright lobes

MZ3

RG
Envelope

Bubble

Star

Disc

Bubble

– –►Unpolarized emission light
——► Polarized scattered light

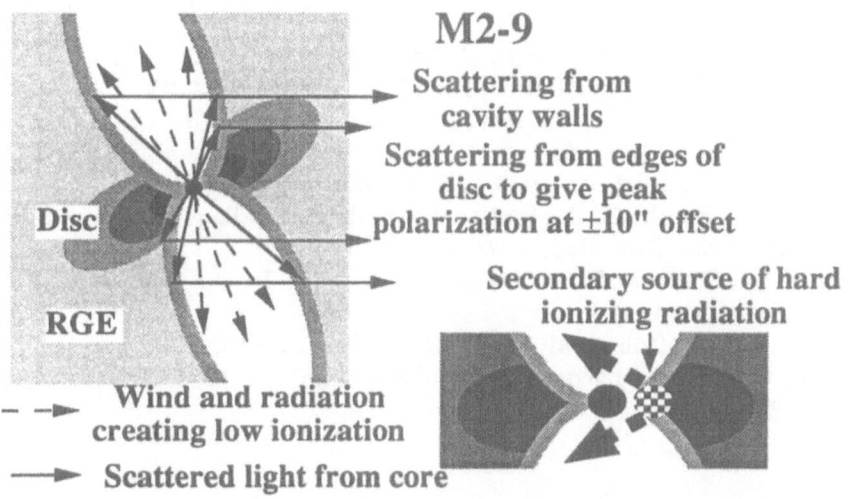

Cavity produced by fast wind

M2-9

Scattering from
cavity walls

Scattering from edges of
disc to give peak
polarization at ±10" offset

Disc

RGE

Secondary source of hard
ionizing radiation

– –► Wind and radiation
creating low ionization

——► Scattered light from core

Fig. 7.   Geometrical models for the three PPNs based on the interpretation of the polarization data. In each case the model shows a cross-section through the centre of the nebula with Earth to the right.
Top left: The Frosty Leo Nebula. In this case the system is illumination bounded and there is no evidence as yet for a secondary wind influencing the structure of the nebula.
Top right: The PPN MZ3. In this case the nebula is seen in a mixture of emission from a hot gas and reflected central starlight scattered in a dusty envelope. The inner emission-line lobes are seen as cavities excavated in the large scale stellar envelope which is seen as a surrounding reflection nebula.
Bottom: The PPN M2-9. In many ways this is similar to MZ3 but the cavities are no longer confined and have broken through the dusty envelope. The insert shows a possible source of the hard ionizing radiation which gives rise to the moving "knots"; this radiation arises from a secondary source which illuminates the cavity walls in such a manner than we get the reflection symmetry between the N and S lobes.

process of excavating bubbles in the preexisting envelope while in M2-9, the most evolved system, the cavities appear to have broken through the envelope and have dispersed it.

Acknowledgments

Observations leading to these data were made on the AAT during 1991/2

References

Allen, D. A. and Swings, J. P.: 1972, *Astrophys. J.* **174**, 583.

Balick, B., Preston, H. L. and Icke, V.: 1987, *Astron. J.*, **94**, 1641.

Carsenty, U. and Solf, J.: 1983, in D. R. Flower (ed), *Proceedings IAU Symposium 103, Planetary Nebulae*, Reidel, Dordrecht, p.510.

Cohen, M., FitzGerald, M. P., Kunkel, W., Lasker, B. M. and Osmer, P. S.: 1978, *Astrophys. J.* **221**, 151.

Cohen, M., Dopita, M. A., Schwartz, R. D. and Tielens, A. G. G. M.: 1985. *Astrophys. J.* **297**, 702.

Forveille, T., Morris, M., Omont, A. and Likkel, L.: 1987, *Astron. Astrophys.*, **176**, L13.

Goodrich, R. W.: 1991, *Astrophys. J.* **366**, 163.

Icke, V., Preston, H. L. and Balick, B.: 1989, *Astrophys. J.* **97**, 462.

King, D. J., Perkins, H. G., Scarrott, S. M. and Taylor, K. N. R.: 1980, *Mon. Not. Roy. Astr. Soc.*, **196**, 45.

Kohoutek, L. and Surdej, J.: 1980, *Astron. Astrophys.* **85**, 167.

Scarrott, S. M., Rolph, C. D., Wolstencroft, R. D., Walker, H. J. and Sechiguchi, K.: 1990, *Mon. Not. Roy. Astr. Soc.* **245**, 484.

Scarrott, S. M., Scarrott, R. M. J. and Wolstencroft, R. D.: 1992, *Mon. Not. Roy. Astr. Soc.* (in press).

Schmidt, D. G. and Cohen, M.: 1981, *Astrophys. J.* **246**, 444.

Rouan, D., Omont, A., Lacombe, F. and Forveille. T.: 1988, *Astron. Astrophys.*, **189**, L3.

IRAS 17423−1755: a BQ[ ] star with a variable velocity outflow

A. Riera
Departament de Física i Enginyeria nuclear, E.U.P.V.G., U.P.C. i Departament d'Astronomia i Meteorologia, Universitat de Barcelona, Av. Diagonal 647, E-08028 Barcelona (Spain)

P. García-Lario
L.A.E.F.F., Villafranca del Castillo, Apdo. de Correos 50727, E-28080 Madrid (Spain)

and

A. Manchado *
Astronomy Department, University of Illinois, 1002 W. Green st., Urbana Il 61801 (USA)

Abstract.
    IRAS 17423−1755 has been identified as a BQ[ ] star. Its more remarkable observed features are large radial velocities, huge line widths and double−peaked profiles arising from its knots, and a high−velocity *jet*. Ejection velocity variation can account for the observational properties detected.

1   Introduction

IRAS 17423−1755 has been identified as a new transition object between the post-AGB phase and the planetary nebula (PN) stage during an observational program of unidentified IRAS sources with far infrared colours similar to those of known PNe. We have performed B, V, R and Hα CCD images, and near-infrared and optical long slit spectroscopy.

    The spectrum of the core corresponds to a B star with strong emission lines of HI Balmer and Paschen series, HeI, FeII, [FeII], OI, CaII, and [CaII]. P-Cygni profiles of HI Balmer lines, HeI and FeII(42) provide direct evidence for a very strong mass loss. IRAS 17423−1755 shares characteristics with BQ[ ] stars (Be stars with forbidden emission lines and strong IR excess), which are believed to be hot post-AGB stars in an early stage of their evolution as proto-PNe.

2   Kinematics of the nebula

IRAS 17423−1755 is a bipolar nebula. Its most conspicuous features are two bow−shaped structures at the ends of the nebula (hereafter called knots) and two collimated structures (*jet* and *counterjet*) detected at a P.A. of 309°. Hα and [NII] emission lines arising from the *jet* are broad and show single−peaked profiles with very high radial velocities, which decrease with distance from the central source (from 870 km s^{-1} to 750 km s^{-1} respect to the central source in 3 arcsec).

    Both knots show large line widths ($\Delta V \sim 265$ km s^{-1}), double−peaked profiles and are moving outward at large radial velocities (~ 425 km s^{-1}), which together

---

* Instituto de Astrofísica de Canarias, E-38200 La Laguna, Tenerife (Spain)

Astrophysics and Space Science **216**: 43–44, 1994.
© 1994 *Kluwer Academic Publishers.*

A. Riera et al.

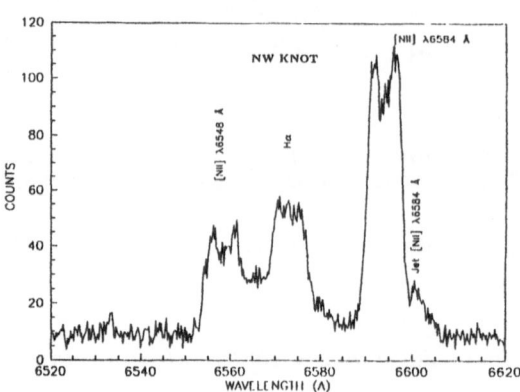

Fig. 1.   Hα and [NII]λ 6548, 6584 Å profiles arising from the NW knot

with the presence of the *jet* are similar to the observational properties of the H–H objects associated to young stars.

Discontinous variations of the radial velocity with position along the jet and the sudden drop in velocity at the knots, indicate that the velocity at which the material is ejected by the source increases. For post-AGB stars the existence of consecutive outflows episodes is suggested. To simplify, we approximate these episodic outflows by a linear increase in the velocity ejection, applying the one-dimensional models developed by Raga *et al.* (1990), who show that the linear increase in the velocity ejection supports the formation of discontinuities which are similar to the working surface of a *jet*. In this scenario, the knots would be the internal working surfaces of the *jet*.

Assuming the source was ejecting material at a constant velocity and that at a certain instant the ejection velocity begins to increase at constant rate, we attempt to fit the velocity–position diagram from the models to the available data.

Acknowledgements

We are grateful to A.C. Raga for his many valuable comments and suggestions. Based on observations collected at the European Southern Observatory (La Silla, Chile) and at the Observatorio del Roque de los Muchachos (La Palma, Spain). A.M. and P.G. have been partially supported by Spanish grant DGICYT PB 90-0570. A.R. has been supported in part by Spanish grant DGICYT PB 87-0371.

References

Raga, A.C., Cantó, J., Binette, L. and Calvet, N.: 1990, 'Stellar jets with intrinsically variable sources', *Astrophys. J.* **364**, 601.

Spectroscopic constraints on outflows from BN-type objects

Janet E. Drew
Department of Physics,
Nuclear Physics Laboratory,
Keble Road,
Oxford,
OX1 3RH.

Abstract.
    New high-quality high spectral resolution observations of the HI line emission from massive young stellar objects are described and discussed. It is proposed that two distinct physical components contribute to the observed emission. One of these is an optically-thick high-velocity stellar wind, the other a more slowly moving optically-thin volume of gas that may, in the case of S106IR at least, be caused by mass loading of the stellar wind. This decomposition is shown to resolve a long-standing problem regarding the relative widths of high and low opacity lines.

Key words: Infrared: stars – stars: early-type, formation – ISM: jets and outflows

1   Introduction

Infrared spectroscopy is now coming of age. The first spectrometers able to operate on astronomical telescopes at wavelengths between 1μ and 5μm were commissioned in a few years around 1980. Their design had to work around the limitation that the area detectors available to optical astronomers had yet to be developed for infrared use – hence the appearance of scanning Fabry-Perot and Fourier Transform spectrometers, alongside grating spectrometers such as UKIRT's CGS2 built around a 7-element detector. This has changed in that many telescopes now have 58×62 IR arrays and are upgrading to 256×256 devices. The possibility of optical-style spectroscopy has brought with it the opportunity for enormous improvements in data quality. High precision at high spectral resolution is now achievable. The subject of this paper, the outflows from deeply embedded massive YSOs, is a re-search topic that stands to gain much from this advance. The observations made of these objects with the first-generation spectrometers during the first half of the 1980s defined the problem – data from the second generation may go along way toward solving them. In this paper, some of the new observations of BN-type objects obtained with Cooled Grating Spectrograph No. 4 (CGS4) on UKIRT are presented and used to motivate a modified and, it is to be hoped, more fruitful interpretation.

    The prototype of the massive YSOs is the Becklin-Neugebauer Object in OMC-1, hence the term 'BN-type object'. These are extremely luminous (L $\gtrsim 10^4 L_\odot$) heavily-obscured (A_V upward of $\sim$15) sources found in regions of massive star formation. So far, none of them has been spatially-resolved at infrared wavelengths. They are just becoming resolvable in the radio domain (e.g. by the recently enhanced *MERLIN* network). An early review of their properties was written by

Astrophysics and Space Science **216**: 45–54, 1994.
© 1994 *Kluwer Academic Publishers.*

Wynn-Williams (1982).

Their most intriguing characteristic is the evidence of outflow at rates over an order of magnitude higher than those sustained by late-O, early-B main sequence stars of comparable luminosity. That this is surprising has much to do with the supposition massive YSOs evolve so rapidly on to the main sequence that it is unlikely any known examples are still approaching it. Compared with normal OB stars, BN-type objects are strong radio sources for their bolometric luminosities but, like normal OB stars, their radio spectral indices fall in the range that is consistent with thermal emission from the $\sim 1/r^2$ density distribution associated with the far reaches of a stellar wind (i.e. $F_\nu \propto \nu^\alpha$ where $\alpha \gtrsim 0.6$). These measurements fix $\dot{M}/v_\infty$, while observed spectral line widths at IR wavelengths set lower limits on v_∞. Mass loss rates derived in this way are of the order of 10^{-6}–10^{-7} $M_\odot\,yr^{-1}$ (Simon *et al.* 1983).

The spectrum of a BN-type object in the infrared is typically made up of a strong highly-reddened continuum and velocity-broadened HI recombination line emission. The contrast of the line emission against the continuum varies greatly from object to object: in some (e.g. GL 2591) the line emission is very difficult to detect while in others (most notably S106IR) the strongest lines have peak fluxes several times that in the adjacent continuum. Another source of diversity within the class is to be found in the estimates of line flux ratios: there are cases where the conclusion that the HI lines are optically-thick is inescapable and examples where Recombination Case B ratios are approached. Chemical species other than hydrogen have been detected in one or two objects at more modest levels of emission (see Scoville *et al.* 1983 and Simon & Cassar 1984). The observed line-widths range from ~ 100 km s^{-1} (FWHM) up to ~ 300 km s^{-1}. If these velocities are representative of the terminal velocities attained, then we have another problem: main sequence OB stars of the same luminosity achieve much higher terminal velocities in the region of 1000 km s^{-1} or more.

The 'surprisingly high' mass loss rates and 'surprisingly low' outflow velocities have provoked two distinct interpretations. On the one hand, line profile synthesis models have been explored that treat the outflows as simple stellar winds from what are, in effect, bloated low-gravity stars (e.g. Felli *et al.* 1984). On the other, remnant protostellar disks have been blamed for the phenomena and analogies have been drawn with the barely less mysterious classical Be stars (e.g. Persson, McGregor & Campbell 1988). In the presence of what can now be seen as primitive data, neither interpretation could move to the point of decisive testing.

A further obscurity of this subject is the relation between BN-type objects ('outflow sources') and ultra-compact HII regions. The latter should distinguish themselves from the former by (i) exhibiting much narrower, optically-thin IR line emission (velocity widths $\lesssim 30$ km s^{-1}), (ii) yielding radio spectral indices characteristic of HII region emission ($F_\nu \propto \nu^\alpha$ where $\alpha \sim 0$ in the optically-thin limit, or $\alpha \sim 2$ for optically-thick emission). This distinction is not always a clean one. For example, the source LkHα 101 (for which A_V happens to be low enough to allow the object to sneak into the Herbig Ae/Be star category) has an IR spectrum compatible with an HII region classification but a radio spectrum that fits well to the outflow model. Is this an outflow source or is it not? Even without this kind of confusion there remains the larger question regarding the evolutionary connection within and between the two groups, if any. Evolution or environment? Nature or nurture?

2   The line width problem

Early in the 1980s, Simon and collaborators set the basis for the interpretation of the IR HI line profiles and flux ratios (Simon *et al.* 1981 and 1983 being the key˙ publications). Some obvious, but important, points were recognised. In particular, it was noted that optically-thicker emission lines must be as broad or broader than less opaque lines if both are produced within an outwardly accelerating radial outflow. For example, Brα should be at least as broad as either Brγ or Pfγ. The IR spectra themselves turned out to be rather uncooperative in this regard. A striking illustration of this is to be found in the observational study of Persson *et al.* (1984, see their Figure 4): they provided examples of comparisons between Brα and Brγ line profiles in the spectra of a number of objects that both conformed with and reversed the expected pattern.

A related problem emerged in attempting to fit observations of the Brα and Pfγ line profiles in the spectrum of S106IR (Bunn & Drew 1992; the observations being due to Garden & Geballe 1986). It was found that the ratio between the line fluxes was very hard to reconcile with the constraint on the mass loss rate provided by radio flux measurements (Felli *et al.* 1984). The observed ratio was suspiciously close to the Recombination Case B value, yet the radio flux indicated an optically-thick wind that should yield relatively suppressed Brα emission. And of course the FWHM of the weaker, less opaque Pfγ line was larger than that of Brα. There seemed to be little reason to imagine that a simple accelerating outflow could give rise to the observed line emission from S106IR.

3   The CGS4 observing programme

The commissioning of CGS4 on UKIRT was an opportunity not to miss. Used in echelle mode this instrument currently offers spectral resolutions of either $\sim$40 km s^{-1} or $\sim$20 km s^{-1} depending on the choice of camera, together with great sensitivity. The statistical noise in much of the data to be discussed here is under 1 percent. Comparable resolving powers have been available before – the advances are in the instrumental line profile and the greatly reduced impact of changing weather. In 1991 a programme of reobservation of the well-known massive YSOs using CGS4 with echelle was started. Its aim is to obtain the high quality well-resolved IR line profiles that can at last, through comparison with model predictions, provide the firm constraints on the outflow kinematics and geometry that have hitherto been beyond reach. Full details of the observations and data extraction are to be found in Bunn (1992). To date, observations of seven massive YSOs have been collected. At the time of writing, a paper on S106IR is in press (Drew, Bunn & Hoare 1993) and another bringing together results on the other sources is in the draft stage (Bunn, Hoare & Drew 1994).

It is inevitable that pride of place first goes to observations of S106IR, the exciting source of the bipolar HII region Sh-2 106. This object appears to be one of those rare gifts of nature that is set up to make things easy for us. Unlike other members of the BN-type class, its orientation is believed to be known. It has been shown by Solf & Carsenty (1982), from optical long-slit spectroscopy of the relatively unobscured HII region, that the ionized gas in either lobe is expanding away from the dust bar containing S106IR at 75 km s^{-1} and that the outflow axis lies almost in the plane of the sky ($i \simeq 75^{O}$). This allows us to assume our sightline to S106IR lies almost in its equatorial plane. Other useful characteristics include the high contrast of its IR line emission relative to the underlying continuum and the independent constraints upon its ionizing flux that have been derived from optical and radio observations of the HII region (O9 is the favoured spectral type, Staude *et al.* 1982; Bally, Snell & Predmore 1983). For this object, a particularly extensive set of observations were made. The line profiles obtained included Brα, Pfγ, Brγ and HeI 2.058μm.

The data on other, perhaps more typical, members of the class is as yet more limited. For most, Brα and Brγ line profiles have been obtained. Observations made at the lower spectral resolution of $\sim$40 km s^{-1} are not as high quality as those made at $\sim$20 km s^{-1}. The reasons for this are poor weather and incomplete removal of a low amplitude ripple due to the order-sorting CVF used with the echelle.

4   Results

4.1   S106IR

In presenting their observations of S106IR, Garden & Geballe (1986) noted that HeI ($5^{3,1}$G–$4^{3,1}$F) 4.049μm emission could be seen shortward of the Brα line. It was this that prompted us to attempt an observation of the HeI (2^1S–2^1P) 2.058μm transition at $\sim$20 km s^{-1} spectral resolution. To our surprise, the line turned out to be a pure absorption feature on-source (Figure 1). Along the slit away from S106IR, narrow nebular emission associated with the HII region replaces the absorption. The on-source profile is clearly skewed toward negative velocities implying formation in an outflow that is most likely to originate from close to the continuum source. This feature is the most direct spectroscopic evidence to date of the stellar wind apparently demanded by the observed radio spectrum.

The lack of significant emission in the on source HeI 2.058μm line suggests that the He$^+$ fraction in the outflow may be small. This is at variance with the observation of HeI emission off-source and indeed with the general excitation of the

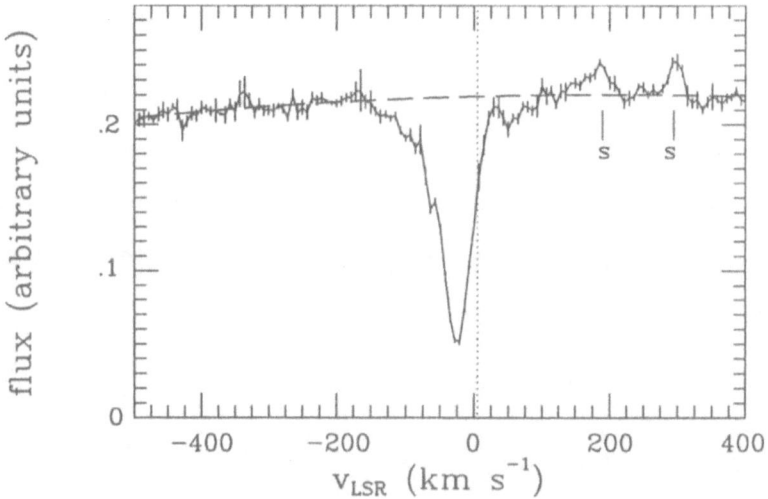

Fig. 1. HeI (2^1S–2^1P) 2.058μm line profile in the spectrum of S106IR. The vertical dotted line marks the systemic velocity of S106IR. The features marked with an 's' are due to uncorrected absorption lines in the standard star spectrum used to remove telluric absorption.

HII region. It seems that the polar and equatorial views of S106IR differ markedly in terms of the apparent ionizing flux.

In section 2, it was noted that it has been found difficult to reconcile the observed HI line profiles with formation in a simple accelerating stellar wind. The new CGS4 data provide a very strong hint as to the cause of the trouble. Earlier observations have for different reasons failed to define the high velocity line wings adequately. Now it is possible to see that the line wings do obey the expected width relation in that the blue wing of Brα extends to higher velocity than it does in Brγ (Brα HWZI = 340$\pm$10 km s^{-1}, Brγ HWZI = 270$\pm$10 km s^{-1}). This is shown in Figure 2 in the form of the Brα/Brγ flux ratio as a function of velocity. The red wings of the two lines are comparable in extent. This can also be reconciled with formation in an accelerating stellar wind: free-free opacity can be significant at $\sim 4\mu$m and thus able to erode the red wing of Brα, while it is negligible at $\sim 2\mu$m, leaving Brγ intact. Indeed the measured radio flux from S106IR can be used to show that it is quantitatively plausible that the free-free optical depth at $\sim 4\mu$m is about unity (for full details see Drew *et al.* 1993).

It is directly apparent from Figure 2 that the Brα/Brγ flux ratio is much higher in the line core than in the wings and indeed it comes close to the optically-thin nebular recombination Case B value for the best estimate of the reddening ($A_V \sim$ 21: Eiroa, Elsasser & Lahulla 1979). To explain the combination of optically-thick line wings (F(Brα)/F(Brγ) $\sim$ 1) and an optically-thin line core (F(Brα)/F(Brα) $\sim$ 2.7), one must appeal either to decelerating outflow or to the superposition of two physically-distinct components. Given the simple 'windy' appearance of the HeI

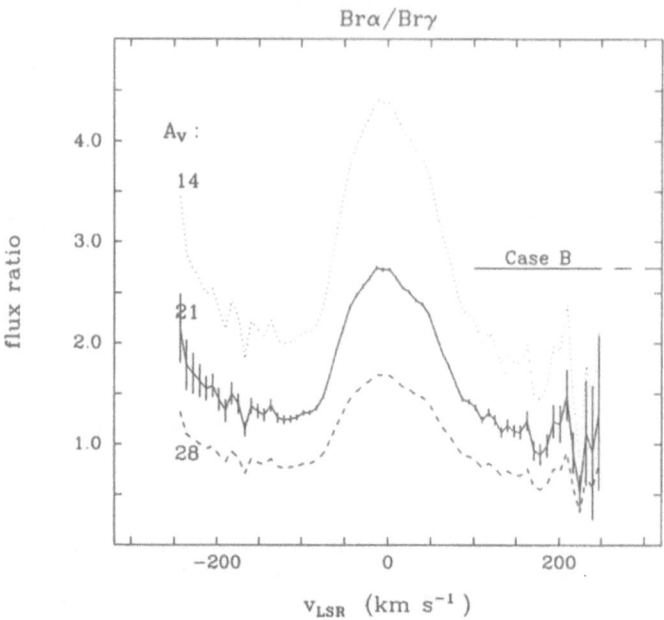

Fig. 2.   The Brα/Brγ flux ratio as a function of velocity derived from the spectrum of S106IR.
The three curves shown were derived for different reddenings. Eiroa, Elsasser & Lahulla (1979)
obtained $A_V = 21$ with a probable uncertainty of ±5 from far-red and near-IR colour measurements. The expected ratio for Recombination Case B (~2.7) is also marked. Note that the
flux ratio for velocities less than -100 km s^{-1} increases with increasingly negative velocity – as
expected for an accelerating stellar wind yielding broader Brα than Brγ.

2.058μm line profile and the existence of the large-scale bipolar flow, the second
option is more appealing. The high-velocity HI profile wings are due to the stellar
wind, while the bright line cores are produced in a 'nebular' region that is the
beginnings of the more sluggish spatially-resolved flow ($v_{exp} \simeq 75$ km s^{-1}, Solf
& Carsenty 1982). This allows us to understand why the Brγ FWHM is greater
than that of Brα – it is a consequence of the greater fraction of the Brγ flux
attributable to the broader stellar wind component (reflecting the fact that the
Brackett line decrement is substantially higher for the nebular component than for
the optically-thick wind).

   The picture of S106IR that emerges from the new data is of an equatorially-
flattened wind driven from the vicinity of the IR continuum source ($\dot{M} \gtrsim 2.7 \times 10^{-6}$
M$_\odot$ yr^{-1} , $v_\infty \gtrsim 340$ km s^{-1}), and a driven HII region expanding at a velocity comparable with that measured from the optically-resolved bipolar flow. Needless to
say, this interpretation raises at least as many questions as it has answered. Here
there is only space to deal with one of the simplest – is S106IR the only example
for which it is necessary to deconvolve the observed HI line spectrum into two
physical components?

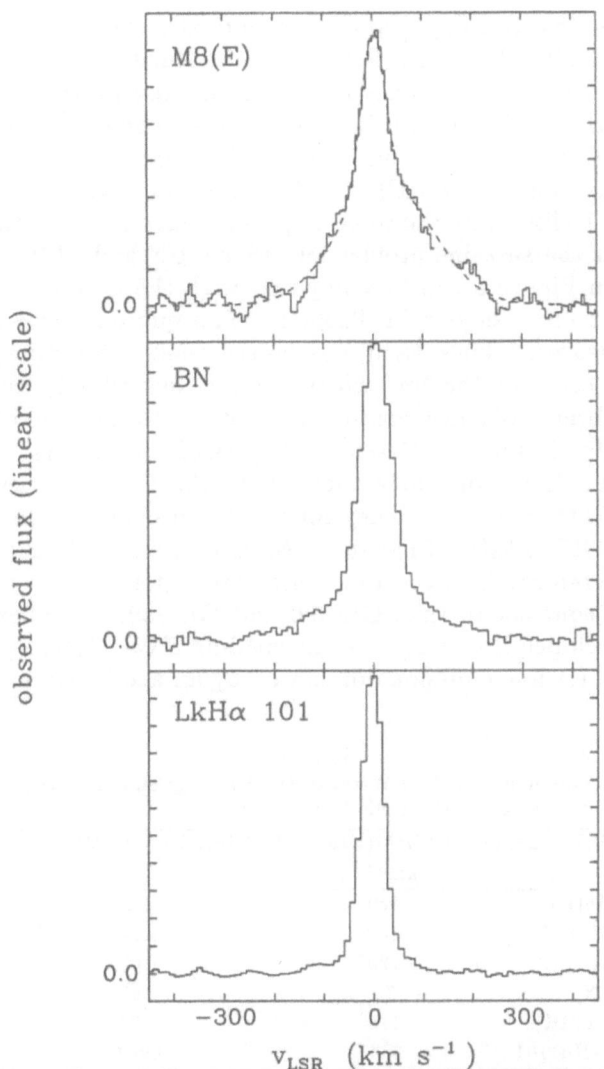

Fig. 3. Observed Brγ line profiles for the sources M8(E), BN and LkHα 101. All were obtained using CGS4 in echelle mode. The spectral resolution is $\sim$20 km s^{-1} for M8(E), and $\sim$40 km s^{-1} for the other two objects. The dashed line superimposed on the M8(E) profile is a multiple gaussian fit included to provide continuity through the gaps caused by deleted telluric features. Note the gradually decreasing contrast of the wing component.

4.2 APPLICATION OF THE TWO-COMPONENT MODEL TO OTHER SOURCES

If it is true that BN-type IR line spectra are composite in the same sense as that of S106IR, we can immediately define two limiting cases. These will be the pure optically-thick stellar wind limit (FWHM[Brα]/FWHM[Brγ] $\geq$ 1), and the opposite extreme of a seemingly unalloyed optically-thin, yet rapidly expanding HII region (FWHM[Brα]/FWHM[Brγ] = 1). Intermediate cases closer to the HII region-dominated limit are the ones for which the line-width ratio will fall below unity, as in S106IR. Of the objects for which CGS4/echelle data have already been obtained, M8(E) provides good examples of line profiles containing an obvious stellar wind component, while LkHα 101 is the most convincing case of an almost pure HII region. To illustrate the trend in profile shape between the limiting cases, Figure 3 contains the Brγ line profiles for M8(E), BN and LkHα 101.

The profiles in Figure 3 can be compared with the data on the Brα and Brγ FWHM measurements set out in Table 1. The quoted line widths have been rounded to the nearest 5 km s^{-1}and may be regarded as no more than 10 percent uncertain. The error in the line width ratio is accordingly no worse than 15 percent. In the absence of a new measurement of Brα in the spectrum of LkHα 101, the measurements of Simon & Cassar (1984, based on 18 km s^{-1}resolution data) are used instead. It is noticeable that where the high velocity line wings are poorly-developed, there is a tendency for the FWHM of Brα to drop below that of Brγ (BN, S106IR, LkHα 101). In these cases we may blame the contrast in Brackett line decrement between the constituent 'stellar wind' and 'HII region' components. In contrast to this, GL 490 and GL 989, resemble M8(E) both in showing well-developed line wings and in yielding FWHM(Brα)/FWHM(Brγ) > 1, as appropriate for line emission dominated by an accelerating outflow.

TABLE I
Measurements of the FWHM of the Brα line and the Brα/Brγ line-width ratio for six BN-type objects

Object name	FWHM(Brα) (km s^{-1})	FWHM(Brα)/FWHM(Brγ)
M8(E)	150	1.5
GL 989	150	1.25
GL 490	120	1.09
BN	70	0.88
S106IR	115	0.77
LkHα 101	38[+]	0.90[+]

[+]The figures for LkHα 101 are from the study of Simon & Cassar (1984).

Only one object for which CGS4 echelle data are available at the time of writing is found to challenge this simple decomposition of the HI line profiles into two components. It is M17(SW) IRS1. Even when the Brα and Brγ line profiles are overplotted on an absolute flux scale, the high velocity wings present in the Brγ profile are wider and stronger than in both Brα and Pfγ (3.75μm). Given that the Pfγ wings are *not* wider than the Brα wings, a likely explanation for the greater strength of the wings in Brγ is reflection by dust. In other words, our sightline into M17(SW) IRS1 is not direct. This is a possibility that it is now feasible to examine by means of high spectral resolution spectropolarimetry.

5   Conclusions

The high-quality echelle spectroscopy that can now be performed in the infrared is beginning to allow full use of observations of emission line profiles for the task of extracting the physical and kinematic information they surely contain. From the data gathered so far, it has been argued that the line emission from the spatially-unresolved luminous sources known as BN-type objects is due to a mix of (i) optically-thick stellar wind and (ii) a more slowly moving optically-thin volume of gas that, in some extreme objects can dominate the HI line emissivity. This is not an entirely new idea – Simon & Cassar (1984) pointed out that two distinct regions of HI line emission were needed in order to produce the observed Brγ emission in LkHα 101 (Simon & Cassar 1984). It is the generalisation of the idea, made possible by the great improvement in data quality, that is the step forward.

It is not immediately obvious just what the optically-thin component must be, since the velocity widths seem to be too large to allow identification with conventional HII regions expanding by means of thermal over-pressure. In the case of S106IR, momentum conservation allows that it may be a mass-loaded flow (cf. Hartquist *et al.* 1986) resulting from the percolation of the stellar wind through interstellar debris left over from the star formation process. It also remains unclear just what the stellar wind really is. In S106IR it is equatorially-flattened and, in objects where radio flux measurements provide mass loss rate estimates, it seems the outflow is much denser than is typical of main sequence stars. Could these outflows be fed from a reservoir of marginally-bound material in a circumstellar disk? A troublesome fact in this and some other respects is the failure so far to find any clear spectroscopic signature of rotational motion.

As well as posing new questions, the two-component model offers the prospect of settling a very basic and thorny one – namely, what is the total extinction towards each object individually? It is usually the case that A_V is highly uncertain. Greater certainty would make it possible to apply a much wider range of diagnostic tools in the analysis of these objects. Interestingly, it can be seen from the line flux ratio plot for S106IR (Figure 2) that the hypothesis the narrower core component is optically-thin would indicate an A_V a little in excess of 21. This happens to be in good agreement with a pre-existing estimate based on a continuum method. The precise value implied by Figure 2 depends on how the Brα/Brγ flux ratio in the stellar wind component is interpolated through line centre. This points the way towards an improved technique for determining reddenings: namely, line profile modelling, aimed at defining the likely characteristics of the stellar wind emission near line centre, followed by deconvolution of the observed velocity-dependent Brα/Brγ flux ratio. In practise, this is likely to be a mildly iterative process.

To make progress on these issues, there is much more that can be done both at the telescope and on the office workstation. The new era in the infrared has only just begun. It is an exciting time to return to the study of this most elusive subset of the young stellar population.

Acknowledgements

The recent work in Oxford directed toward understanding massive YSOs has been very much a team effort. Accordingly, it is a pleasure to thank Melvin Hoare for the regular exchange of ideas, comment and criticism over the past few years and Jenny Bunn for her expertise and tireless efforts at the CGS4 data face. All the new data discussed in this article were obtained at the United Kingdom Infrared Telescope. JED presently holds a Science & Engineering Research Council Advanced Fellowship.

References

Bally, J., Snell. R.L., Predmore, R., 1983, *Ap. J.*, **272**, 154
Bunn, J.C., D.Phil.Thesis, Oxford University
Bunn, J.C., Drew, J.E., *Mon. Not. R. astr. Soc.*, **255**, 449
Drew, J.E., Bunn, J.C., Hoare, M.G., 1993, *Mon. Not. R. astr. Soc.*, in press
Eiroa, C., Elsasser, H., Lahulla, J.F., 1979, *Astr. Ap.*, **74**, 89
Felli, M., Staude, H.J., Reddmann, T., Massi, M., Eiroa, C., Hefele, H., Neckel, T., Panagia, N. 1984, *Astr. Ap.*, **135**, 261
Garden, R.P., Geballe, T.R., *Mon. Not. R. astr. Soc.*, **220**, 611
Hartquist, T.W., Dyson, J.E., Pettini, M., Smith, L.J., 1986, *Mon. Not. R. astr. Soc.*, **221**, 715
Persson, S.E., Geballe, T.R., McGregor, P.J., Edwards, S., Lonsdale, C.J., 1984, *Ap. J.*, **286**, 289
Persson, S.E., McGregor, P.J., Campbell, B., 1988, *Ap. J.*, **326**, 339
Scoville, N., Kleinmann, S.G., Hall, D., Ridgway, S.T., 1983, *Ap. J.*, **275**, 201
Simon, M., Cassar, L., 1984, *Ap. J.*, **283**, 179
Simon, M., Felli, M., Cassar, L., Fischer, J., Massi, M., 1983, *Ap. J.*, **266**, 623
Simon, M., Righini-Cohen, G., Fischer, J., Cassar, L., 1981, *Ap. J.*, **251**, 552
Solf, J., Carsenty, U., 1982, *Astr. Ap.*, **113**, 142
Staude, H.J., Lenzen, R., Dyck, H.M., Schmidt, G.D., 1982, *Ap. J.*, **255**, 95
Wynn-Williams, C., 1982, *Ann. Rev. Astr. Astrophys.*, **20**, 587

First Wavelet Analysis of Emission Line Variations in Wolf-Rayet Stars
Turbulence in Hot-Star Outflows

A.F.J. Moffat and S. Lépine
Département de Physique, Université de Montréal, C. P. 6128, Succ. A , Montréal, QC, H3C 3J7, Canada, and Observatoire du mont Mégantic

R.N. Henriksen
Department of Physics, Queen's University, Kingston, ON, K7L 3N6, Canada

and

C. Robert
Space Telescope Science Institute, 3700 San Martin Dr., Baltimore, MD, 21218, USA

Abstract. The quantification of stochastic substructures seen propagating away from the centers of emission lines of Wolf-Rayet (WR) stars is extended using the powerful, objective technique of wavelet analysis. Results for the substructures in one WR star so far show that the scaling laws between (a) flux and velocity dispersion and (b) lifetime and flux, combined with (c) their mass spectrum, strongly support the hypothesis that we are seeing the high mass tail-end distribution of full-scale supersonic compressible turbulence in the winds. This turbulence sets in beyond a critical radius from the star and shows remarkable similarity to the hierarchy of cloudlets seen in giant molecular clouds and other components of the ISM.

The velocity dispersion is larger on average for substructures (interpreted as density enhanced turbulent eddies) propagating towards or away from the observer, suggesting that the turbulence is anisotropic. This is not surprising, since the most likely force which drives the wind *and* the ensuing turbulence alike, radiation pressure, is directed outwards in all directions from the star. It is likely that a similar kind of turbulence prevails in the winds of all hot stars, of which those of WR stars are the most extreme.

The consequences of clumping in winds are numerous. One of the most important is the necessary reduction in the estimate of the mass-loss rates compared to smooth outflow models.

Key words: Stars: Hot, Wolf-Rayet — Turbulence

1   Introduction

Wolf-Rayet stars are the epitome of hot, massive stars with sustained winds. With mass-loss rates $\dot{M} > 10$ times those of their O-star progenitors, the slower inner parts of WR winds are optically thick in all but some of the cooler WN sequence stars. This circumstance explains why WR stars, unlike O-stars, exhibit strong emission lines in the visible part of the electromagnetic spectrum. Thus, the properties of WR winds can be readily studied in the *visible* using modern, high spectral resolution, high S/N techniques.

Until the launching of the Hubble Space Telescope, UV spectroscopy of the very strong resonance emission lines, even in O-stars, was limited to low S/N using IUE. Nevertheless, IUE UV spectroscopy has been very useful in revealing strong variability mainly of the *absorption* components of P Cygni lines formed in the intervening column of the wind between the observer and the projected stellar

Astrophysics and Space Science **216**: 55–65, 1994.
© 1994 *Kluwer Academic Publishers.*

disk (St-Louis 1990).

Optical spectroscopy of various subordinate emission lines in WR stars allows a more global view of the wind variability. This is important in order to distinguish between shells and clumps and to make meaningful statistical studies. This paper is a first report on the application of the new technique of wavelet transforms to quantify this variability. It represents an extension of an initial spectroscopic study of WR wind variability by Moffat *et al.* (1988), carried further by Robert (1992; cf. also Moffat & Robert 1992).

2   Data Base

As described elsewhere (Robert 1992; Moffat & Robert 1992), we have at our disposal some 20–30 regularly repeated spectra on 3–4 consecutive nights for each of 9 WR stars of various subclasses in both the WN and WC sequences. These optical spectra were obtained in the yellow at CFHT (coudé) in 1987 and 1988, and ESO (Cassegrain échelle) in 1989. They have a 2-pixel resolution of ~ 0.2Å, S/N ~ 300 and are spaced out typically in ~ 1 hour intervals for each star. We have concentrated our studies so far mainly on the best isolated emission lines of HeII 5412 in the WN stars and CIII 5696 in WC. Fig. 1 shows an example montage of one night's data for the first star selected for wavelet analysis in this paper: WR 135 (subtype WC8).

3   Multi-Gaussian Analysis

A first attempt to quantify the emission–line substructures seen in HeII 5412 (WN) and CIII 5696 (WC) was made using multi-gaussian fits — up to some 20 components simultaneously— to each line, after subtracting off the minimum base profile (Robert 1992). Each substructure (cf. Fig.1) is characterized by three parameters: f, its flux expressed in continuum units times kms^{-1}; σ_v , its velocity dispersion (= HWHM of the fitted gaussian); and v_c , its central Doppler velocity. (Later, for the wavelet analysis, a fourth parameter was added: the lifetime τ .) The main results of this first study (cf. Moffat & Robert 1992), typical for each of the 9 WR stars, are as follows:

- Subpeaks appear at random (spatially i.e. in Doppler velocity, and temporally) and are seen at all times.

- Subpeaks move away from line center (thus accelerate) with time; allowing for projection angles, this behavior is similar to the discrete absorption components (DAC) seen propagating in UV P Cygni absorption edges.

- The emission line variability at various positions on a line $\sigma(\lambda)$, is proportional to the mean line profile $I_l(\lambda)$.

- The absorption components of P Cygni lines are more variable than the emission components.

- The relative global line variability increases for lines of lower ionization level; the lines of highest ionisation show no detectable variations.

- The average number of detected subpeaks for a given line is proportional to the wind terminal velocity, i.e. WR stars with faster winds tend to show proportionately more subpeaks. (Perversely, this makes WR stars with faster winds less variable in global line profile as well as continuum photometry and polarisation, due to Poisson statistics.)

- The velocity widths are larger for subpeaks that appear at either edge of the emission line ($\sigma_v \sim 50 km s^{-1}$), compared to those at line center ($\sigma_v \sim 150 km s^{-1}$). (We no longer believe that this is due to the radial velocity gradient of the wind alone, since unrealistic stretching would be required.)

- Down to the noise limit, the subpeak flux frequency spectrum shows a power law with rapidly increasing numbers of subpeaks with smaller flux: $N(f)df \sim f^{-2.4\pm0.2}df$ on average for 7 WR stars with reliable power-law slopes.

These findings are compatible with anisotropic, supersonic turbulence of a compressible medium (Moffat & Robert 1992) — although they do not prove it. As such, the observed subpeaks would represent individual eddies propagating outwards in the wind. It is possible that, beyond a certain critical radius ($\sim$ several stellar radii), the whole wind is in a state of full-scale turbulence!

4 Wavelet Analysis

The discrete, isolated nature of the line subpeaks lends itself in a natural way to analysis by wavelet transforms (cf. Farge 1992) as opposed to Fourier transforms, which are more appropriate for periodic phenomena. This has been done successfully for the first time in the case of cloudlets in a GMC by Gill & Henriksen (1990: hereafter GH): they carried out a wavelet analysis of a velocity contour map in ^{13}CO for the star-forming region L1551.

It is expected that, compared to multi-gaussian analysis, the convolution of the WR spectral data by an appropriate wavelet function should lead to greater objectivity and detectability, hence more robust statistics of the subpeaks. We follow GH's example in choosing as wavelet transform function a so-called Mexican Hat form, which is essentially the second derivative of a Gaussian. Such a function is simple, well-behaved, roughly similar in shape to the subpeaks being analyzed and satisfies the admissibility criterion of wavelets, namely its average over velocity is zero. The normalized Mexican Hat transformation (convolution) of the original function $I(x)$ is written as (cf. Daubechies 1992):

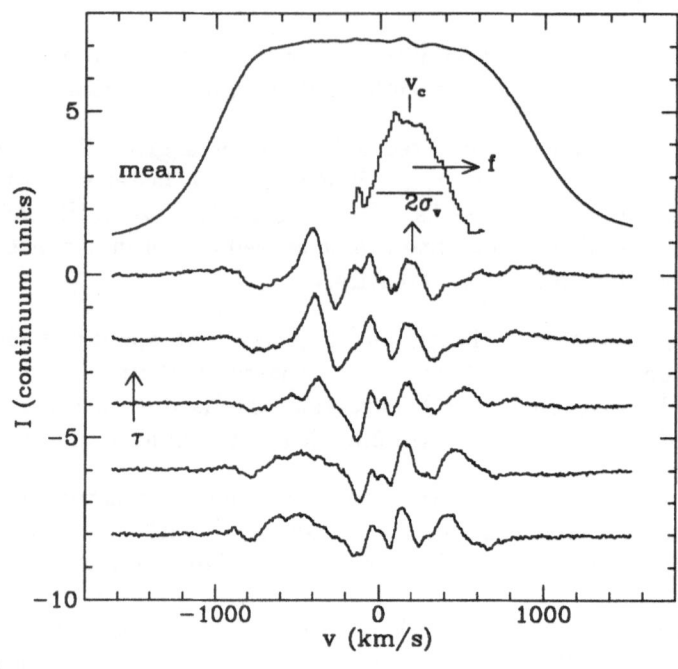

s

Fig. 1.   Montage of one night of spectra for the CIII 5696 line of WR135. All 26 spectra are combined in the mean, with scale at left. The individual differences from the mean spectra are expanded in vertical scale by a factor two. The time difference between adjacet spectra is ~1 hour ±5 %. A zoom of one feature is shown with its derived parameters.

$$(T^{wav}I)(\sigma_v, v_c) = \sigma_v^{-1/2} \int g[(v - v_c)/\sigma_v]I(v)dv,$$

where

$$g(x) = (1 - x^2)exp(-x^2/2).$$

Note that the Mexican Hat function $g(x)$ is non-orthogonal, i.e. some cross-talk will prevail between nearby subpeaks of different width. Nevertheless, as noted by Langer *et al.* (1993), such non-orthogonal functions are often better suited than orthogonal ones.

So far, we have applied the wavelet technique to only one WR star, WR135 = HD192103 (WC8), in which the strong, variable line of CIII 5696Å is most suited. Fig.2 illustrates the wavelet transforms for one spectrum of this star after subtracting off the 4-night mean profile. Note that for a given σ_v, the wavelet

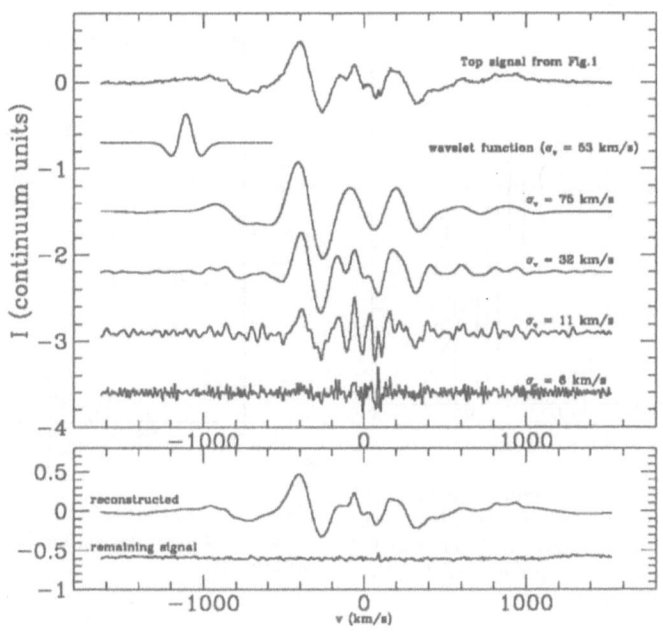

Fig. 2.   Illustration of the wavelet analysis on the top WR135 spectrum in Fig.1 . The wavelet transform is shown for four values of σ_v.

function acts like a filter, sensitive only to subpeaks of nearly the same width as the wavelet itself. Individual subpeaks are then extracted one by one, by taking the values of σ_v, f and v_c where the transform amplitude is a maximum. This was done iteratively, i.e. the extracted wavelet was subtracted from the original profile and the process repeated, until a uniform residual noise level remained (cf. Fig.2). In this way, we recovered ~twice as many subpeaks in the same spectra as previously using multi-gaussian fitting.

Applied to all 26 spectra of WR135, we identified some 1000 individual sub-peaks. Many of these, especially the bigger ones, are seen only slightly modified in sequential spectra, leaving a total of 108 identified unique subpeaks over the consecutive 4 nights. Based on these 108 subpeaks, we find the following results:

- A flux power law spectrum, $N(f) \sim f^{-1.8\pm0.1}$ (cf. Fig.3) (With the same data, the multi-gaussian analysis of Robert (1992) gave $N(f) \sim f^{-1.7\pm0.5}$.)

- Scaling law between flux and velocity width: $f \sim \sigma_v^{2.0\pm0.4}$ (cf. Fig.4).

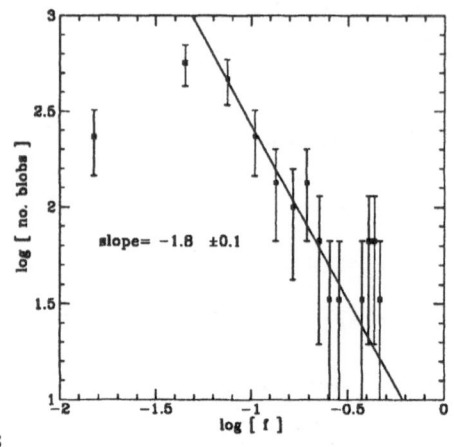

S

Fig. 3.   Flux power spectrum for WR135's subpeaks. The straight line is a weighted least squares fit, ignoring the first two points at low flux. Open symbols refer to points omitted from the fit due to biased statistics.

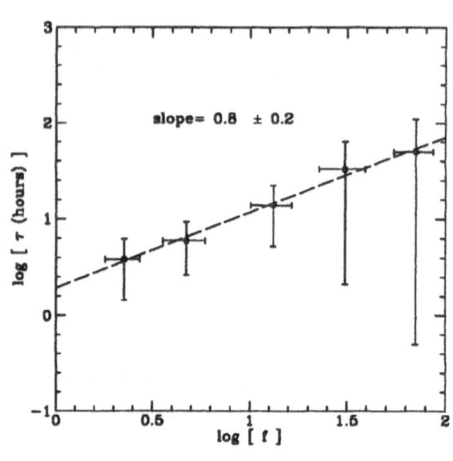

S

Fig. 5.   Lifetime vs. flux for WR135's subpeaks.   The straight line is a weighted least squares fit.

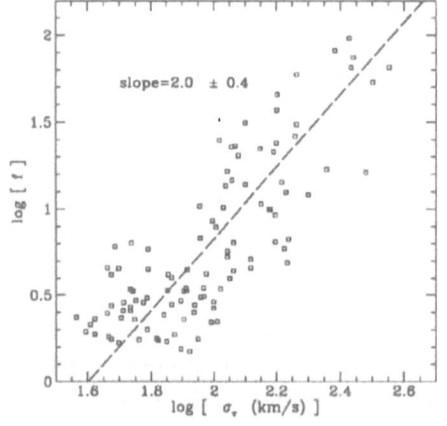

S

Fig. 4.   Flux vs. velocity dispersion for WR135's subpeaks. The straight line is an unweighted least squares fit.

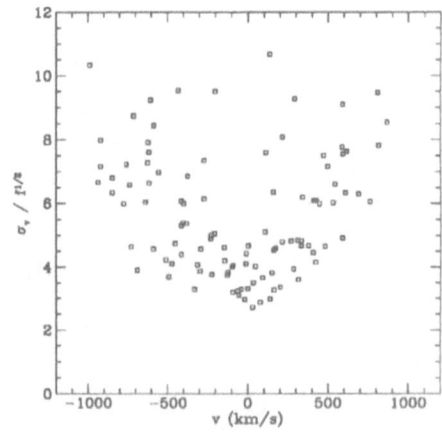

S

Fig. 6.   Variation of normalized velocity dispersion with position on the line.

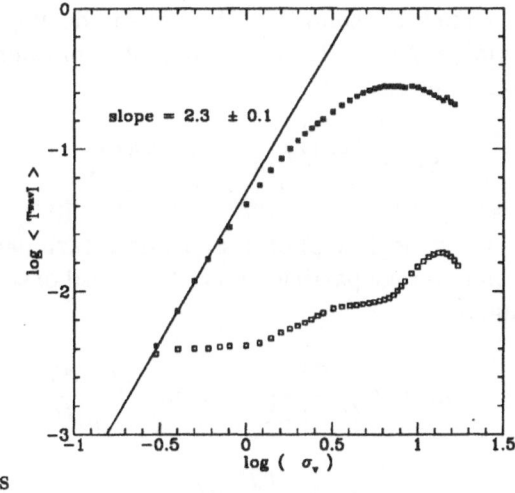

Fig. 7.   Squared amplitude averaged over position of the wavelet transform (filled symbols) as a function of velocity dispersion after subtracting off the background signal (open symbols) based on analysis of the continuum. The asymptotic slope is the fractal dimension of the signal.

- Scaling law between lifetime and flux: $\tau \sim f^{0.8\pm0.2}$ (cf. Fig.5). Although subpeak lifetimes are generally longer than one night's data, we can estimate $\tau(f)$ from the total number of subpeaks and the number of times one sees subpeaks appear or disappear (assumed to occur with constant frequency) for a given range of flux. This law is likely to have introduced overestimates of small subpeaks in the flux power law spectra since our statistics are based on many hours of observation (instead of a "snapshot view"). On the other hand, the number of short-lived subpeaks (thus small ones) will have been underestimated, since we have only used those that appeared in at least two consecutive spectra.

- Variation of velocity width (σ_v) with position on the line (v), after allowing for the above f–σ_v scaling law: cf. log–log plot of $\sigma_v/f^{1/2}$ vs. v in Fig.6. This comfirms and improves the previous analysis using multi-gaussian fitting.

- A log–log plot of total flux in a given range of velocity width(σ_v) vs. σ_v (Fig.7) leads to an asymptotic slope at the low end of 2.3 ± 0.1, which can be identified as the fractal dimension (cf. GH) of the wind clumps.

5   Interpretation

A necessary (but not sufficient) condition that one is dealing with turbulence is that the Reynolds number $Re \gg 1$. In a wind of expansion speed $v_w(r)$ at a distance r from the star, one has:

$$Re \approx \frac{r v_w(r)}{\nu_{thermal}} \approx \frac{r}{l_{mfp}} \cdot \frac{v_w(r)}{c_s},$$

where ν is the viscosity (we neglect photon and micro-turbulent viscosities), l_{mfp} the mean free path of the average particle and c_s the speed of sound in the medium. Furthermore, in cgs units:

$$l_{mfp} \approx 3.2 \; 10^6 \frac{T^2}{Z^2 n_e ln\Lambda_c}, \quad \Lambda_c = \frac{1.3 \; 10^4 T^{3/2}}{n_e^{1/2}} \; ;$$

$$c_s = (\frac{kT}{\mu m_h})^{1/2}.$$

Thus, with typical WR wind values where the observed lines form : $r \sim 5R_*(R_* \sim 5R_\odot), v_w \approx 2000 km s^{-1}, T \sim 30000K, n_e \approx 10^{10} cm^{-3}$ and He dominates, one finds $l_{mfp} \sim 30m$ (short!) and $Re \sim 10^{10}$. Thus, turbulence is likely to exist if there is a driving force.

A critical test for turbulence is to look for scaling laws as a result of energy cascading and dissipation from large to small eddies (Larson 1981; Henriksen 1991; Wolfire *et al.* 1992), among l (size), ρ (density) and σ_v. In the case of compressible turbulence, one has no *a priori* knowledge of $\rho(l)$ or $\sigma_v(l)$, as in the case of incompressible turbulence, where $\rho \sim const.$ and $\sigma_v \sim l^{1/3}$ (the famous Kolmogoroff relation) from conservation as energy is transferred between different scales. One thus needs two conserved quantities (i.e. over the cascades from large to small scales). For example, in GMC's, where virial equillibrium is assumed to prevail and gravity is the driving force, one has:

- (a) constant pressure, i.e. $\rho \sigma_v^2 \sim const.$, and

- (b) lifetime $\tau \sim \frac{l}{\sigma_v} \sim \frac{1}{\sqrt{G\rho}}$.

These lead to $\sigma_v \sim l^{1/2}$ and $\rho \sim l^{-1}$. The same relations prevail when (b) is replaced by a constant column density (Larson 1981): (c) $\rho l \sim const.$

In the case of a WR wind, it is likely that (a) also prevails, although little justification can be made for (b). However, if (c) is valid, we arrive at the same scaling laws as in GMC. Despite this uncertainty, we adopt (c) as a working model. In any case, it is well known that large changes in the *physics* often yield small changes in the turbulence scaling laws.

Unfortunately, we cannot observe l and ρ directly (σ_v *is* however observed). We use f and τ instead, which are related to ρ, l and σ_v by: $f \sim \rho^2 l^3$ for optically thin recombination emission lines, neglecting variations in temperature (this is reasonable, because f does not vary with viewing angle and cooling times for the dense clumps to the ambient temperatures are very short; also, Falgarone *et al.* (1991) found in GMC's that the excitation of ^{13}CO 3–2 vs. 2–1 remains constant over at least an order of magnitude in line intensity) and $\tau = \frac{l}{\sigma_v}$. We thus predict for compressible turbulence:

$$f \sim \sigma_v^2 \; ; \; \tau \sim f^{1/2}.$$

The *observations* discussed above for WR 135 give:

$$f \sim \sigma_v^{2.0\pm0.4} \; ; \; \tau \sim f^{0.8\pm0.2}.$$

These are remarkably similar!

Furthermore, conversion to clump mass via $m(l) \sim \rho l^3$, thus $f \sim m^{1/2}$, yields the mass frequency spectrum:

$$n(m)dm = N(f)df \sim f^\alpha df \; (observed \; form),$$

from which:

$$n(m) \sim m^\gamma \; ; \; with \; \gamma = (\alpha - 1)/2.$$

Thus, with $\alpha = -1.8 \pm 0.1$ for WR135 (or -2.4 ± 0.2 for 7 WR stars on average), one finds $\gamma = -1.4 \pm 0.05(-1.7 \pm 0.1)$. This is again remarkably close to the power law index found in GMCs : mean $\gamma \approx -1.5$, range -1 (young) to -2 (older) (Williams & Blitz 1993).

Another interesting analogy with GMC's is anisotropy. In WR winds, we find a mean true axis ratio:

$$\frac{l_{transversal}}{l_{radial}} \sim [\frac{\sigma_v(trans)}{\sigma_v(radial)}]^2 \sim (0.3)^2 \sim 0.1.$$

In GMC's, observed projected axial ratios (but random orientation!) lie in the range 0.1—1 leading to *true* axial ratios that are typically $\gtrsim$ 30 % smaller (Fleck 1992). In GMC's, the stochastically oriented elongations are believed to be caused by anisotropic gravitational field fluctuations whereas in WR winds, the elongations are all radially oriented, defined by the direction of radiation pressure and gravity. Again, the similarity is remarkable, apart from the orientation statistics.

Finally, we note that the fractal dimension of the hierarchical WR clumps discussed above (2.3 ± 0.1) is very similar to the fractal dimension derived from ^{13}CO in the L1551 outflow region: 2.35 ± 0.01 (see GH).

Taken together, the above evidence strongly favors anisotropic, supersonic, compressible turbulence in WR135's wind. Since all WR stars observed intensively so far do behave similarly, and WR stars are extreme manifestations of winds in hot, luminous stars, it is possible, or even likely that all hot–star winds show the same basic phenomenon. This being the case, the observed velocity dispersion is interpreted as eddy rotation. It is quite plausible that the observed spectral sub-peaks represent recombination radiation from eddies, driven by radiatively induced stochastic wind instabilities (e.g. Owocki et al. 1988). This occurs preferentially in the radial direction away from the star. If the observed scaling laws can be extrapolated to predict rapidly increasing numbers of small, unresolved clumps, then it is clear that we are dealing with full–scale turbulence, beyond a critical radius from the star (several R_*), where large $v_w(r)$ drives Re to large values. It is interesting to note that full–scale turbulence in GMC's produces relatively smooth ^{12}CO profiles when many clumps are seen simultaneously in the telescope beam, even at very high signal–to–noise (S/N $\sim$ 350): (cf. Wolfire et al. 1992). The same situation may prevail for many WR emission lines.

6   Consequences of Turbulent Clumping

It appears very likely that one will have to abandon the paradigm of *smooth* outflows from hot–stars. The effects of clumping are numerous. We list below some of those that occur to us, without elaborating (this will be done in a forthcoming paper by Moffat 1993).

- Hot–star winds (especially those of WR stars) are ideal, unique cosmic laboratories in which to observe and study for its own sake time–dependent compressible turbulence essentially in 3-D (via projected Doppler separation). In GMC's the time scales are long; in stellar photospheres, one essentially has only a 2-D view.

- $\dot{M}$'s based on radio/IR free-free emission or recombination line fluxes and the assumption of smooth flows may have to be revised downwards, typically by a factor $\sim$ 3 (Moffat & Robert 1993).

- Lower $\dot{M}$'s will lead to lower initial stellar masses on average for massive stars, derived from present masses.

- If full–scale turbulence really does prevail, the discrete clumps will be usable to trace the wind velocity law $v(r)$ (this has been done by Robert (1992), with the result that wind acceleration is much lower than predicted by the usually assumed β-law, with $\beta \sim 1$) and shed light on the possibility of rotating winds.

- If clumping increases outwards in the temperature–stratified winds of WR stars, then lines of lower ionisation will be enhanced, so that wind ionisation temperatures will be underestimated and possibly even abundances falsified.

7   Remaining Questions

We list below some interesting problems which remain to be tackled:

- What is the detailed form of the clumps (e.g. absolute sizes and masses; internal temperature; velocity and density profiles)? One will have to study the exact spectral profiles of individual clumps in lines of different ionisation to probe this.

- Does the mass–spectrum continue to rise dramatically to lower (presently unobservable) clump masses? More observations at even higher S/N and spectral resolution with shorter time resolution will be necessary.

- What are the details concerning growth, lifetime and decay of clumps? Longer, continuous observing runs will be required to answer this.

- Do large eddies fragment into shorter lived smaller ones?

- What are the details concerning the actual transfer of the radiative driver to produce turbulence? Present 1-D models will have to be expanded to 3-D to do this properly.

References

Daubechies, I. 1992, *Ten Lectures on Wavelets*, SIAM, Philadelphia

Falgarone, E., Phillips, T.G. & Walker, C.K. 1991, *ApJ*, **378**, 186

Farge, M. 1992, *Ann.Rev.Fluid Mech.*, **24**, 395

Fleck, R.C. 1992, *ApJ*, **401**, 146

Gill, A.G., Henriksen, R.N. 1990, *ApJ*, **365**, L27

Henrisksen, R.N. 1991, *ApJ*, **377**, 500

Langer, W.D., Wilson, R.W. & Anderson, C.H. 1993, *ApJ*, **408**, L45

Larson, R.B. 1981, *MNRAS*, **194**, 809

Moffat, A.F.J., Drissen, L., Lamontagne, R. & Robert, C. 1988, *ApJ*, **334**, 1038

Moffat, A.F.J. & Robert, C. 1992, *ASPC*, **22**, 203

Moffat, A.F.J. & Robert, C. 1993, *ApJ, submitted*

Moffat, A.F.J. 1993, *in preparation*

Owocki, S.P., Castor, J.I. & Rybicki, G.B. 1988, *ApJ*, **335**, 914

Robert, C. 1992, *Ph.D. thesis, Univ. de Montréal*

St-Louis, N. 1990, *Ph.D. thesis, Univ. College London*

Williams, J.P. & Blitz, L. 1993, *ApJ*, **405**, L75

Wolfire, M.G., Hollenbach, D. & Trelens, A.G.G.M. 1992, *ApJ*, **402**, 195

Complex Structure Associated with the Wolf-Rayet Star WR147

R. J. Davis
N.R.A.L., Jodrell Bank, Macclesfield, Cheshire SK11 9DL, U.K.

The Wolf-Rayet star WR147 is known to be bright at X-ray and radio wavelengths (van der Hucht *et al.*, 1981, Caillaut *et al.*, 1985, Abbot *et al.*, 1986).

Observations have been made with old MERLIN at 5GHz on the Wolf-Rayet star WR147. The object is resolved into two radio sources, separated by 0.6 arcsec, corresponding to a spatial separation of 1100 AU at an assumed distance of 1.9kpc (Moran *et al.*, 1989). The computed brightness of the two sources, $\sim$ 10000K, is typical of optically thick thermal emission from a stellar wind, but compared with other observations suggests that some part of the emission is non-thermal.

Optical CCD images of WR147 were obtained on WHT on La Palma. No secondary source was evident within 0.6 arcsec north or south of the source with $\Delta m < 1.9$ min, confirming the optical speckle results (Röser and Bastian, 1988).

Ed Churchwell *et al.*, (1992) have now shown this source to vary, particularly at L Band. They are unable to resolve the components at this frequency with the VLA. Thus to study this interesting complex Wolf-Rayet star system which has bipolar flow, thermal and non-thermal compact radio emission and optical emission coincident with the one of the radio sources, MERLIN observations are essential.

The flux density and position of the source were accurately related to that of the phase reference. The flux density of the reference source, 2005+403, was determined to be 4.4Jy using a comparison observation of the calibration source 3C286, the flux density of which was assumed to be 7.4Jy at 5GHz.

Accurate measurement of the optical position has shown that the stellar source is associated with the southerly component of the radio emission. Combining results at 1.6GHz and 5GHz gives a spectrum for the southerly source consistent with optically thick thermal emission. The northerly source appears to be non-thermal and is modelled in terms of a shock in the stellar wind. The 1.6GHz map also shows radio emission on both sides of the star with a spectral index of -0.6. This emission is typical of bipolar outflow and if confirmed is the first bipolar outflow found in a Wolf-Rayet star.

Since the expected radio flux density was low, a phase referencing technique (Peckham, 1973) was used in which twelve minute observations of AS 431 were interspersed with three minute observations of a phase reference source. The data were edited, vector-averaged and calibrated using routines in the OLAF[1] software package and the visibility data were transformed to an intensity map using the Astronomical Image Processing System (AIPS). Initially the phase reference source

---

[1] Off Line Analysis Format - An astronomical image processing package that is designed specifically to process MERLIN or VLBI data.

Astrophysics and Space Science **216**: 67–68, 1994.
© 1994 *Kluwer Academic Publishers.*

was mapped by the standard self calibration technique (Hogbom, 1974). The derived telescope gain solutions were then applied to AS431 and the resulting data were further processed by Fourier and CLEAN techniques (Lortet et al., 1987).

We therefore suggest that the southernmost radio source originates from thermal emission from a stellar wind. Assuming that this is the source of all the thermal emission in AS 431, the spectral index from 5GHz to 25μm (the latter being from the IRAS point source catalogue with stellar flux subtracted Stickland et al., 1985) would be $\alpha = 0.6$. This is the canonical value for thermal bremsstrahlung emission from a constant outflow optically thick ionised wind.

We may then calculate the required mass loss rate for such a process from this component, assuming an optical depth of unity and a uniform isotropic flow from equation 20 of Wright and Barlow (1975), i.e.

$$\dot{M} = C \ V_t \ S_\nu^{3/4} \ D^{3/2} \ M_\odot \ yr^{-1}$$

where V_t is the terminal velocity in the wind in km s^{-1}, S_ν the observed flux density in Jy, D the distance in kpc and C a constant which is sensitive only to the ionic composition of the wind.

For Wolf-Rayets it appears that C is between 0.7×10^{-6} and 2.1×10^{-6} (Abbott et al., 1986). Thus, taking $C = 1.4\times10^{-6}$, $D = 1.9$ kpc and $V_t = 1000$ kms^{-1} and a flux density of 21.4mJy we arrive at a mass loss rate of $2.1\times10^{-4}M_\odot yr^{-1}$. This is still several times greater than typical mass loss rates calculated for single Wolf-Rayet stars. It should be noted however that $\dot{M}$ is very sensitive to D and the quoted distance is uncertain by ± 1 kpc.

The old MERLIN map at 5GHz gives an indication that the northerly component is flattened perpendicular to the jet direction. New measurements with MERLIN clearly show the flattening and also a shell-like structure around the actual star. At 5GHz the jet appears to extend only to the north.

References

Abbott, D. C., Bieging, J. H., Churchwell, E., and Torres, A. V.: 1986, *Astrophys. J.* **303**, 239–261.

Caillault, J. P., Chanan, G. A., Helfand, D. A., Patterson, J., Nousek, J. A., Takalo, L. O., Bothun, G. D., and Becker, R. H.: 1985, *Nature* **313**, 376–378.

Churchwell E., Bieging, K. A., van der Hucht, K. A., Williams P.M., Spoelstra T.A. and Abbott, D. C.: 1992, *Astrophys. J.* **393**, 329–340.

Hogbom, J. A., *Astron. Astrophys. Suppl.* **15**, 417–426.

Lortet, M. C., Blazit, A., Bonneau, D. and Foy, R.: 1987, *Astron. Astrophys.* **180**, 111–113.

Moran, J. P., Davis, R. J., Bode M. F., Taylor A. R., Spencer, R. E., Argue A. E., Irwin M. J. and Skanklin J. D.: *Nature* **340**,449–450.

Peckham, R. J.: 1973, *Mon. Not. Roy. Astr. Soc.* **165**, 25–38.

Röser, S. and Bastian, U.: 1988, *Astron. Astrophys. Suppl.* **74**, 449–451.

Stickland, D. J., Lloyd, C. and Willis, A. J.: 1985, *Astron. Astrophys.* **150**, L9–L11.

Ulvestad, J. S. and Wilson, A. S.: 1989, *Astrophys. J.* **343**, 659–671.

van der Hucht, K. A., Conti, P. S., Lundstrom, I. and Stenholm, B., *Space Sci. Rev.* **28**, 227–306.

Wright, A. E. and Barlow, M. J.: 1975, *Mon. Not. Roy. Astr. Soc.* **170**, 41–51.

The Importance of Continuum Radiation for the Stellar Wind Hydrodynamics of Hot Stars

M. Runacres
Astrofysisch Instituut, Vrije Universiteit Brussel, Pleinlaan 2, B-1050 Brussels, Belgium

and

R. Blomme
Royal Observatory of Belgium, Ringlaan 3, B-1180 Brussels, Belgium

Abstract. The contribution of bound-free and free-free processes to the outward acceleration of ζ Pup is studied and is found to be negligible.

Key words: Hydrodynamics, Radiative Transfer, Early-Type Stars

1 Introduction

Early-type stars (of spectral type O, B or Wolf-Rayet stars) continuously eject material and are therefore surrounded by a stellar wind. The material is driven away from the star by radiation pressure: when a photon is scattered by an ion in a spectral line, momentum is transferred from the photospheric radiation field to the stellar wind.

2 Model

In order to model the stellar wind, we have to solve the equations of hydrodynamics. Traditionally the acceleration due to electron scattering and line scattering is included, but not the effect of bound-free and free-free scattering. We have studied their importance by first constructing such a traditional model and then calculating the acceleration due to bound-free and free-free scattering.

General assumptions include spherical symmetry, time independence and a constant temperature. The wind is allowed to expand freely to infinity and the accelerations included are due to gas pressure, electron scattering and the line scattering of some 200,000 spectral lines. Solving the equations of hydrodynamics, we find the density and velocity structure. Using this density we then alternatingly solve the non-LTE equations of radiative transfer and statistical equilibrium for a Hydrogen + Helium mixture, until convergence is reached.

The additional acceleration due to bound-free and free-free processes is given by:

$$\frac{2\pi}{c\,\rho(r)} \int_0^{+\infty} d\nu \int_{-1}^{+1} \mu d\mu\, \chi_\nu(r)\, I_\nu(r,\mu)$$

where χ_ν is the opacity due to bound-free and free-free scattering and I_ν is the specific intensity. The integrals can easily be evaluated using the physical quantities (opacity, specific intensity, mass density) which have already been calculated in our model.

3   Results and Discussion

We have applied our model to the star ζ Pup (mass loss = 4.5 10^{-6} $M_\odot$/yr). The different contributions to the total outward acceleration are plotted on Fig. 1 as a function of distance. One can see that electron scattering is the most important continuum contributor to the outward acceleration. Our study shows that bound-free and free-free processes are negligible. However the possibility remains that for higher mass loss rates (e.g. for Wolf-Rayet stars) these processes are significant.

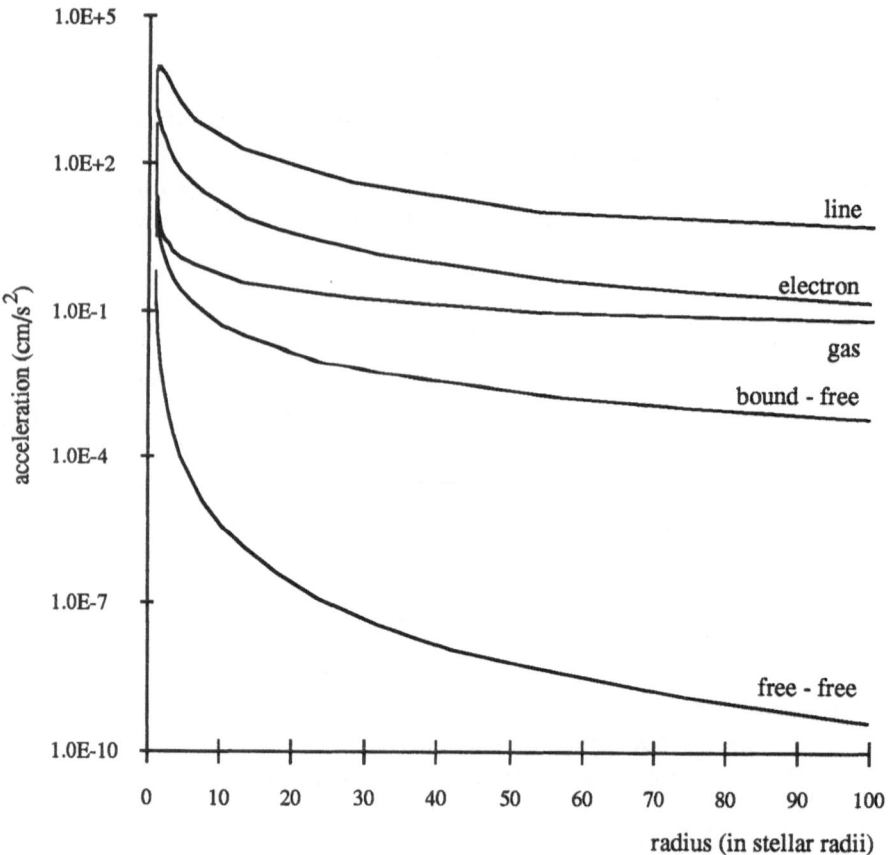

Fig. 1.   The different contributions to the total outward acceleration as a function of distance

Herbig Ae/Be Stars

T. P. Ray
School of Cosmic Physics, Dublin Institute for Advanced Studies,
5 Merrion Square, Dublin 2, Ireland

Abstract. Herbig Ae/Be stars are the higher mass counterparts of the T Tauri stars. In comparison with the latter, however, relatively little is known about them. After a historical introduction, we briefly review their optical and UV spectroscopic properties. We consider the evidence for and against disks around Herbig Ae/Be stars; the existence of which remains highly controversial. We also examine in-depth their interaction with the surrounding medium as manifested through optical outflows. It is shown that although there are similarities with analogous outflows from lower mass young stars, those from Herbig Ae/Be stars may be more poorly collimated. Jets, however, are found in at least some cases.

Key words: pre-main sequence stars, disks, outflows

1 Historical Perspective

It is over 30 years since George Herbig proposed that a certain group of stars (later to be known as Herbig Ae/Be stars and hereafter referred to as HAEBE stars) were the higher mass counterparts of the T Tauri stars (TTSs). In a milestone paper Herbig (1960) defined HAEBE stars by four main properties:

(i) Strong emission lines (by analogy with the TTSs), particularly those of the Balmer series.

(ii) Spectral class earlier than F0, thereby ensuring that the star, when it eventually settled on the main sequence, would be either an A or B-type star.

(iii) A location (at least on the sky) within the boundaries of a molecular cloud.

(iv) The star illuminated a reflection nebula.

This last criterion, in conjunction with the previous one, virtually ensured that such a star is associated with a molecular cloud as opposed to being just along our line of sight to it. Herbig (1960) originally proposed just over two dozen stars as members of this group, although his survey was far from complete even in terms of sky coverage. It was over two decades before Finkenzeller and Mundt (1984) extended this list to nearly 60 stars, but additional examples have been discovered in the meantime using, for example, the IRAS database (Hu *et al.*, 1989). Strom *et al.*(1972) determined the effective temperature and surface gravity of 14 stars from Herbig's original list and placed them in the HR diagram for the first time. They established that, like the TTSs, these stars lay well above the main sequence. A subsequent study by Cohen and Kuhi (1979) came to the same conclusion.

Further evidence for the pre-main sequence nature of HAEBE stars came from studies in the near-infrared with most stars showing an excess (see, for example

Astrophysics and Space Science **216**: 71–86, 1994.
© 1994 *Kluwer Academic Publishers.*

Cohen, 1980, Finkenzeller and Mundt, 1984) and some even exhibiting the 9.7μm silicate feature (Cohen, 1980). While it has been suggested that this near-infrared excess could be understood in terms of free-free emission, as with the classical Be stars (e.g. Lorenzetti, Saraceno and Stratfella, 1983), such a model does not explain the presence of the silicate feature. Moreover, the spectral energy distributions of most HAEBE stars show large mid- and far-infrared excesses, in some cases even at 100 μm (Hu *et al.*, 1989); thus one has to come to the conclusion that they are surrounded by dust like the TTSs (see Bertout, 1989 for a review of TTS properties).

Finkenzeller (1985) studied the projected rotational velocities, i.e. vsini values, for a large sample of HAEBE stars and discovered that they rotated at higher velocities than the TTSs. Their rotational velocities are, however, lower than those of Be stars and their distribution of vsini's is very different from that of normal B-type field stars. Finkenzeller (1985) concluded that this was further evidence for their youth.

In summary there can be little doubt now that HAEBE stars are pre-main sequence objects, a view that will be reinforced by the more recent work reviewed here. The break-up of the paper is as follows: after surveying their UV and optical properties we go on to discuss the controversy as to whether these stars are surrounded by disks. We then consider how these stars interact with their environment through optical outflows. Finally we examine, from a theoretical standpoint, the evolutionary status of these stars and the origin of their winds and outflows.

2   Optical and UV Observations

Spectral studies, such as those of Finkenzeller and Mundt (1984), allow us to classify HAEBE stars according to their Hα profiles. There are essentially three types of profile: P Cygni, single peak, and double peak. Of the three, double peaked Hα is the most common ($\approx$ 50%). Evidently P Cygni profiles imply the presence of winds but the other types of profiles can be interpreted in various ways. [BTo complicate matters further the Hα profiles of at least some stars are found to change over a number of years (see, for example, Catala, 1989).

A comparison of the emission line properties of HAEBE stars with classical T Tauri stars (CTTSs) has been made in the optical and near-infrared range by Hamann and Persson (1992). Obviously, as one might expect, HAEBE stars have hotter emitting envelopes and so lines of neutral species such as FeI or CaI are either very weak or, in most cases, absent from their spectra. Hamann and Persson (1992) do, however, point out that there are many spectral similarities between these and the CTTSs. For example, the presence of the CaII triplet near-red emission is observed in over 80% of the HAEBE stars, and while this is almost always seen in the CTTSs, it occurs less frequently in the classical Be stars (20%). Hamann and Persson (1992) suggest that the spectral similarities between HAEBE stars and CTTSs could be explained if the emission line activity of many HAEBE stars is likewise a direct result of accretion at a star/disk boundary (see §3).

IUE observations of the brightest HAEBE stars show emission lines of both high and low excitation species in their SWP spectra, these species include OI, CII, SiIV and CIV. Lyman α is seen in emission in at least the Herbig Ae stars but, since *it is redshifted* and does not vary in phase with the other UV emission lines, it is presumably formed in a different region (Blondel, Talavera and Tjin A Djie, 1993). The shape of the Lyman α line is certainly inconsistent with it arising from a chromosphere and it has been suggested that its redshift arises from the recombination of infalling matter i.e. accretion (Blondel *et al.*, 1993).

One important difference between CTTSs and HAEBE stars is that whereas CTTSs often show a "UV bump" attributable to accretion (see below), HAEBE stars in contrast show strong UV depletion (Catala, 1989). As we shall see, this presents a problem for disk accretion models since one expects a significant fraction of the accretion energy to be released at these wavelengths. However, as suggested by Catala (1989), one way around this difficulty, at least in the case of the more embedded HAEBE stars, may be to suppose that there is additional extinction along the line of sight to the star provided by a dust shell. This is in line with some recent findings on HAEBE stars (see below and Natta *et al.*, 1993).

3 Evidence for Disks Around Herbig Ae/Be Stars

As will be shown there is considerable controversy as to whether HAEBE stars have disks like those that almost certainly surround CTTSs. It is clear that both groups of stars have many properties in common e.g. spectral characteristics, infrared excesses and mm emission. The idea that disks surround HAEBE stars is therefore virtually an extension of the canonical models for CTTSs, (see, for example, Basri and Bertout, 1993). Nevertheless it should be emphasized that alternative possibilities have been invoked to explain the observed properties of HAEBE stars in which the star itself, rather than a disk, plays the crucial role (see, for example Catala, 1989).

As in the case of the CTTSs most of the evidence for disks is circumstantial but still quite compelling. Hillenbrand *et al.*(1992) have shown that the spectral energy distributions of HAEBE stars can be divided into 3 main groups depending on spectral index (see Fig. 1). Group I have a spectral index $\alpha \approx -\frac{4}{3}$ with α defined by $\lambda F_\lambda \approx \lambda^\alpha$ (for the infrared spectrum longward of the K band). Group II have either flat or rising spectra while Group III emit virtually like blackbodies. Group I is by far the largest category but more importantly the observed infrared spectrum is exactly the type one would expect from either a passive (i.e. reprocessing) or active accretion disk. Unfortunately it is not possible to distinguish between these two possibilities on the basis of the spectral energy distribution alone without accurately knowning the stellar luminosity (Hillenbrand *et al.*, 1992).

Further indirect evidence for disks exists. For example, as with the CTTSs, one

T. P. Ray

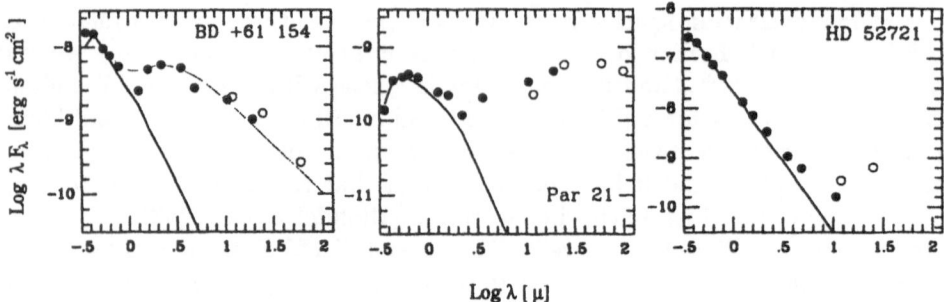

Fig. 1. Classification of HAEBE star spectral energy distributions after Hillenbrand *et al.*(1992).
Spectra have been corrected for extinction. Examples are given of Group I (left), Group II
(middle) and Group III (right) sources. The meaning of the symbols and lines are as follows: open
circles are IRAS Point Source Catalogue data, solid lines are scaled spectral energy distributions
appropriate to corresponding main sequence stars of identical spectral type and the dashed line
shown for the Group I source BD +61°154 represents the best fit disk model.

could argue that the optical extinction towards these stars is only a few magnitudes
at most but any reasonable *spherical* distribution of dust around them that could
produce the observed far-infrared spectrum would at the same time give rise to
many orders of magnitude greater optical extinction than seen. We can conclude
that at least some of the dust must not be distributed spherically but instead
in another distribution (e.g. a disk) which allows light to preferentially escape in
certain directions. That is not to say that a combination of disk plus a dusty halo
may be present as has been been argued by Natta *et al.*(1993).

Moreover, the presence of disks is also suggested by spectroscopy: in those
rare cases where high velocity forbidden line emission is seen in HAEBE stars it
is always blueshifted. As with the TTSs (see, for example Bertout, 1989) this
is taken to mean that any redshifted high velocity component is obscured by an
occulting disk. An example of this effect is shown in Fig. 2 (Corcoran and Ray,
1994a).

Optical spectroscopy also indicates that accretion may be occurring in at least
some HAEBE stars. Initial studies by Corcoran and Ray (1994a) have found
indications of "veiling[1]"(see Fig. 3) as in CTTSs. This phenomenon is normally
attributed to the presence of an additional hot continuum source, apart from the
photosphere, perhaps generated by the boundary layer of an accretion disk.

Estimates of the amount of material in the putative disks can be derived from
mm continuum observations assuming the material is optically thin at these wave-

---

[1] i.e. the photospheric absorption lines are not as deep, with respect to the continuum, as in a
main sequence star of the same spectral type.

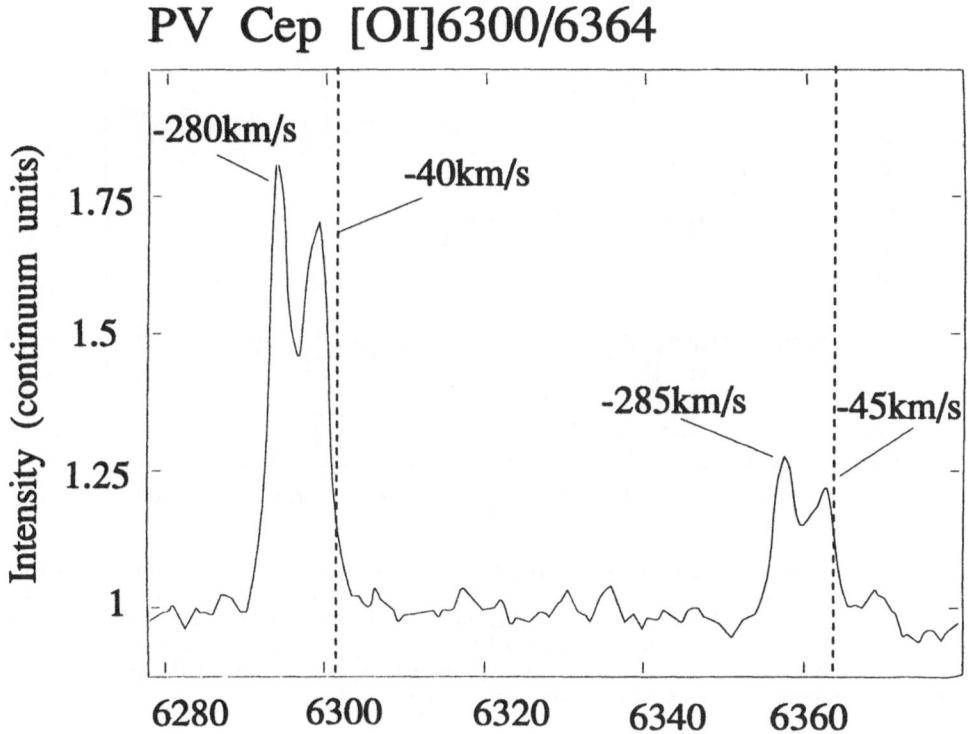

Fig. 2.   The [OI]λλ6300,6364 doublet region of PV Cep's spectrum. The dashed vertical line corresponds to the systemic velocity of this HAEBE star. Note how the forbidden lines are blueshifted with a velocity (for the higher velocity component) typical of optical outflows. Any redshifted component is presumably hidden behind PV Cep's disk.

lengths. Calculated masses are in the range $10^{-2} - 10M_{\odot}$, i.e. some ten times higher than is the case of CTTSs. It has been argued by Hillenbrand *et al.*(1992) and Natta *et al.*(1993) that if we combine such mass estimates with derived accretion rates then the estimated disk lifetimes are relatively short (at most about 1 Myr). This may have repercussions for the formation of large grains around these stars and ultimately any larger bodies. One has to, however, be cautious since the derived accretion rates, which are based on near-infrared excesses, appear anomalously high and could at least in some cases be partly attributed to incorrectly estimated extinction rates (Ray, 1993).

KK Oph and HR 8443 (A3)

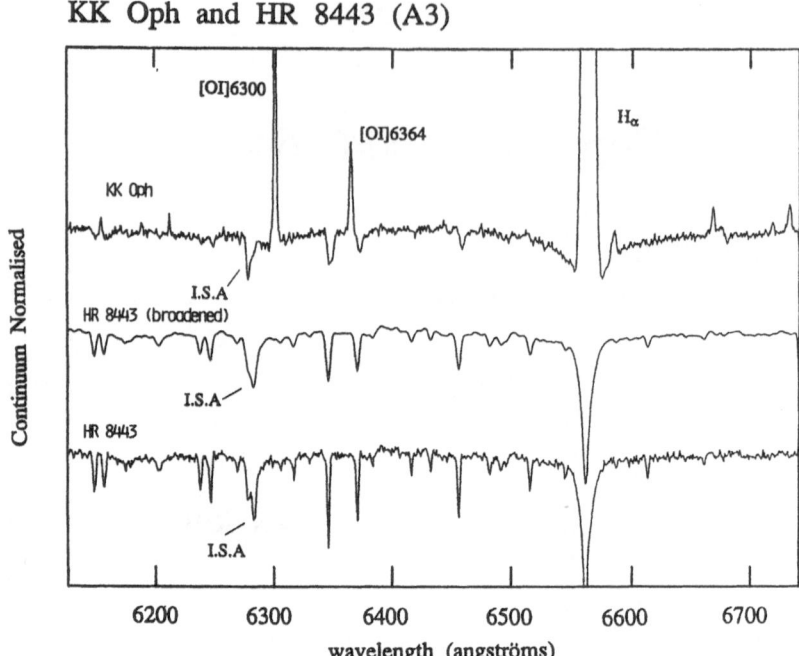

Fig. 3. The spectrum of the HAEBE star KK Oph (top) is compared with a main sequence star of the same spectral type (HR 8443, bottom) broadened in velocity to match KK Oph. All spectra have their continua normalized to unity and then offset in the flux direction for clarity. It can be readily seen that the absorption lines in the HAEBE star are less deep than in the corresponding main sequence star, an effect known as "veiling" (see text). It is due to the presence of an additional continuum source apart from the photosphere.

The disk hypothesis for HAEBE stars has not, however, gained universal acceptance and it is clear that there are some problems. For example, one intriguing feature of the spectral energy distributions of HAEBE stars is the often observed dip around 2-3 μm (see Hillenbrand *et al.*, 1992 and Fig. 4). This has been variously ascribed to the destruction of dust grains or magnetospheric accretion. According to the first model, grains are thought to be destroyed at temperatures above 2000K i.e. at distances less than 5-10R_* from a HAEBE star. If this dust is contained within a disk, we would then expect, at such distances, a drastic drop in the disk's opacity, effectively giving rise to an opacity hole. In turn this would cause a dip in the spectral energy distribution in the near-infrared. The magnetic accretion model, originally proposed by Camenzind (1990) in the context of CTTSs and by Königl (1991), is that the magnetic field of the star permeates the disk out to a distance of several stellar radii. Accretion then occurs either through the magnetosphere (i.e. along the magnetic field lines) onto the magnetic poles or perhaps via magnetic accretion columns (Königl, 1991). Here an actual physical "hole" develops in the disk but the result is as before and we expect a dip in the

Spectral Energy Distribution of V1685 Cyg

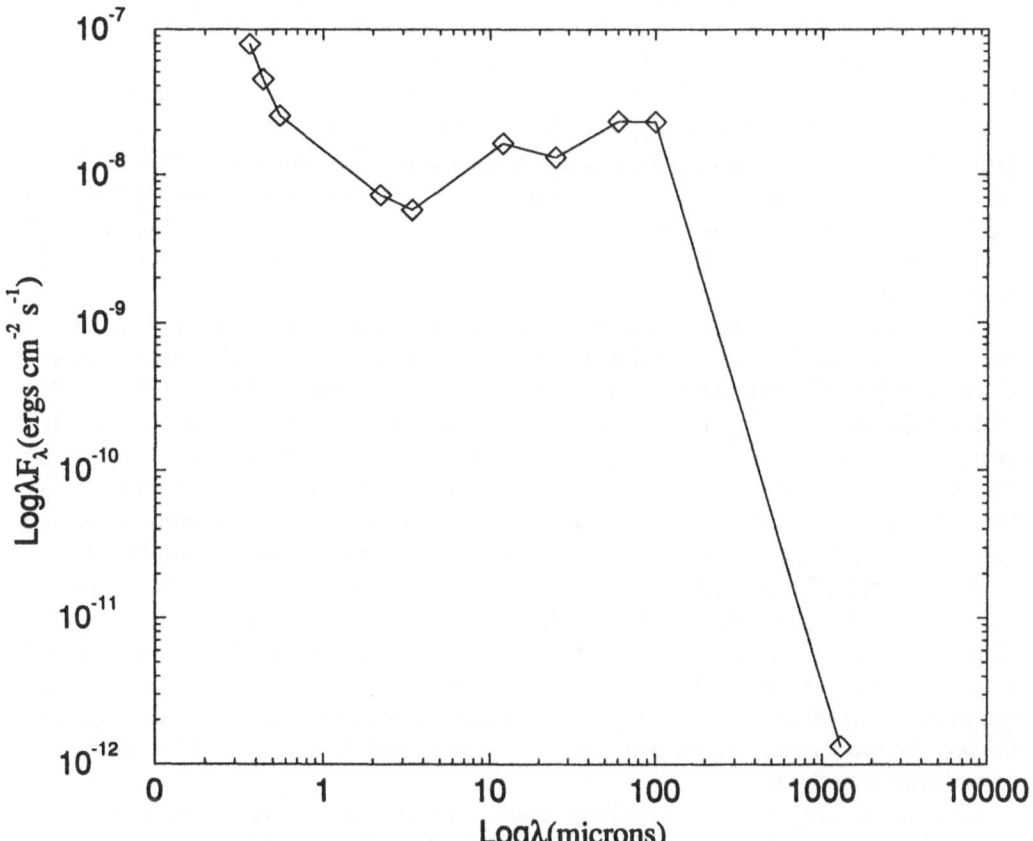

Fig. 4. The spectral energy distribution of the HAEBE star V1685 Cyg. Note the dip around 3μm which has been attributed to such diverse causes as grain destruction, magnetospheric accretion or small dust grains transiently heated by UV photons (see text)

near-infrared spectrum. Magnetically guided accretion, which one might expect to occur at velocities close to the escape velocity of the star, has the further advantage that it could explain the redshifted line profiles observed in, for example, Lyman α (see above). Hartmann, Kenyon and Calvet (1993) have shown, however, that there are problems with both scenarios. Firstly in connection with the grain destruction model, standard disk theory shows that, given the high rates of accretion for HAEBE stars, the disk will remain *optically thick* even if the grains are destroyed. The Hartmann *et al.*(1993) objection to the magnetic accretion model is that unless conditions, such as magnetic field strength, are similar in all HAEBE stars, we expect the effective hole to vary enormously in size and this is not observed. Hartmann *et al.*(1993) propose the 3μm feature is due to transient heating of small dust grains by UV photons (see also Siebenmorgen *et al.*, 1992) and furthermore *that there is no need to invoke disk models* to explain the spectral energy distribution of HAEBE stars. Moreover accretion disks, Hartmann *et al.*(1993) argue, suffer the problem that, ultimately, energy has to be released at UV wavelengths and UV excesses are not seen. We caution, however, that even amongst the limited sample of HAEBE stars with IUE data investigated by Hartmann *et al.*(1993), some appear to have additional UV emission such as T Ori (*ibid.* Fig. 2). Note also that any additional UV absorption in the surrounding medium, due to for example a dust halo, has been ignored. It certainly seems premature to dismiss the disk hypothesis yet *for at least some members* of the HAEBE class.

The point was brought up above that in addition to the Group I sources a number of HAEBE stars have flat or rising spectral energy distributions. Many of these Group II objects are optical outflow sources (e.g. LkHα 198, R Mon, R CrA and see §4) and so may represent less evolved HAEBE stars. Here the spectral energy distribution is best modeled by a combination of a star+disk+envelope (Natta *et al.*, 1993). The envelopes themselves have been seen directly using the KAO by Natta *et al.*(1993) who found that several of the Group II objects imaged by them could be resolved at 50 and 100 μm. Natta *et al.*(1993) were unable to explain how such large scale structures (of order $5 \times 10^3 - 10^5$ AU) could emit at such wavelengths in terms of a simple star+envelope model; some degree of reprocessing of the stellar light by a disk, or generation of longer wavelength radiation by accretion, had to occur as well. In some cases the envelopes had steep gradients (with density $\rho \approx \rho^{-2}$) while in others the density distribution appeared flatter. Obviously the steep gradient in the ambient density could be attributed to ongoing accretion.

Finally we should say something about the HAEBE stars with little or no infrared excess i.e. the Group III sources. Hillenbrand *et al.*(1992) speculated that these may be the HAEBE equivalent of the weak line T Tauri stars (Bertout, 1989) which seem to be disk-less versions of the CTTSs. Group III sources are relatively rare: for example only six (10%) are listed in the original Hillenbrand *et al.*(1992) sample. It is interesting to note that two of these (HD 52721 and HD 53367) were in fact thought to be classical Be stars by Hamann and Persson (1992). Despite this it seems unlikely that all Group III sources are classical Be stars given their association with star forming regions. Obviously Group III stars, about which very little is known, require further attention.

4   Optical Outflows from Herbig Ae/Be Stars

Although Herbig-Haro (HH) outflows have been known for several decades, and HH jets for at least ten years, most effort has concentrated on understanding this phenomenon amongst low mass stars. Up to relatively recently, few intermediate stars were known to have optical outflows. As explained by Corcoran, Ray and Mundt (1993a) the reasons for this are partly historical and partly pragmatic. Early studies tended to concentrate on nearby low mass star forming regions, like Taurus, ignoring the more distant clouds where higher mass stars form. An additional problem is that HII regions are sometimes present (Poetzel, Mundt and Ray, 1992), making it difficult to discern HH from photo-dissociated emission.

Despite these difficulties, an increasingly large number of optical outflows from intermediate mass stars are now known. Some "old" examples in the literature include V645 Cyg (Goodrich, 1986; Hamann and Persson, 1989; Zou, 1989), R Mon (Brugel et al., 1984), Cepheus A (Lenzen, 1988) and HH 80/81 (Rodríguez and Reipurth, 1989; Martí, Rodríguez and Reipurth, 1993). In recent years several studies have been undertaken of optical outflows from intermediate mass pre-main sequence stars including both embedded infra-red sources (EIRS) and optically visible HAEBE stars. Several new outflows have been discovered including those associated with AFGL 2591 and MWC 1080 (Poetzel et al., 1992). In many cases the optical outflow appears poorly collimated, a point we shall return to later, but jets have been seen from LkHα 234, AFGL 4029 (Ray et al., 1990) and Z CMa (Poetzel, Mundt and Ray, 1989). We note however that Z CMa, although originally listed as a HAEBE star, is most probably an FU Orionis object.

Goodrich (1993) has found four new possible HH objects in the vicinity of the HAEBE stars, HK Ori, BD +46° 3471, BD +41° 3731 and LkHα 198, but the nature of several of these objects has yet to be confirmed. In even more recent work, new outflows associated with Cep A (Corcoran et al., 1993a), LkHα 198 (Corcoran, Ray and Bastien, 1994a), V380 Ori (Corcoran and Ray, 1994b), VdBH 65B, R CrA and T CrA (Eislöffel and Ray, 1994) have been discovered. We will concentrate here on briefly discussing two examples, namely V380 Ori and LkHα 198, to illustrate the nature of HAEBE optical outflows.

V380 Ori is located in the NGC 1999 reflection nebula close to the well-known optical outflow HH1/2. There is a triangular Bok globule near this star and HH35 extends roughly in the direction of V380 Ori. A few years ago Strom et al.(1986) discovered some additional faint HH objects to the south-west but the connection with V380 Ori remained unclear. The deeper CCD images of Corcoran and Ray (1994b) show that both these objects and HH35 appear to be part of a emission loop structure (Fig. 5) which may in turn delineate a wind blown cavity associated with V380 Ori. Because V380 Ori is located at the edge of a molecular cloud, optically we see only one half of the outflow, although the counterflow is visible in CO (Corcoran and Ray, 1994b). The nature of the emission loop is somewhat obscure but a similar loop has been seen in the case of Cep A (Corcoran et al., 1993a) and here it is almost certainly shocked (HH) emission.

LkHα 198, like V380 Ori, is one of the stars originally listed by Herbig (1960). Close by is another Herbig Ae/Be star V376 Cas (see Fig. 6). Strom et al. (1986) discovered one HH object close to LkHα 198 which they found to be of low radial velocity and which they associated with it. A closer study (Corcoran et al., 1993b;

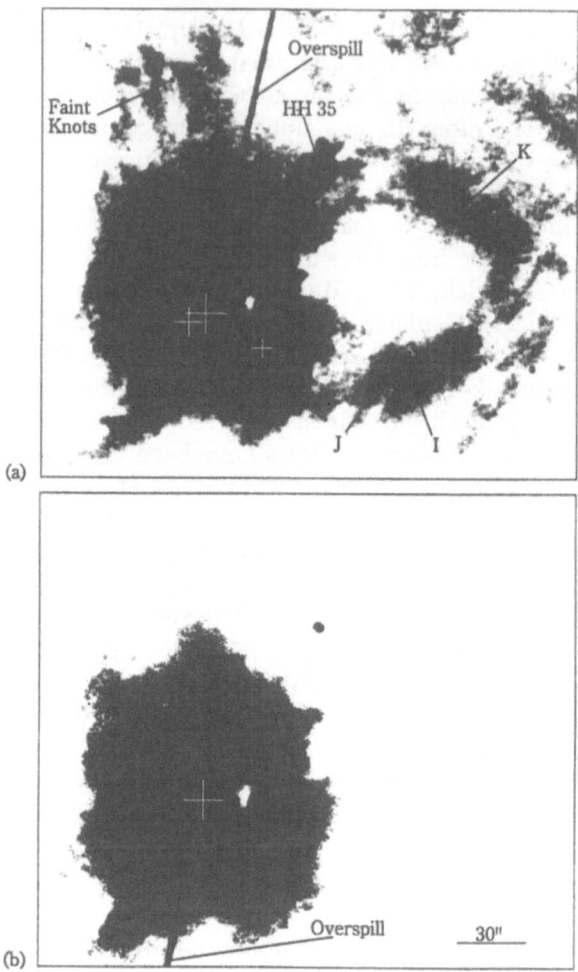

Fig. 5.   The V380 Ori Region. (a) A Hα image of NGC 1999 and the region to its west. The
plus signs, in order of decreasing size, mark the positions of V380 Ori and the K band sources
V380 Ori-B and V380 Ori-C respectively. V380 Ori-B is certainly a star but V380 Ori-C appears
to be shocked molecular hydrogen emission and part of the V380 Ori outflow. Prominent within
the image is a "loop" feature composed of several sub-condensations: these include HH35, and a
number of newly-discovered objects, labeled I, J and K. To the north of NGC 1999, some faint
knots are seen at the ends of "finger-like" emission. (b) The same area as in (a) but through a
narrowband red continuum filter *excluding* HH emission lines. Note the disappearance of both
the "loop" feature and the emission knots to the north. In all frames (i.e. Figs. 5 and 6) north is
to the top and east is to the left. From Corcoran and Ray (1994b)

1994a) however reveals that this HH emission comes not from LkHα 198 but from
a nearby "companion" which is referred to as LkHα 198B. This star appears to
be responsible not only for the HH emission but illuminating the two large scale

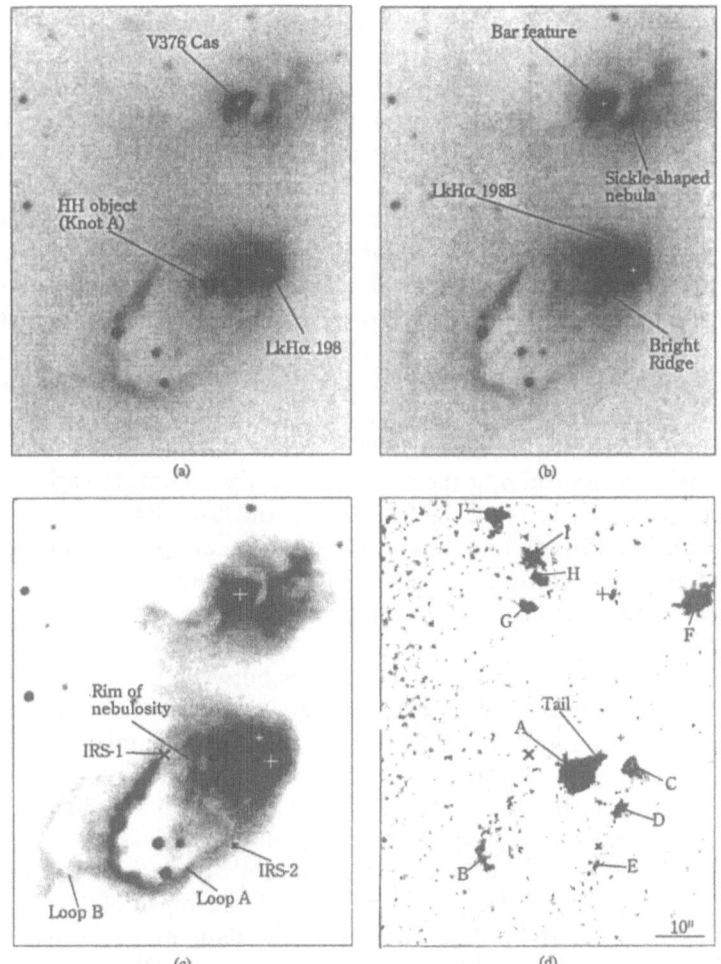

Fig. 6.   The optical outflows in the LkHα 198 region. (a) A [SII] image of the area around LkHα 198 and V376 Cas with the previously known HH object (knot A) to the southwest of LkHα 198. (b) The same area as in (a) but through a nearby red continuum filter as in Fig. 5. Note that the HH object disappears. Note also the optically faint star (LkHα 198B) situated about 5" north of LkHα 198. (c) As in (b) but at higher contrast revealing the presence of an additional faint reflection loop (Loop B). Also marked are the positions of two K-band sources IRS-1 and IRS-2 (see Corcoran *et al.*1994a for details). (d) A [SII]–red continuum difference image (as in Fig. 5) showing virtually pure HH emission. LkHα 198 is seen to be associated with a jet at a PA ≈ 160 degrees. The HH object knot A has a tail pointing back to LkHα 198B and is part of an outflow from this star and is not associated with LkHα 198 as previously thought.

reflection loops seen in Fig. 6.   It is faintly visible in the optical, despite being immersed in reflection nebulosity, but is clearly seen in the near infrared (Corcoran *et al.*, 1994a). At 10μm it is almost of comparable luminosity to LkHα 198 (Lagage *et al.*, 1993). Whether LkHα 198 or LkHα 198B is responsible for driving the molecular outflow in this region is unknown. The latter source however is the more likely candidate as it is more embedded. Turning to LkHα 198 itself, this also has a collimated optical outflow, i.e. a HH jet, traces of which can be seen to the west of the innermost brightest reflection loop (Fig. 6) at a P.A. ≈ 160 degrees. An additional HH object (at a projected separation of 81" from LkHα 198) has also been found (Goodrich, 1993) directly along the LkHα 198 jet axis. Part of a counterflow may also be visible (i.e. knot F in Fig. 6).

There could be a tendency for optical outflows from intermediate mass stars to be more poorly-collimated than their lower mass counterparts (see, for example, Poetzel *et al.*, 1992). A classic example, although not strictly a HAEBE star since it is embedded, is Cep A. To the west of this source lies GGD 37, a large complex of arc-shaped HH objects which have been individually modeled as bow shocks (Hartigan, Raymond and Hartmann, 1987). The GGD 37 HH objects are widely scattered in position angle giving the outflow a poorly-collimated appearance. Such a morphology contrasts sharply with that of the outflows from low mass stars, like HH34-IRS (Bührke, Mundt and Ray, 1988), which are almost invariably highly collimated, at least far from the source (Edwards, Ray and Mundt, 1993). The eastern optical outflow from Cep A is similarly observed to be poorly collimated (Corcoran *et al.*, 1993a).

There are few other morphological differences between the optical outflows from low and intermediate mass stars although the outflows from intermediate mass stars are larger, as one might expect. With regard to kinematics, optical outflows from HAEBE stars tend to have (radial) velocities which are typically 2-3 times higher than those from low mass stars. Such a growth in velocity is more or less in line with the increase in the escape velocity of the star.

Of particular interest is the fact that mass loss rates of optical outflows scale in the same way with luminosity, irrespective of the mass of the star (Edwards *et al.*, 1993). This clearly points to one mechanism being responsible for the outflow phenomenon in both T Tauri stars and Herbig Ae/Be stars, although what that mechanism might be is far from certain. It should also be mentioned that radio studies, in addition to optical observations, can give us valuable information regarding mass loss from HAEBE stars. As in the case of CTTSs the fluxes are typically very low (usually a few hundred microJanskys) and the emission is quite compact (scales of a few arcseconds at most) so that only instruments like the VLA can carry out this type of observation. Recently Skinner, Brown and Stewart (1993) have finished an extensive survey of radio continuum emission from HAEBE stars. They detected approximately one quarter of them and find that the emission is predominantly thermal and in many cases appears to be wind-related. There is a clear deficit of A-type stars amongst those detected probably because of the decreasing levels of ionization in their winds. Both the optical and the radio data suggest mass loss scales with the luminosity of the star like $L_*^{0.6}$.

5   Some Theoretical Comments

We make only a few salient points here regarding the evolution of HAEBE stars and the formation of their optical outflows. The reader is referred to Palla and Stahler (1990; 1992; 1993) for more details of the former topic and Ray and Mundt (1994) for the latter.

Low mass stars and intermediate mass stars were thought originally to have evolve differently onto the main sequence. In the case of intermediate mass stars, early estimates (e.g. Larson, 1972) implied that the accretion times were longer than the Kelvin-Helmholtz time; such stars would thus contract and begin burning hydrogen *while still accreting*. Intermediate mass young stellar objects were thus expected to join the main sequence as protostars and thus any subsequent evolution would be *along, and not towards, the main sequence*. Based on such calculations, we would not have expected to observe pre-main sequence stars with masses greater than say 2-3 $M_\odot$. This, of course, flies in the face of observational reality given the large number of HAEBE stars known!

The principal problem with the earlier protostar calculations (such as those made for example by Iben, 1965) is that they neglected the effects of deuterium burning. Palla and Stahler (1990; 1992; 1993) have extended spherical accretion models to include its effects. Briefly they find that for intermediate masses, deuterium burning occurs in an outer sub-surface shell causing a swelling in the protostar radius. This increase in the radius lowers the temperature of the inner core, thereby allowing accretion to higher masses take place before hydrogen ignites.

As with CTTSs, such protostellar collapse calculations allow one to establish an initial mass-radius relationship for intermediate mass stars, i.e. a "birthline". As can be seen from Fig. 7, there is remarkable agreement between the limits set by the birthline and the positions of the HAEBE stars in the HR diagram. While such agreement is encouraging, we should caution however that the positioning of these stars *depends not only on the luminosity of the star, but also on its accretion luminosity,* which is not allowed for. Moreover, as we have already mentioned, extinction values towards some of these stars are unreliable and so there are corresponding further uncertainties in their positions in the HR diagram.

Turning now to the optical outflows seen from some HAEBE stars, the question naturally arises as to what drives them? Edwards *et al.*(1993) have claimed that it is the disk, rather than the star, that is ultimately responsible for producing an outflow from a low mass young star. They argue this on several grounds, including the lack of any outflows associated with weak-line T Tauri stars which in other respects resemble the CTTSs. Since however (see §4) it is likely that the same mechanism is responsible for outflows from both low and intermediate mass young stars, disks would have to surround at least some HAEBE stars. Apart from the presence of HH flows there are other observational characteristics that point to a close link between the outflows from CTTSs and some HAEBE stars. In particular,

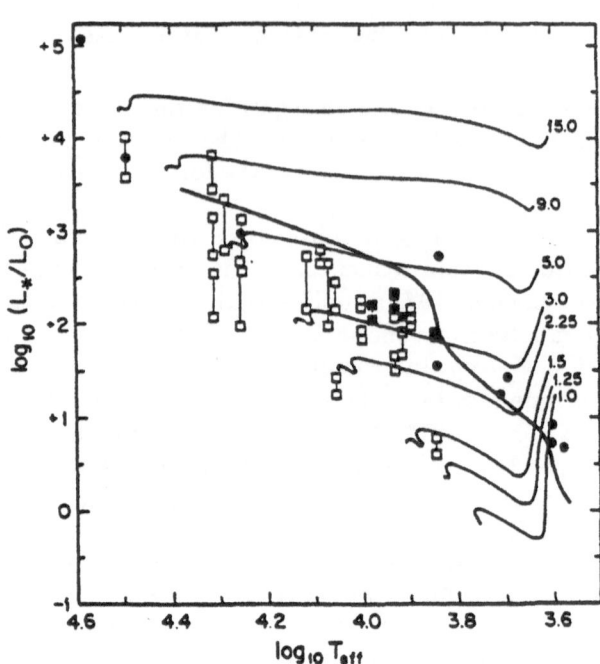

Fig. 7.   Evolutionary tracks and the location of the stellar birthline according to Palla and
Stahler (1990). The birthline is seen to transverse the evolutionary tracks of Iben (1965) for
stars of various masses. The empty squares represent HAEBE stars and the filled symbols are
molecular outflow sources. Note the good agreement of the upper envelope of these stars with
the calculated birthline when one allows for deuterium burning.

the strength of the Hα emission in HAEBE stars scales with the excess luminosity
(Hillenbrand *et al.*1992) in the same way as in CTTSs (see, for example, Cabrit *et
al.*, 1990). Moreover there is a smooth transition in such properties between the
two groups of stars. This argues strongly, as with mass loss rates (see §4), *for one
physical mechanism* being responsible for outflows in low and intermediate mass
stars.

    In summary Herbig Ae/Be stars resemble their lower mass counterparts, the
T Tauri stars. This is true not only for the stars themselves but also for large
scale phenomenona like outflows. It seems highly plausible that disks surround at
least some Herbig Ae/Be stars but it remains to be seen whether such disks are
common.

Acknowledgements

TR would like to thank Jochen Eislöffel and Dave Corcoran for critically reading
this manuscript. He also appreciates John Dyson's patience in awaiting delivery!

References

Basri, G. and Bertout, C.: 1993, in *Protostars and Planets III*, eds. E. Levy and J. Lunine (University of Arizona Press), p. 543.
Bertout, C.: 1989, *Ann. Rev. Astron. Astrophys.* **27**, 351.
Blondel, P.F.C., Talavera, A. and Tjin A Djie, H.R.E.: 1993, *Astron. Astrophys.* **268**, 624.
Brugel, E.W., Mundt, R. and Bührke, T.: 1984, *Astrophys. J. Lett.* **287**, L73.
Bührke, T., Mundt, R. and Ray, T.P.: 1988, *Astron. Astrophys.* **200**, 99.
Cabrit, S., Edwards, S., Strom, S.E. and Strom, K.M.: 1990, *Astrophys. J.* **354**, 687.
Camenzind, M. 1990, *Rev. Modern Astron.* **3**, 234.
Catala, C.: 1989, in *Low Mass Star Formation and Pre-Main-Sequence Evolution*, ed. B. Reipurth, (ESO, Garching), p. 471.
Cohen, M.: 1980, *Mon. Not. Roy. Astr. Soc.* **191**, 499.
Cohen, M. and Kuhi, L.V.: 1979, *Astrophys. J. Suppl.* **41**, 743.
Corcoran, D. and Ray, T.P.: 1994b, in preparation.
Corcoran, D., Ray, T.P. and Bastien, P.: 1994a, *Astron. Astrophys.*, submitted.
Corcoran, D., Ray, T.P. and Mundt, R.: 1993a: *Astron. Astrophys.* **279**, 206.
Corcoran, D., Ray, T.P. and Mundt, R., Poetzel, R.: 1993b, in *Stellar Jets and Bipolar Outflows*: Proceedings of the 6th International Workshop of the OAC (1991), eds. L. Errico and A. Vittone (Dordrecht, Kluwer), in press.
Corcoran, M. and Ray, T.P.: 1994a, in *The Nature and Evolutionary Status of Herbig Ae/Be Stars*, eds. P.S. Thé, M.R. Perez and E.P.J. van den Heuvel, P.A.S.P. Conf. Ser., in press.
Edwards, S., Ray, T.P. and Mundt, R.: 1993, in *Protostars and Planets III*, eds. E.H. Levy and J. Lunine, (University of Arizona press), p. 567.
Eislöffel, J. and Ray, T.P.: 1994, in *Nature and Evolutionary Status of Herbig Ae/Be Stars*, ed. P.S. Thé, M.R. Perez and E.P.J. van den Heuvel, P.A.S.P. Conf. Ser., in press.
Finkenzeller, U.: 1985, *Astron. Astrophys.* **151**, 340.
Finkenzeller, U. and Mundt, R., 1984, *Astron. Astrophys. Suppl.* **55**, 109.
Goodrich, R.W.: 1986, *Astrophys. J.* **311**, 882.
Goodrich, R.W.: 1993, *Astrophys. J. Suppl.* **86**, 499.
Hamann, F. and Persson, S.E.: 1989, *Astrophys. J.* **339**, 1078.
Hamann, F. and Persson, S.E.: 1992, *Astrophys. J. Suppl.* **82**, 285.
Herbig, G.: 1960, *Astrophys. J. Suppl.* **4**, 337.
Hartigan, P. Raymond, J.C. and Hartmann, L.: 1987, *Astrophys. J.* **316**, 323.
Hartmann, L., Kenyon, S.J. and Calvet, N.: 1993, *Astrophys. J.* **407**, 219.
Hillenbrand, L.A., Strom, S.E., Vrba, F.J. and Keene, J.: 1992, *Astrophys. J.* **397**, 613.
Hu, J.Y., Thé, P.S. and de Winter, D.: 1989, *Astron. Astrophys.* **208**, 213.
Iben, I.: 1965, *Astrophys. J.* **141**, 993.
Königl, A. 1991, *Astrophys. J. Lett.* **370**, L39.
Lagage, P.O., Olofsson, G., Cabrit, S., Cesarsky, C.J., Nordh, L. and Rodriguez Espinosa, J.M.: 1993, *Astrophys. J.* **417**, L79.
Larson, R.B.: 1972, *Mon. Not. Roy. Astr. Soc.* **157**, 121.
Lenzen, R.: 1988, *Astron. Astrophys.* **190**, 269.
Lorenzetti, D., Saraceno, P. and Stratfella, F.: 1983, *Astrophys. J.* **264**, 554.
Martí, Rodríguez, L.F. and Reipurth, B.: 1993, *Astrophys. J.* **416**, 208.
Natta , A., Palla, F., Butner, H.M., Evans, N.J. and Harvey, P.M.: 1993, *Astrophys. J.* **406**, 674.
Palla, P. and Stahler, S.W.: 1990, *Astrophys. J. Lett.* **360**, L47.
Palla, P. and Stahler, S.W.: 1992, *Astrophys. J.* **392**, 667.
Palla, P. and Stahler, S.W.: 1993, *Astrophys. J.* **418**, 414.
Poetzel, R., Mundt, R. and Ray, T.P.: 1989, *Astron. Astrophys.* **224**, L13.
Poetzel, R., Mundt, R. and Ray, T.P.: 1992, *Astron. Astrophys.* **262**, 229.
Ray., T.P., 1993, in *The Cold Universe*, Proceedings of the XIIIth Moriond Astrophysics Meeting, eds. T. Montmerle, C. Lada and J. Tran Thanh Van, Editions Frontières, in press.
Ray, T.P. and Mundt, R.: 1994, in *The Nature and Evolutionary Status of Herbig Ae/Be Stars*, eds. P.S. Thé, M.R. Perez and E.P.J. van den Heuvel, P.A.S.P. Conf. Ser., in press.
Ray, T.P., Poetzel, R., Solf, J. and Mundt, R.: 1990 *Astrophys. J. Lett.* **357**, L45.

Rodríguez, L.F. and Reipurth, B.: 1989, *Rev. Mex. Astr. Astrophys.* **17**, 59.

Siebenmorgen, R., Krügel, E. and Mathis, J.S.: 1992 *Astron. Astrophys.* **266**, 501.

Skinner, S.L., Brown, A., Stewart, R.T.: 1993, *Astrophys. J. Suppl.*, submitted.

Strom, K.M., Strom, S.E., Wolff, S.C., Morgan, J. and Wenz, M.: 1986, *Astrophys. J. Suppl.* **62**, 39.

Strom, S.E., Strom, K.M., Yost, J., Carrasco, L. and Grasdalen, G.: 1972, *Astrophys. J.* **173**, 353.

Zou, H.: 1989, Ph.D. Thesis, University of Heidelberg.

3-D radiative line transfer for Be star envelopes

W. Hummel
Astronomisches Institut, Ruhr-Universität Bochum, Postfach 10 21 48, W-4630 Bochum 1, Germany

Abstract. Based on numerical three-dimensional radiative line transfer calculations Hα emission line profiles of circumstellar Be star envelopes have been derived. The results show that the so-called winebottle-type emission line profiles can be explained by the combination of rotational broadening and non-coherent scattering in optically thick Keplerian disks. In a further calculation the stellar wind model of Be star envelopes has been re-investigated assuming an additional expansion component in the velocity field. The resulting asymmetric winebottle-type profiles and asymmetric shell-type emission lines with blue-shifted central depressions are in contradiction with the observed line shapes. It is concluded that isotropic stationary outflows are not suitable to explain observed asymmetric emission line profiles of Be stars.

Key words: Be stars, line transfer, emission lines

1 Introduction

Be stars are defined as early-type stars of spectral class B and luminosity classes III-V which at least sometimes exhibit emission-lines in their spectra. These emission line profiles (hereafter ELPs) are generated in circumstellar gaseous envelopes (Struve, 1931). Be stars are fast rotators and are variable on timescales between days and years in most spectral regions. The origin and evolution of Be star circumstellar envelopes (hereafter CEs) is still unknown.

Observed high-resolution Hα-ELPs of Be stars can be divided into four groups (Dachs, 1987; Hanuschik et al., 1988): (a) double-peak profiles, (b) winebottle-type profiles, (c) shell profiles with a deep central depression below the stellar continuum and (d) asymmetric profiles. Double-peak profiles are due to kinematic broadening (Smak, 1969); shell profiles can be understood by shear broadening (Horne & Marsh, 1986). The so-called winebottle-type structure can be explained by the combined influence of kinematic broadening and non-coherent scattering broadening in a homogeneous disk (Hummel & Dachs, 1992).

Asymmetric ELPs of Be stars are supposed, according to Marlborough (1969) to arise from a radially symmetric expansion in the CE, caused by a radiatively driven stellar wind. Alternatively eccentric trajectories like elliptical rings (Huang, 1973), elliptical disks (Křiž, 1979) or disk perturbations (Kato, 1983) are discussed.

Based on 3-D radiative line transfer calculations, we present theoretical Hα ELPs for two Be star model CEs: a Keplerian disk with an inhomogeneous density distribution (model A) and a Keplerian disk with an additional expansion velocity (model B) in order to test the stellar wind model of Be stars.

Astrophysics and Space Science **216**: 87–91, 1994.
© 1994 *Kluwer Academic Publishers.*

2   The model

The two models consist of a central B3III star, represented by a sphere including
limb-darkening. The isothermal, purely hydrogenic CE extends up to $R_d = 10R_*$
and the vertical thickness of the disk is controlled by the scale-height $H(r)$. The
envelope is in hydrostatic equlibrium in the z-direction, while in r-direction a
power-law ($\alpha = 2$) is assumed (Pringle, 1981):

$$N_H = N_0 \left(\frac{r}{R_*}\right)^{-\alpha} \exp\left[-\frac{1}{2}\left(\frac{z}{H(r)}\right)^2\right]$$

3   3-D radiative line transfer

The radiative line transfer equation is solved by an implicit first order finite volume
element method for a two-level atom with complete redistribution in the observer's
frame in three spatial dimensions (Adam *et al.*, 1989). We used 51 spatial gridpoints in each dimension of the cartesian grid and 80 angle gridpoints for $\frac{1}{8}$ of
the solid angle 4π. The wavelength step-size $\Delta\lambda = 0.36$Å is equal to the thermal
width $\Delta\lambda_{th} = 0.4$Å of the intrinsic profile function (at $T = 20000$K for hydrogen).
After the convergence of the Λ-iteration the emission line profiles F_λ are calculated
for four inclinations i, where i is the angle between the disc rotation axis and the
observer.

4   Model A: Keplerian Disk

The resulting ELPs are shown in Fig.1 at four different inclinations i for a Keplerian
disk which is optically thin (lower curves, $N_0 = 10^{12}$cm^{-3}) and for another one
that is optically thick (upper curves, $N_0 = 10^{14}$cm^{-3}) for vertical line radiation.

4.1   OPTICALLY THIN KEPLERIAN DISK

For $i = 0^0$ (pole-on view) neither kinematics nor scattering influences the line
radiation and the emergent intensity is the intrinsic profile function. For higher
values of i the rotational broadening distribution for Keplerian disks broadens the
ELP yielding a double-peak profile (Smak, 1969). At edge-on view ($i = 90^0$), the
disk is becoming optically thick and shell profiles arise as a consequence of shear
broadening (Rybicky & Hummer, 1983; Horne & Marsh, 1986).

4.2   OPTICALLY THICK KEPLERIAN DISK

For $N_0 = 10^{14}$cm^{-3} the envelope is optically thick already for line radiation in the
vertical direction. At $i = 0^0$ the typical double-peak profile of an optically thick
finite slab is visible and can be understood by non-coherent scattering broadening:
Line photons which are absorbed and re-emitted several times scatter from the
line center into the line wings, where they can already escape from deeper regions

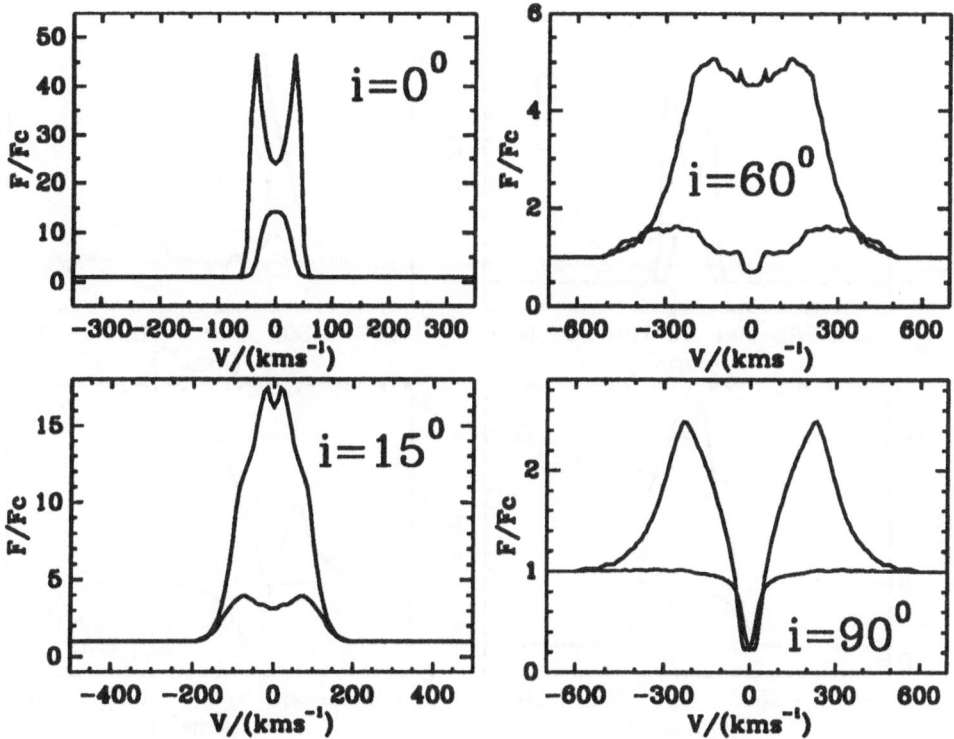

Fig. 1.   Emission line profiles of a Keplerian disk (model A) for different inclination angles i. Upper profiles: $N_0 = 10^{14} \mathrm{cm}^{-3}$. Lower profiles: $N_0 = 10^{12} \mathrm{cm}^{-3}$, enlarged by a factor of 10, except at $i = 90^0$.

of the slab. In other words: In the line wings, the observer looks deeper into the disk where the source function is larger.

Obviously, for any inclination $i > 0^0$ in the optically thick case each peak of the rotationally broadened double-peak profile (lower curves) will split into two peaks due to non-coherent scattering, producing a wine-bottle-type ELP. Usually, among the four secondary peaks, only those with smallest radial velocities appear as true peaks, while the other two peaks appear as flank inflections. For edge-on view, this fine structure vanishes and shear broadening is dominant resulting into a shell-type profile. This model explains all types of observed symmetric Hα ELPs of Be stars in the full inclination range $i = [0^0...90^0]$.

5   Model B: Expanding Keplerian Disk

In order to re-investigate a frequently discussed Be star model assuming radial outflows across the disk-shaped CE (Marlborough, 1969), an additional linear expansion velocity: $V_{ex} = \Delta V_{th} * (\tilde{r}/R_*)$ is superimposed to the Keplerian rotation

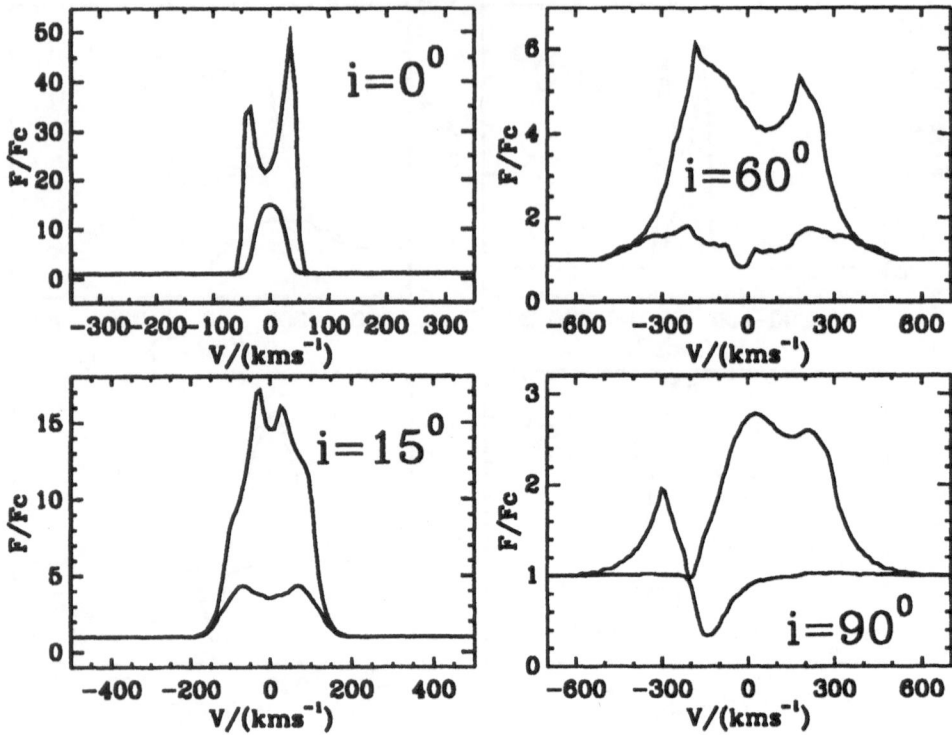

Fig. 2.   Emission line profiles of a slowly expanding Keplerian disk (model B) for different inclination angles i. Upper profiles: $N_0 = 10^{14}\,\mathrm{cm}^{-3}$. Lower profiles: $N_0 = 10^{12}\,\mathrm{cm}^{-3}$, enlarged by a factor of 10, except at $i = 90^0$.

of model A, where $\tilde{r} = \sqrt{r^2 + z^2}$. Fig.2 shows the resulting ELPs of model B for four inclination angles and the footpoint densities $N_0 = 10^{12}\mathrm{cm}^{-3}$ (lower curves) and $N_0 = 10^{14}\mathrm{cm}^{-3}$ (upper curves).

5.1   OPTICALLY THIN EXPANDING KEPLERIAN DISK

The resulting ELPs (Fig.2, lower profiles) are symmetric up to $i \simeq 45^0$. At edge-on view ($i = 90^0$) an asymmetric and blue-shifted depression occurs as a consequence of the expansion velocity of the gas along the line of sight.

5.2   OPTICALLY THICK EXPANDING KEPLERIAN DISK

For $i = 0^0$, the vertical component of the radially symmetric expansion velocity V_{ex} is responsible for the red-side dominated double peak profile. Asymmetric winebottle-type ELPs with the opposite sign of asymmetry ($F_{\mathrm{Violet}}/F_{\mathrm{Red}} > 1$; short: $V/R > 1$) are produced in the inclination range $15^0 \leq i \leq 60^0$. For nearly edge-on view the expansion velocity dominates the ELP formation in the CE. The blue-shifted line depression at $V_{\mathrm{rad}} = -180\mathrm{kms}^{-1}$ for $i = 90^0$ corresponds to the maximum expansion velocity at the outer rim of the disk ($R_\mathrm{d} = 10R_*$) in front of the star. The resulting ELPs are similar to those obtained in the calculations of Poeckert & Marlborough (1978) and Waters & Marlborough (1992).

Comparison with the observations (Dachs et al., 1992; Hanuschik et al., 1988) shows that both signs of asymmetry (V>R as well as V<R) are observed in Hα winebottle-type ELPs. The large negative radial velocity of the depression of calculated profiles (Fig.2, $i = 90^0$) indicating the expansion velocity at $r = R_\mathrm{d}$ is clearly at variance with typical values of $V_\mathrm{rad} \simeq \pm 40\mathrm{kms}^{-1}$ in observed Hα shell profiles. If this observed small velocity shift V_rad is interpreted in terms of maximum expansion velocities then no significant ELP asymmetry should be expected at small inclinations (Fig.2, $i = 15^0$). Furthermore, this model B does not explain the cyclic V/R variability observed in Hα ELPs (Cowley & Gugula, 1973; Hanuschik, 1993).

6   Conclusions

The present model of an inhomogeneous Keplerian disk with large vertical optical thickness for line radiation can explain all types of symmetric Hα emission line profiles observed in Be star spectra.

Line profiles calculated under the assumption of an additional radially symmetric expansion velocity are in contradiction with observed asymmetric Hα line profiles. Therefore there is no evidence that the formation of Be star envelopes is based on a stationary mass loss caused by a radiatively driven stellar wind. Elliptical trajectories in the disk could be more suitable to understand the properties of asymmetric Be star emission line profiles (Hanuschik, 1993).

References

Adam, J., Innes, D.E., Shaviv, G., Störzer, H., Wehrse, R.: 1989, *Theory of Accretion Discs*, (eds. Meyer, F., Duschl, W.J., Frank, U., Meyer-Hofmeister, E.) NATO ASI C **290**, 403
Cowley, A., Gugula, E.: 1973, *Astron. Astrophys.* **22**, 203
Dachs, J.: 1987, in *'Physics of Be stars'*, IAU Coll. **92**, 149
Dachs, J., Hummel, W., Hanuschik, R.W.: 1992, *Astron. Astrophys. Suppl.* **95**, 437
Hanuschik, R.W.: 1993, *these proceedings*
Hanuschik, R.W., Kozok, J.R., Kaiser, D.: 1988, *Astron. Astrophys.* **189**, 147
Horne, K., Marsh, T.: 1986, *MNRAS* **218**, 761
Huang, S.S: 1973, *Astrophys. J.* **183**, 541
Hummel, W., Dachs, J.: 1992, *Astron. Astrophys.* **262**, L17
Kato, S.: 1983, *Publ. Astron. Soc. Japan* **35**, 24
Kříž, S.: 1979, *Bull. Astron. Soc. Czes.* **30**, 951
Marlborough, J.M.: 1969, *Astrophys. J.* **156**, 575
Poeckert, R., Marlborough, J.M.: 1978, *Astrophys. J. Suppl.* **38**, 229
Pringle, J.: 1981, *Ann. Rev. Astron. Astrophys.* **14**, 137
Rybicky, G.B., Hummer, D.G.: 1983, *Astrophys. J.* **274**, 380
Smak, J.: 1969, *Acta Astron.* **19**, 155
Struve, O.: 1931, *Astrophys. J.* **73**, 94
Waters, L.B.F.M., Marlborough, J.M.: 1992, *Astron. Astrophys.* **253**, L25

Radiatively Driven Winds Using Lagrangian Hydrodynamics

John M. Porter
Astrophysics,
Nuclear and Astrophysics laboratory,
Keble Road,
Oxford,
OX1 3RH.

Abstract. The study of radiatively driven winds in O and B stars has been achieved mainly through analytic solutions of the equation of motion. These solutions can only be applied to simple geometries. Presented here are preliminary results of a particle based Lagrangian code (applied to a simple 1D radiatively driven wind test case), which will be able to resolve structure down to the smallest scales without becoming temporally prohibitive. This approach to modelling radiatively driven winds can be applied to the dispersal of massive young stellar objects' natal circumstellar material, Be star wind structure and CV disc winds.

Key words: stellar winds, hydrodynamics

1   Introduction

The model of an ionised wind being driven predominantly by spectral lines from the central star is very successful in explaining the overall features of massive OB star winds. The basic steady state model for radiation driven mass loss was derived by Castor, Abbott and Klein, 1975. However, linear stability analyses (Owocki and Rybicki, 1986) have showed that line driven winds are extremely unstable and thus nonlinear growth of instabilities may give the flow considerable structure (see also Owocki, Castor and Rybicki, 1988). In order to examine fully the temporal and spatial characteristics of radiatively driven flows, the equation of motion needs to be solved using a hydrodynamic code.

2   1D radiatively driven wind simulation

Lagrangian codes represent the fluid as a set of discrete particles. In order to solve the flow, internal (pressure) and external (gravity) forces are calculated on each individual particle and it is then moved in space under the action of these forces. The most well known technique of this sort is that of smoothed particle hydrodynamics (SPH). This is not used here though - a Lagrangian grid is used. This is a grid where the nodes are coincident with the particle positions, and which will change, as the particles move, to ensure monotonicity - the particles r and s at physical positions (r_x, r_y, r_z) and (s_x, s_y, s_z) will have positions (i, j, k) and (m, n, o) respectively in the grid, where $r_x < s_x$ if $i < m$, $r_y < s_y$ if $j < n$ and $r_z < s_z$ if $k < o$. This grid gives immediate information on a particles' nearest neighbours - something which SPH codes do not provide.

   In this implementation (written by Rob Whitehurst), a standard force multpiler

Astrophysics and Space Science **216**: 93–94, 1994.
© 1994 *Kluwer Academic Publishers.*

is used, of the form $M = kt^{-\alpha}$, where k and α have the values 1/30 and 0.7 respectively, to represent the line force on the gas. Figure 1 shows the output from the simulation of a wind from a $60M_\odot$, $9.66\times10^5 L_\odot$ star (the mass loss rate is $\sim6.6\times10^{-6} \dot{M}_\odot \, yr^{-1}$ and the terminal velocity is $\sim2000 km \, s^{-1}$).

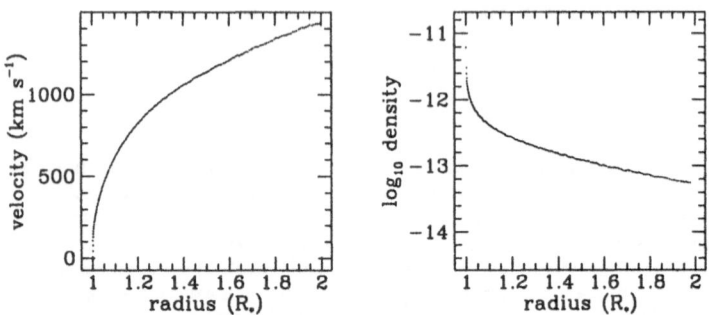

Fig. 1.   Final velocity and density profiles of a $60M_\odot$ 1D wind simulation

3   Applications

It is highly probable that late in the obscured stage of massive young star evolution, it is encircled by a disc like structure. In high mass young stellar objects (eg. the exciting source of the S106 nebula), millimeter molecular line observations suggest elongated structures, which may be large scale discs. These structures are undetectable in MS stars of the same spectral type. The effect of radiation pressure from a central star, on a disc like structure needs to be examined to ascertain if it is important in disc dispersal. Analysis suggests that this effect may be too weak to produce such a dispersion. However, potentially unphysical assumptions used in the analysis may be avoided with the implementation of this Lagrangian mesh code. Also, the mechanical effects of a central stellar wind impacting on the circumstellar material could, in principle, be included in the code.

    Bjorkman and Cassinelli (1993, preprint) have proposed a model for both the equatorial structure and the stellar wind of Be stars. The model predicts that a disc would form in the star's equatorial plane due to streamline convergence of an initially isotropic radiatively driven wind. The large contrast between the outflow velocities at the poles and the equator from observations would then be explained. A hydrodynamical investigation of this model would also be possible with this code, and would give insight into the stability of this "compression" disc.

References

Castor, J. I., Abbott, D. C. and Klein, R. I.: 1975, *Astrophys. J.***195**, 157.
Owocki, S. P., Castor, J. I. and Rybicki, G. B.: 1988, *Astrophys. J.***335**, 914.
Owocki, S. P. and Rybicki, G. B.: 1986, *Astrophys. J.***309**, 127

Parametric determination of the inclination of Keplerian circumstellar discs from spectropolarimetric profiles of scattered lines

Kenneth Wood
Department of Physics and Astronomy, The University of Glasgow

March 1993

Abstract. Polarimetric line profiles arising from the Doppler redistribution of monochromatic stellar line radiation, Thomson scattered in a Keplerian rotating circumstellar disc are presented. It is shown that analysis of the scattered line profiles at different wavelengths which, due to Doppler redistribution, sample different disc regions allows the disc inclination to be determined.

Key words: Thomson scattering, lines, spectropolarimetry

1 Introduction

The spectral shape of the Stokes fluxes resulting from Thomson scattering of stellar line radiation in a moving planar disc are determined by considering the direction of the maximum scattered electric vector (polarisation direction) and also the Doppler shift of scattered radiation at each point on the disc, the latter being illustrated by isowavelength–shift contours which determine the relative wavelength shift due to scattering from different disc regions, Wood et al (1993). Assuming the initial stellar line is monochromatic and adopting certain parametrisations of the disc velocity and density distributions yields analytic forms for the scattered Stokes fluxes, thus allowing most of the disc parameters to be determined from analysis of spectropolarimetric data of scattered lines, Wood & Brown (1993).

The spectropolarimetric line profiles arising from scattering in a rotating disc and the method for determining the inclination and maximum rotational speed are presented below.

2 Keplerian circumstellar disc

The Stokes fluxes arising from scattering of stellar line radiation in a Keplerian rotating disc (with $\beta_0 c$ being the rotational speed at the stellar surface) are shown in Fig. 1, where $\alpha = (\lambda - \lambda_0)/\lambda_0$ is the dimensionless wavelength shift from line centre. Due to the symmetry of the isowavelength–shift contours between the front and rear halves of the rotating disc (Fig. 2) the U component of the flux cancels out on summation over each contour, hence the position angle is constant across the line. At line centre (the $\alpha = 0$ contour), the Q flux is negative, as is evident from the scattered polarisation direction along this contour. Moving away from line centre, more positive Q contributions raise the Q flux to a maximum positive value, beyond which the shape of the contours yield more negative contributions, thus reducing the Q flux to zero at $|\alpha| = \beta_0 \sin i$ where there is no scattered flux

Astrophysics and Space Science **216**: 95–97, 1994.
© 1994 *Kluwer Academic Publishers.*

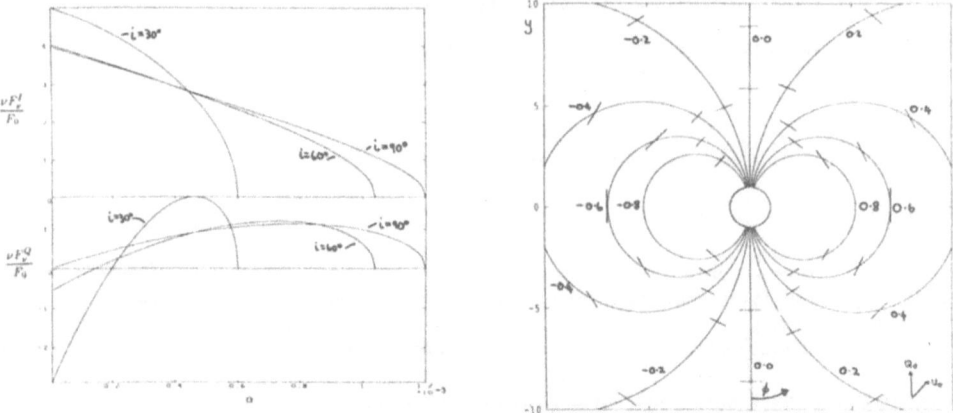

Fig. 1.   (a) The resulting spectral shapes of the Stokes fluxes after scattering monochromatic stellar line radiation in a rotating disc whose velocity at the stellar surface is 450 km.s^{-1}. The profiles are symmetric about $\alpha = 0$ so only one half is shown for inclinations of 30°, 60° and 90°. (b) Isowavelength–shift contours, $\alpha \times 10^3$ = Constant, on a rotating disc viewed by an observer inclined at 60° to the rotation axis and positioned at infinity in the plane containing the rotation axis and the line $\phi = 0$. The axes are labelled in units of stellar radii. The short bars are the projection on the sky of the local polarisation vector for scattering of stellar radiation from different disc regions. The positive Stokes Q and U directions, (Q_0, U_0), for the entire system are also shown.

since no material is moving fast enough to give the required Doppler redistribution of the scattered radiation.

Use of a scattered spectral line profile, as opposed to broad–band measurements – cf Brown & McLean (1977), allows different parts of the disc to be picked out as stated above. Setting $\alpha \approx 0$ (i.e. close to the centre of the scattered line, but outwith the stellar line) picks out the isowavelength–shift contour elements at $\phi = 0, \pi$ where the scattering angles are $\pi/2 + i$ and $\pi/2 - i$, giving different values of the scattered to polarised fluxes. This then allows the inclination to be determined from solution of,

$$\frac{F_\nu^Q(\alpha \approx 0)}{F_\nu^I(\alpha \approx 0)} = \frac{-\cos^2 i}{1 + \sin^2 i} \, . \tag{1}$$

The parameter β_0 is then obtained from the maximum wavelength shift of the scattered flux profiles ($\alpha_{max} = \beta_0 \sin i$) using the inclination value determined from Eq. 1.

3   Conclusions

This analysis has demonstrated how wavelength dependent polarimetric line profiles may arise through scattering alone without having to appeal to other opacity sources, as had previously been the case (Poeckert and Marlborough, 1978). For the case presented here of a Keplerian rotating disc the position angle is constant across the scattered line. However, when the disc velocity is a combination of rotation *and* expansion position angle changes *do* occur due to the symmetry breaking of the isowavelength–shift contours (Wood et al, 1993).

The work presented is applicable to the scattering of *any* stellar absorption or emission lines in rotating discs and illustrates that analysis of the resulting scattered polarimetric line profiles yield disc parameters, in particular the inclination, which are not obtainable from continuum spectroscopy or photometry. Forthcoming advances in high resolution CCD polarimeters and the further development of this theoretical framework for interpreting polarimetric line profiles will yield a novel method for determining the structure of stellar winds.

References

Brown J.C., Mclean I.S., 1977, *Astron. Astrophys.*, **57**, 141
Poeckert, R., Marlborough, J.M. : 1978, *Ap. J.*, **220**, 940.
Wood, K., Brown, J.C., Fox, G.K. : 1993, *Astron. Astrophys.*, **271**, 492.
Wood, K., Brown, J.C. : 1993, *Astron. Astrophys.*, submitted.

Observational Evidence for Global Oscillations in Be Star Disks

R. W. Hanuschik

Astronomisches Institut, Ruhr-Universität Bochum, D-44780 Bochum, Germany

Abstract. From a long-term spectroscopic observing campaign of Be stars we have found that their emission line profiles separate into two classes, the second of which is very asymmetric. The sign of asymmetry changes on a timescale of a few years. We show that these line profiles and the well-known enigmatic V/R variations observed in some Be stars are two aspects of the same phenomenon, one-armed global oscillations in a Keplerian disk. We present model profiles and a line fit to support this hypothesis and find good overall agreement.

Key words: Be stars, emission lines, circumstellar matter

1 Introduction: Emission lines from Be star disks

1.1 BE STARS

Be stars are early-type stars surrounded by an extended circumstellar envelope, the origin of which is still unclear. We have started an extensive observing programme of emission lines in Be stars in 1982 (Hanuschik, 1986, 1987, Hanuschik *et al.*, 1988, Dachs *et al.*, 1992) using ESO's 1.4m Coudé Auxiliary Telescope and the Coudé Echelle Spectrograph. Profiles have been measured at high resolution ($R \geq 50000$, $\Delta v \leq 6$ km s^{-1}) and high SNR (100 – 1000). We have focused our survey on a comparative study of optically thick (Hα, Hβ) and optically thin (mainly Fe II λ5317) emission lines. The latter are mainly kinematically broadened and offer insight into the kinematic and density structure of the circumstellar envelope. The intrinsic resolution of this structure is limited by the thermal width of Fe lines at 10^4 K, 3 km s^{-1}.

The optically thick Balmer lines are broadened in addition by radiative transfer effects [winebottle-type profile (Hummel & Dachs, 1992; Hummel, these proceedings), electron scattering wings] which dominate over kinematic broadening. Therefore these lines mainly provide gross information on parameters like total emission strength.

1.2 EMISSION LINE PROFILES

We have found that virtually all Fe II emission line profiles can be divided into two classes (cf. Fig. 1): *class 1* profiles showing symmetric, twin-peak profiles, with $V \approx R$ (the violet and red peak flux, resp.), and *class 2* profiles, i.e. asymmetric profiles ($V \neq R$) with often only a single, sometimes very sharp peak. Prototype examples of these "steeple-type" profiles are those of δ Cen (Fig. 1). About 2/3 of our programme stars show class 1 profiles. In addition, a subvariety exists, with a deep, sharp central absorption, the *shell profiles*. These profiles arise in case of inclination $i \approx 90°$ when part of the stellar photosphere becomes occulted by the

Astrophysics and Space Science 216: 99–103, 1994.

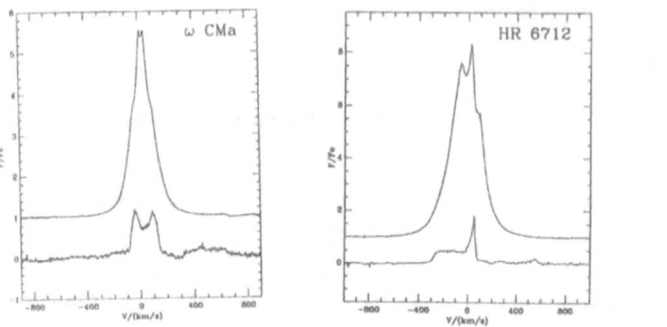 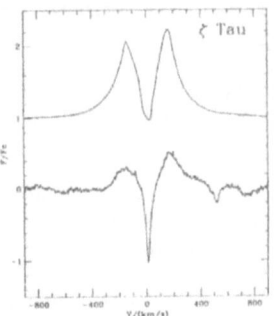

Fig. 1. Typical Hα (top) and Fe II λ5317 (bottom) emission line profiles. From left to right: class 1, class 2, shell profile (from Dachs *et al.*, 1992). Flux scale for Fe II line is 10 times enlarged

circumstellar disk.

A similar classification scheme can be defined for the Hα profiles; however, differences in profile shape are less pronounced for these profiles as their kinematical information is washed out, and we prefer to investigate Fe II line profiles whenever they are available.

For the class 1 profiles it has been shown (e.g., Hanuschik, 1989) that Be star envelopes essentially are Keplerian disks, with a structure similar to classical accretion disks. Typical disk radii are less than 10 stellar radii.

2   Class 2 profiles

Observed properties of Fe II profiles of class 2 are in puzzling contrast to these findings. Most important properties are:

1. They are constant in shape and overall intensity on timescales shorter than a few months.

2. On timescales of a year or more we observe quite frequently a change of symmetry (Fig. 2) with a quasi-cyclic behaviour.

3. Though rather complex, *the profile shape looks very similar in different stars.* The most asymmetric shape is characterized by a steep dominant peak on one side of the profile, a high-velocity tail on the same side, and a plateau, sometimes with a faint secondary peak, extending to the other side.

4. The observed cycle time of V/R change is $\geq$ 7 years.

The observed line profile asymmetry cannot be caused by a small-scale density inhomogeneity (e.g. a clump), because then the enhanced emission feature should migrate with time across the line profile (S-wave). This is not observed here: we see strong asymmetry at certain phases of the cycle, but symmetry ("fake class 1") in between (Fig. 2). Such a pattern can only be produced by a *large-scale* density inhomogeneity. In addition, such feature cannot be co-rotating (orbital period is

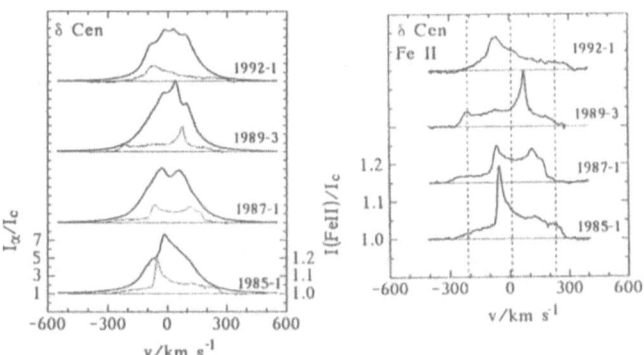

Fig. 2.   Class 2 emission line profiles of δ Cen over 7 years. Left: Hα (solid) and Fe II (dotted line); flux scale for Fe II is 10 times enlarged. Right: Close-up of Fe II line

of the order of days), but must instead be an almost stationary pattern.

3   Global oscillations

3.1   THE MODEL

A purely geometrical model capable of explaining many features of class 2 profiles is the elliptical disk model of Huang (1973). It proposes an elliptical rather
than circular shape of the disk, with uniform eccentricity of orbits. Despite its
merits (model line profiles show some similarity with class 2 profiles), this model
is unable to explain the required stability of orbits which are expected to precess
differentially (due to the flattened central star) and circularize quickly.

A more promising approach has been devised by Kato (1983) and more recently
by Okazaki (1991) and Papaloizou et al. (1992): thin, non-self gravitating Keplerian disks can become subject to global $m = 1$ oscillations. The latter authors
include the effect of the non-spherical gravitational potential of the central star
on the disk which results in a slow precession of the whole density wave (which
would be stationary in case of a perfectly spherical star). The oscillation period
(= precession rate at the outer disk radius) is of the order of 10 years.

3.2 V/R CYCLES

There are two predictions of the global oscillation hypothesis which can be tested
empirically: the line profile shape, and the V/R cycle time. In fact, V/R variability
in Be stars has been known for a long time (e.g. for β^1 Mon: Cowley & Gugula,
1973; γ Cas, Doazan et al., 1987; ζ Tau, Mon et al., 1992), from a long time base
(70 years in case of γ Cas) with however mostly noisy or unpublished line profiles.
Observed cycle times are about 5–10 years, in agreement with the predictions of
the global oscillation model.

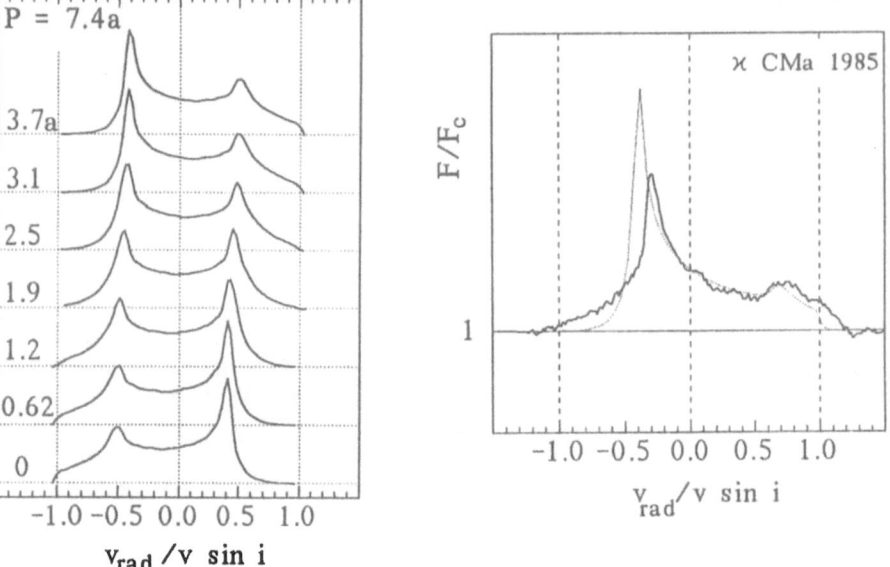

Fig. 3. **a** (left): Model line profiles for Okazaki's (1991) perturbed r^{-2} density law; time elapsed is marked ($T = 0$ corresponds to minimum V/R). **b** (right): Measured class 2 profile in κ CMa, 1985 (solid line), and model line profile (dotted line)

3.3 LINE PROFILES

Our fully resolved optically thin line profiles over about one cycle (Fig. 2) provide a second, more stringent test. In order to test the predictions of the oscillation model, we have performed simple line profile calculations based on Okazaki's (1991) predicted density perturbation superposed on a r^{-2} density law. As in his model, the outer disk radius is five stellar radii. The optically thin emission line profile is obtained by integrating the (vertically averaged) density distribution along radial velocity contours. The results for a half-cycle are shown in Fig. 3a (full cycle period is 7.4 years). A qualitative fit of a model profile to the observed Fe II $\lambda 5317$ emission in κ CMa in 1985 is shown in Fig. 3b.

We note the following results:

1. In the model profiles, the higher peak is always shifted to lower (absolute) radial velocities, and vice versa. This is also well established observationally.

2. In the course of the slow precession, the observed line asymmetry (= log V/R) changes sign, with a short period of $V = R$. Again, this is well established observationally (Fig. 2).

3. The tentative profile fit to the 1985 κ CMa emission line (Fig. 3b) reproduces all main features well, with the exception of the high-velocity tail at negative radial velocities.

4 Conclusions

We conclude that

1. observed class 1 profiles arise from circumstellar Be star disks without perturbation (classical accretion disk); class 2 profiles arise from those disks which are perturbed by a one-armed global oscillation;

2. long-term, hitherto enigmatic V/R variations and class 2 profiles are two manifestations of the same phenomenon;

3. the observed cyclic rather than periodic V/R variability is due to slow changes in density structure or disk radius.

Long-term observations of V/R behaviour furthermore show that V/R oscillations may suddenly set in, as well as may occasionally die out. While the latter could be due to viscous damping, the former behaviour could offer the extremely interesting opportunity to watch the excitation mechanism for the oscillation. This mechanism could be related, or be identical, to the still unknown mechanism giving birth to the envelope itself and thereby to the Be phenomenon.

Acknowledgements

The author acknowledges stimulating discussions with Drs. Dietrich Baade, Joachim Dachs and Wolfgang Hummel, and thanks Oliver Dietle for reduction of the 1992 measurements.

References

Cowley, A., Gugula, E.: 1973, *Astron. Astrophys.* **22**, 203

Dachs, J., Hummel, W., Hanuschik, R.W.: 1992, *Astron. Astrophys. Suppl.* **95**, 437

Doazan, V., Rusoni, L., Sedmak, G., Thomas, R.N., Bourdonneau, B.: 1987, *Astron. Astrophys.* **182**, L25

Hanuschik, R. W.: 1986, *Astron. Astrophys.* **166**, 185

Hanuschik, R. W.: 1987, *Astron. Astrophys.* **173**, 299

Hanuschik, R.W.: 1989, *Astrophys. Space Sci.* **161**, 61

Hanuschik, R.W., Kozok, J.R., Kaiser, D.: 1988, *Astron. Astrophys.* **189**, 147

Huang, S.S: 1973, *Astrophys. J.* **183**, 541

Hummel, W., Dachs, J.: 1992, *Astron. Astrophys.* **262**, L17

Kato, S.: 1983, *Publ. Astron. Soc. Japan* **35**, 24

Mon, M., Kogure, T., Suzuki, M., Singh, M.: 1992, *Publ. Astron. Soc. Japan* **44**, 73

Okazaki, T.: 1991, *Publ. Astron. Soc. Japan* **43**, 75

Papaloizou, J.C., Savonije, G.J., Henrichs, H.F.: 1992, *Astron. Astrophys.* **265**, L45

Coupled Stellar Jet/Molecular Outflow Models

A. C. Raga
Mathematics Department, UMIST, P.O. Box 88, Manchester, M60 1QD, U.K.

Abstract.
Molecular outflows can be modelled as environmental material entrained into high velocity stellar jets. Even though models of this entrainment process are at this time quite uncertain, a few preliminary theoretical efforts have been made. Three different models are discussed, in which the molecular outflows are identified with the turbulent mixing layer at the edge of a jet, with a turbulent envelope (driven by a large number of internal working surfaces in the jet) and with the wake of one working surface.

1 Introduction

Young stellar objects are observed to eject both highly collimated Herbig-Haro (HH) jets and less well collimated molecular outflows. This situation has given rise to two categories of models :
- "two-wind models", in which the HH objects (or jets) and the molecular outflows are assumed to come from two dynamically different flows (e. g., from a stellar wind and from a disk wind, respectively, Pudritz, 1988),
- "unified models" , in which the HH jets and the molecular outflows are assumed to be different manifestations of the same flow.

In this second category of models, the most attractive possibility appears to be to model the fast, partially ionized HH objects as a collimated wind ejected from the young stellar objects, and to identify the molecular outflows with molecular environmental material entrained into or swept up by this fast wind. For such a model to be applicable, the momentum rate associated with the HH objects has to be large enough to drive the molecular outflow. This "momentum rate problem" has been the focus of extended arguments as to whether or not the molecular outflows might be driven by the optically detected HH objects. The resolution of this question appears to be quite difficult due to the fact that the determination of the momentum rates of both the molecular outflows (Cabrit, 1994) and the HH objects (Raga, 1991) is highly model dependent, which results in very large uncertainties. However, the general consensus now seems to be that HH objects do have enough momentum to drive the molecular outflows, in agreement with the original suggestion of Snell *et al.* (1985) (though it should be noted that in the intervening period the opposite view was generally held, see Mundt *et al.*, 1987).

The present paper discusses three possible ways of modelling molecular outflows as environmental material entrained into a fast, collimated wind. The possibility of modelling molecular outflows as a turbulent mixing layer around a laminar jet is discussed in §2. A more complex model, in which the turbulent mixing layer is modified by the presence of a number of "internal working surfaces" (resulting

Astrophysics and Space Science **216**: 105–112, 1994.
© 1994 *Kluwer Academic Publishers.*

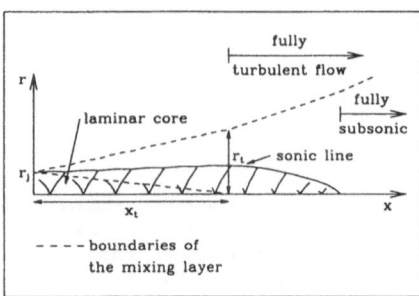

Fig. 1. Schematic diagram showing the transition from a laminar to a fully turbulent jet (taken from Cantó and Raga, 1991). The dashed lines indicate the inner and outer boundaries of the turbulent mixing layer. The jet becomes fully turbulent at $x = x_t$, where the inner boundary of the mixing layer reaches the symmetry axis. The crossing shocks in the jet beam extend out to the sonic line, which lies inside the mixing layer. When the jet becomes fully turbulent, the sonic line curves towards, and eventually reaches the symmetry axis. At larger distances from the source, the jet flow is fully subsonic.

from a source velocity time variability), is presented in §3. Finally, the entrainment of environmental material at one working surface is discussed in §4.

2   Turbulent Mixing Layer

Supersonic, laminar laboratory jets develop a turbulent mixing layer along the jet/environment boundary. This mixing layer is very thin at the injection point, and grows both into the surrounding environment and into the jet, eventually reaching the symmetry axis at a distance x_t from the source. For distances $x > x_t$ the jet is fully turbulent. This is shown in a schematic way in figure 1.

For $x < x_t$, the situation is as follows. The material in the inner part of the mixing layer has a velocity equal to the jet velocity v_j, as it is in contact with the laminar core (which preserves the full initial jet velocity, as it does not "feel" the drag from the surrounding environment). The outer part is almost at rest, as it is in contact with the undisturbed environment (which is slowly being dragged into the mixing layer). The velocity profile across the width of the mixing layer is approximately linear (Cantó and Raga, 1991), resembling the well known Couette flow (= the viscous flow between a stationary and a moving infinite, plane plate). So, at any given distance from the source, the mixing layer has a velocity profile (across the outflow axis) with all velocities ranging from 0 to v_j. From this it is evident that somewhere in the cross section of the mixing layer there has to be a sonic line (see figure 1).

For $x > x_t$, the flow is fully turbulent. Now, the whole cross section of the jet is coupled to the surrounding environment by the turbulent viscosity. Because of this, the velocity across the whole cross section of the jet starts to decrease with increasing x (in contrast, for $x < x_t$, the central, laminar core preserves the initial jet velocity, see above). At large enough distances from the source, the flow slows down enough so that it becomes fully subsonic (see figure 1).

Cantó and Raga (1991) have derived a simple expression for the distance x_t at which the jet becomes fully turbulent :

$$x_t = 16.9 M_j r_j = 0.55 \, \text{pc} \left(\frac{M_j}{20}\right) \left(\frac{r_j}{5 \times 10^{15} \text{cm}}\right) , \qquad (1)$$

where M_j is the Mach number of the jet and r_j is its initial radius. For the second equality, typical values of these parameters for jets from young stars have been used. This distance is of the same order as the observed lenghts of molecular outflows (Lada, 1985) and somewhat larger than the typical lengths of HH jets (Mundt et al., 1987). Because of this, it is tempting to identify the observed molecular emission as coming from either the turbulent mixing layer, or from the fully turbulent flow found farther downstream (Cantó and Raga, 1991; Stahler, 1994).

However, it appears that the observational properties of molecular outflows cannot be explained convincingly as emission from a turbulent boundary layer. The main problem of this interpretation of molecular outflows is that the predicted length-to-width ratios appear to be too long. Cantó and Raga (1991) find that the outer radius r_t of the mixing layer at a distance x_t from the source (i.e., at the point where the flow becomes fully turbulent, see figure 1) is given by :

$$r_t = 3 \left(\frac{c_l}{c_j}\right)^2 r_j , \qquad (2)$$

where c_j is the sound speed of the jet (supposedly of the order of 10 km s^{-1}, as the jet has to have a temperature $\sim 10^4$ K in order to produce the observed optical emission), c_l is the sound speed of the material in the mixing layer, and r_j is the initial jet radius. From equations (1) and (2) we would then conclude that the collimation ratio $q \equiv$ length/width of the observed outflow is approximately given by :

$$q \approx \frac{x_t}{2\, r_t} \approx 2.8 \, M_j \left(\frac{c_j}{c_l}\right)^2 . \qquad (3)$$

Now, if we want to model molecular outflows as the emission from a mixing layer, it is clearly necessary for the mixing layer gas to be quite cool (otherwise, it would not be molecular). From this we would conclude that we need to look at models with $c_j/c_l > 1$. For a jet Mach number $M_j = 20$ we would then have $q > 50$, in other words, a length-to-width ratio similar to the value found for the HH jets (Mundt et al., 1987. Such a value of the collimation ratio seems to be in clear disagreement with the $q \sim 2$ values measured from observations of molecular outflows (Lada, 1985).

The remaining possibility would be that the molecular outflow corresponds to the emission from the fully turbulent flow region (i.e., the $x > x_t$ region, see figure 1). This seems to be in principle possible at least for some outflows from young stars, in which the optically detected outflow (which we would identify with the laminar core present in the $x < x_t$ region) is observed close to the source, and the molecular outflow (which we would identify with the fully turbulent, $x > x_t$ region) extends to larger distances from the source. However, in other cases (e. g., in HH 46/47 and HH 111) the molecular outflows and the HH jets have similar spatial extensions.

To explain these objects, we need a model in which a narrow jet and a much wider molecular outflow are observed to coexist at similar distances from the source. An attempt at producing such a model is described in the following section.

3   Turbulent Envelope

A time-dependence in the ejection velocity results in the formation of two-shock "internal working surfaces" that travel down the beam of the jet (Wilson, 1984; Raga *et al.*, 1990). These working surfaces intercept material from the beam of the jet, and eject it sideways into the surrounding environment (see figure 2 and Raga and Kofman, 1992).

This sideways ejection of material by the internal working surfaces forces jet material into the turbulent mixing layer which surrounds the jet. This has the effect of "puffing up" the mixing layer into a broader "turbulent envelope" (see figure 3). Raga *et al.* (1993) have modelled such turbulent envelopes, and find that they have widths ~ 10 times larger than the width of the jets (see figure 4). This results in values of the length-to-width ratio $q \sim 10$, which, though on the high side, are in better agreement with the $q \sim 2$ values measured for molecular outflows (see §2 and Lada, 1985).

Also, from these turbulent envelope models it is possible to obtain predictions of velocity channel maps, similar to the ones obtained in molecular line observations (see figure 3). Future comparisons between such predictions and observations of molecular outflows should be useful for deciding the possible relevance of the turbulent envelope models.

4   Working Surface

Finally, we discuss a model in which the molecular outflows are identified with the material in the "wake" of a working surface of an HH jet. The situation is shown in the schematic diagram of figure 4. An internal working surface in the jet flow ejects material sideways. This material interacts with the surrounding environment, forming a two-shock structure, with a bow shock being pushed into the surrounding environment.

In between the two shocks, a turbulent mixing layer is formed, in which the

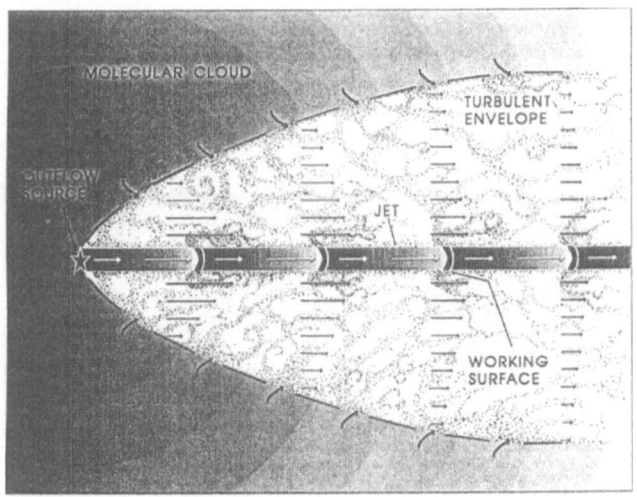

Fig. 2.   Schematic diagram of a turbulent envelope driven by the passage of many internal working surfaces (taken from Raga *et al.*, 1993). The internal working surfaces eject material sideways out of the jet beam. This material couples with the surrounding environment, leading to the formation of a turbulent envelope around the jet.

material ejected sideways by the working surface mixes with the post-bowshock environmental gas. As a result of the turbulent dissipation, this turbulent mixing layer is warm, with temperatures $\sim 10^3$–10^4 K.

Once the internal working surface has passed by, the warm material in the turbulent mixing layer starts to expand, filling in the cavity left behind by the working surface. This refilled cavity (see figure 4) is then identified with the observed molecular outflows. Raga and Cabrit (1993) have studied such turbulent wakes, and find that their kinematical properties are similar to the ones of high velocity (Richer *et al.*, 1992) or of "bullet like" (Bachiller and Gómez-Gonxáles, 1992) molecular outflows.

An alternative model in which molecular outflows are also identified with environmental material pushed by a working surface of a jet has been proposed by Masson and Chernin (1993) (also see the contribution of Chernin in this volume). In this model, it is assumed that the environmental material that goes through the bowshock remains confined in a narrow, high density layer, and does not re-expand to fill in the cavity left behind by the working surface.

It is at this time somewhat hard to say whether the scenario of Raga and Cabrit (1993) (in which the swept-up environmental material re-expands to fill in the cavity) or the one of Masson and Chernin (1993) (in which the swept-up

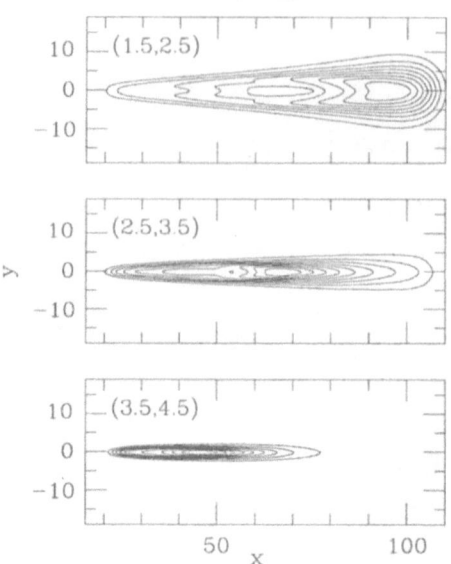

Fig. 3.   [Velocity channel intensity]Velocity channel intensity maps predicted from the turbulent envelope model (taken from Raga and Cabrit, 1993). A high (bottom), an intermediate (center) and a low velocity (top) map are shown. The emission is assumed to be optically thin.

material remains in a narrow shell) is correct. In order to resolve this question, it will be necessary to carry out high resolution, axisymmetric numerical simulations with realistic parametrizations of the atomic and molecular cooling and of the turbulent dissipation.

It should also be possible to resolve the question of which of these two scenarios is correct by comparison with observations. For example, observations of the Orion B jet (Richer *et al.*, 1992) seem to indicate a "filled in cavity" morphology, in better agreement with the model of Raga and Cabrit (1992). However, it is at this time unclear which of the two pictures better fits the characteristics of molecular outflows in general.

5   Conclusions

Three different possible models for the production of molecular outflows as environmental material entrained into a high velocity jet have been presented. The first possibility is that molecular outflows might correspond either to the turbulent mixing layer at the edge of a jet, or to the fully turbulent region farther downstream (where the turbulent mixing layer has eroded all of the laminar core of the jet). As the turbulent mixing layer is very narrow, the emission from this region would have very high length-to-width ratios, in apparent disagreement with observations of molecular outflows.

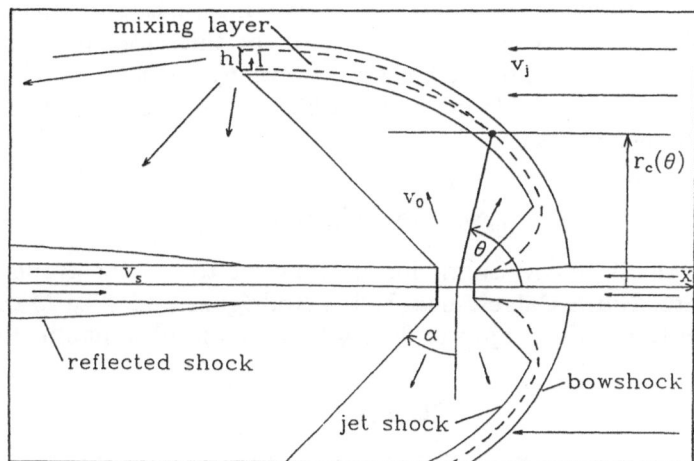

Fig. 4. Schematic diagram showing the interaction between an internal working surface and the surrounding environment (taken from Raga and Cabrit, 1993). The material ejected sideways by the internal working surface interacts with the surrounding environment, forming a two-shock "snowplow" structure. The velocity shear across the contact discontinuity between the two shocks results in the formation of a turbulent mixing layer. Once the working surface has gone by, this warm turbulent mixing layer expands to fill in the cavity which has been left behind by the working surface.

The fully turbulent region farther downstream can in principle be broader, in better agreement with observations. However, if the observed molecular outflows do correspond to this region of the flow, we would expect to observe sources with relatively short HH jets, and with molecular outflow lobes which extend to much larger distances. This situation apparently is not found in many outflows (see, e. g., Reipurth and Olberg, 1991; Richer et al., 1992).

This leads us to the two other models discussed above. These models stress the importance of the jet working surfaces on the entrainment of environmental material. In the "turbulent envelope" model, the average effect of the passage of many internal working surfaces is considered. In the "working surface" model, the flow of environmental material induced by the passage of a single working surface is considered.

Depending on the state of evolution of the outflow system, either one or the other (or both) of these pictures might be applicable. One would be tempted to think that while young outflows might be better modelled in terms of the interaction of a single working surface with an undisturbed environment (the "working surface" model), more evolved outflow systems might be in the "turbulent envelope" regime. Obviously, more work is necessary in order to clarify the possible relevance of this conjecture.

To finalize, one should point out again that an alternative model for the entrainment of environmental material at the working surface of the jet has been proposed by Masson and Chernin (1993) (also see this volume). Also, Padman (in this volume) has presented a kinematic working surface model, which produces predictions of observable quantities which seem to be in surprisingly good agreement with observations.

Acknowledgments

I would like to thank Sylvie Cabrit for allowing me to use the graph of figure 4, and Jorge Cantó, Nuria Calvet, Luis Felipe Rodríguez and José María Torrelles for allowing me to use the graphs of figures 2 and 3 ahead of publication.

References

Bachiller, R. and Gómez-Gonzáles, J.: 1992, *Astron. Astrophys. Rev.* **3**, 257.
Cabrit, S.: 1994, in *Stellar Jets and Bipolar Outflows*, eds. L. Errico and A. A. Vittone, Kluwer, Dordrecht, in press.
Cantó, J. and Raga, A. C.: 1991, *Astrophys. J.* **372**, 646.
Lada, C.: 1985, *Ann. Rev. Astron. Astrophys.* **23**, 267.
Masson, C. R. and Chernin, L. M.: 1993, *Astrophys. J.* **414**, 230.
Mundt, R., Brugel, E. W. and Bührke, T.: 1987, *Astrophys. J.* **319**, 275.
Pudritz, R.: 1988, in *Galactic and Extragalactic Star Formation*, eds. R. Pudritz and M. Fich, NATO ASI Series, Kluwer, Dordrecht, p. 135.
Raga, A. C.: 1991, *Astron. J.* **101**, 1472.
Raga, A. C. and Cabrit, S.: 1993, *Astron. Astrophys.* **278**, 267.
Raga, A. C., Cantó, J., Calvet, N., Rodríguez, L. F. and Torrelles, J. M.: 1993, *Astron. Astrophys.* **276**, 539.
Raga, A. C., Cantó, J., Binette, L. and Calvet, N.: 1990, *Astrophys. J.* **364**, 601.
Raga, A. C. and Kofman, L.: 1992, *Astrophys. J.* **386**, 222.
Reipurth, B. and Olberg, M.: 1991, *Astron. Astrophys.* **246**, 535.
Richer, J. S., Hills, R. E. and Padman, R.: 1992, *Mon. Not. Roy. Astr. Soc.* **254**, 535.
Snell, R. L., Bally, J., Strom, S. E. and Strom, K. M.: 1985, *Astrophys. J.* **290**, 587.
Stahler, S. W.: 1994, in *Astrophysical Jets*, ed. M. Livio, C. O'Dea and D. Burgarella, Cambridge Univ. Press, in press.
Wilson, M.: 1984, *Mon. Not. Roy. Astr. Soc.* **209**, 923.

Modelling Jet-Driven Molecular Outflows

L. M. Chernin* and C. R. Masson
Center for Astrophysics, 60 Garden St., Cambridge MA 02138, USA

June 17, 1993

Abstract.
    We have investigated the basic physical properties of the outflow that is created by a supersonic jet in a dense molecular cloud. We show that the dynamics of the interaction is strongly controlled by the rapid cooling of the post-shock gas at the head of the jet. The velocity of the gas is high in the vicinity of the jet head, but decreases rapidly as more material is swept-up. This type of outflow produces extremely high velocity clumps of post-shock gas which resemble the features seen in outflows. We also show that momentum transfer in bow shocks is more important than entrainment in high Mach number jets, as found in the protostellar environment.

Key words: hydrodynamics, protostars, jets, entrainment.

1   Introduction

Although over 150 molecular outflows from protostars have been discovered (Fukui 1989) there is yet to be a consensus on how these flows are driven. Molecular outflows are often associated with Herbig- Haro (HH) objects and highly collimated optical jets (e.g. Mundt, Brugel and Bührke 1987, hereafter MBB) with velocities of 100-400 km s^{-1}. On the other hand, molecular outflows are poorly collimated and have much lower (< 20 km s^{-1}) velocities. It has been suggested that these types of flows may be dynamically related (Padman these proceedings; Masson and Chernin 1993, hereafter MC) and several authors have begun to explore various models of outflows driven by jets (Raga these proceedings; MC). In this paper, we investigate a supersonic jet interacting with a dense quiescent medium and derive the properties of the swept-up gas.

2   Momentum Transfer by Bow Shocks

The working surface at the head of a jet contains a jet shock where the jet material is decelerated and a bow shock in which the ambient material is accelerated. The resultant evolution of the post-shock gas depends on the strength of the cooling in the shocked gas. Since gas in molecular outflows has typical densities of 1000 cm^{-3} and the bow shock velocities are a few hundred km s^{-1}, the cooling distance of the shocked ambient gas is much less than a jet radius (Chernin *et al.* 1993). Thus, the shocked ambient gas is initially driven away from the jet axis by the high thermal pressure at the apex of the shock. At later times, after the rapid cooling, the outflow evolves in a momentum conserving snowplough fashion. We have calculated the appearance of a jet-driven molecular outflow by making

---

* Current address: Astronomy department, U. California, Berkeley CA 94720, USA

Astrophysics and Space Science **216**: 113–117, 1994.
© 1994 *Kluwer Academic Publishers.*

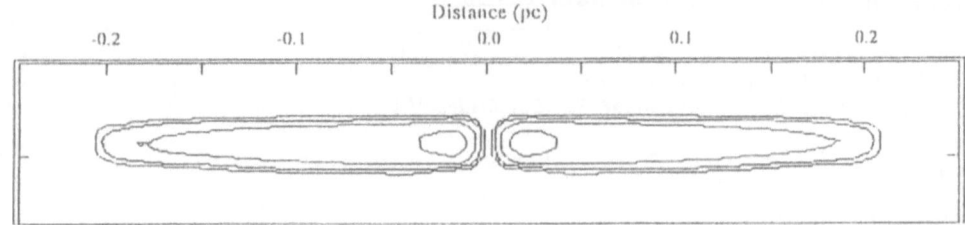

Fig. 1.   Column density map for the model outflow as observed at an angle of 45 degrees. The contours are logarithmic in column density

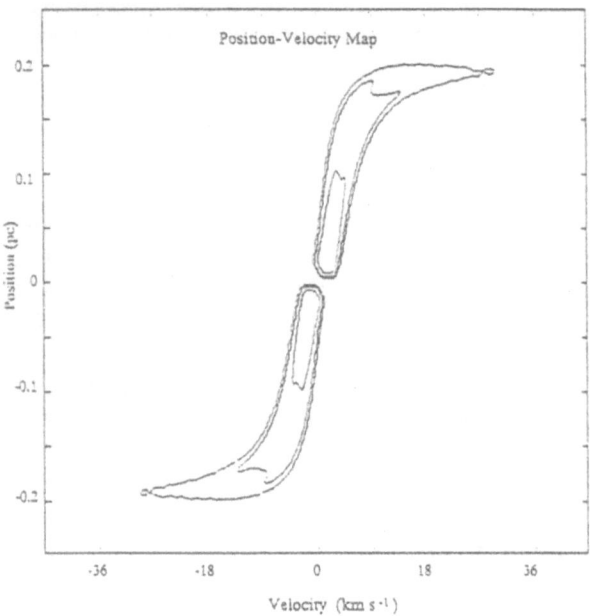

Fig. 2.   Position-velocity diagram for the model outflow as observed at an angle of 45 degrees. The contours are logarithmic in column density.

a simple approximation for the cooling. Before the shocked gas cools, we assume that the thermal pressure imparts a small perpendicular component, in addition to the longitudinal component required by momentum conservation (MC). The jet creates a shell of swept-up material that is roughly cylindrical. The shape of the outflow in Figure 1 is similar to that seen in collimated outflows such as NGC 2024 (Richer *et al.* 1992). A wider outflow can only be created if the jet direction changes (MC). The position-velocity diagram shown in Figure 2 shows a linear increase in velocity with distance which is due to the density gradient, as well as a sharp bend (or hook) at the highest velocities which is due to the bow shock. The

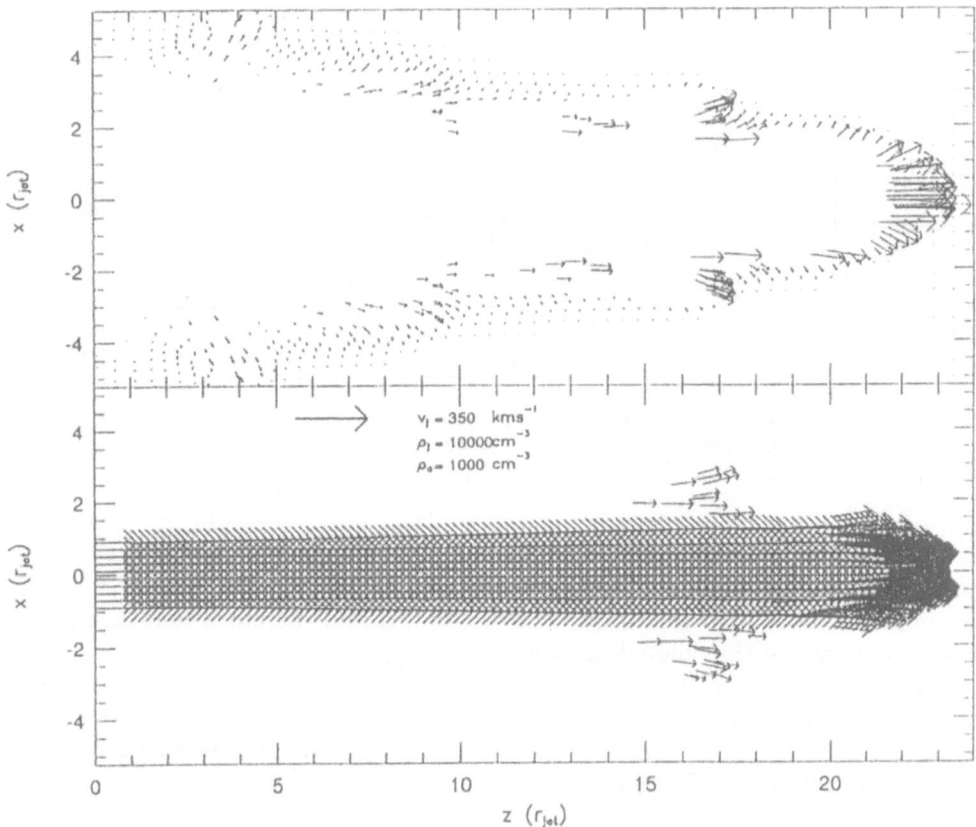

Fig. 3.   Velocities of sph particles within $y = \pm 0.25 r_j$ of y=0. The upper panel shows the ambient particles and the lower panel shows the jet particles.

presence of such high velocities at large distances from the star agrees well with observations, but is hard to obtain with poorly collimated winds (Shu *et al.* 1988) or with turbulent jet models (e.g. Raga, these proceedings).

The above results are based on a simple assumption about the behaviour of the shocked gas, but we have also used smoothed particle hydrodynamical (sph) simulations in order to determine the characteristics of a jet/cloud interaction in more detail (Chernin *et al.* 1993). Figure 3 shows an example of a simulation with a dense jet, $n_j = 10^4$ cm^{-3}, travelling at 350 km s^{-1} into a quiescent medium of $n_{amb} = 1000$ cm^{-3}. The ambient gas (top panel) is swept-up into a dense shroud which we identify with the molecular outflow. Since this simulation was done in a uniform density medium, the shroud increases in width behind the bow shock, unlike the shape shown in Figure 1. The velocities of the ambient particles are only high near the jet head and are much lower in the shroud. The shocked gas

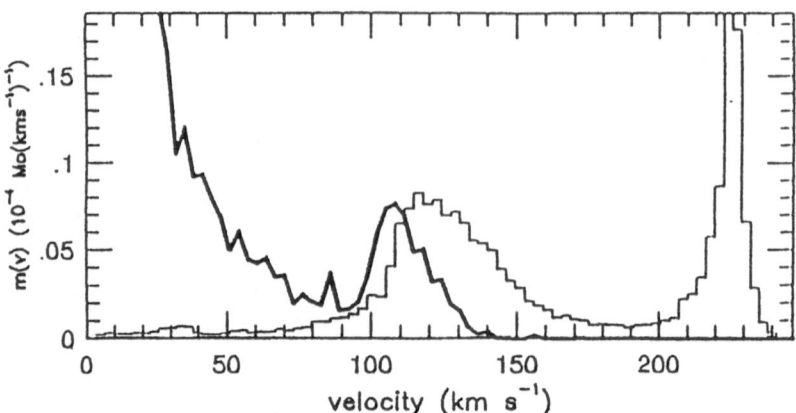

Fig. 4.   Mass-velocity profiles from the sph simulations. The histogram shows the jet particles and the heavy solid line shows the ambient particles. Note the two concentrations of the jet particles corresponding to the pre- and post-shock gas respectively.

has cooled to less than 10^4 degrees within a distance of a jet radius. There is a low density region, or cocoon, between the jet and the shroud. These properties are similar to those with the simple analytic model and therefore confirm that the basic ingredients of that model are correct.

Figure 4 shows the distribution of sph particles as a function of velocity. Note that the post shock gas piles up in the working surface forming a clump which we identify with the so called extremely high velocity CO features as seen in outflows (see MC and references therein).

3   Momentum Transfer by Turbulent Mixing

Besides the bow shock momentum transfer described above, jets can also transfer momentum through turbulent mixing at a velocity shear. This process has been called steady-state entrainment, since it occurs all the way along the length of the jet (De Young 1986). Figure 5 shows an example of this in the case of a Mach three jet. We have found that this type of entrainment only occurs in low Mach number jets as they have weak bow shocks and do not have a cocoon separating the jet beam from the ambient gas (Chernin et al. 1993). Therefore steady state entrainment cannot be important in protostellar jets in dense clouds because of the high Mach numbers, i.e. $10 < M < 40$ (MBB).

4   Conclusions

A simple jet-driven outflow model can account for the major properties of outflows. The strong cooling of the bow shock results in a momentum conserving snowplough of ambient gas. A highly supersonic jet transfers momentum to the ambient gas

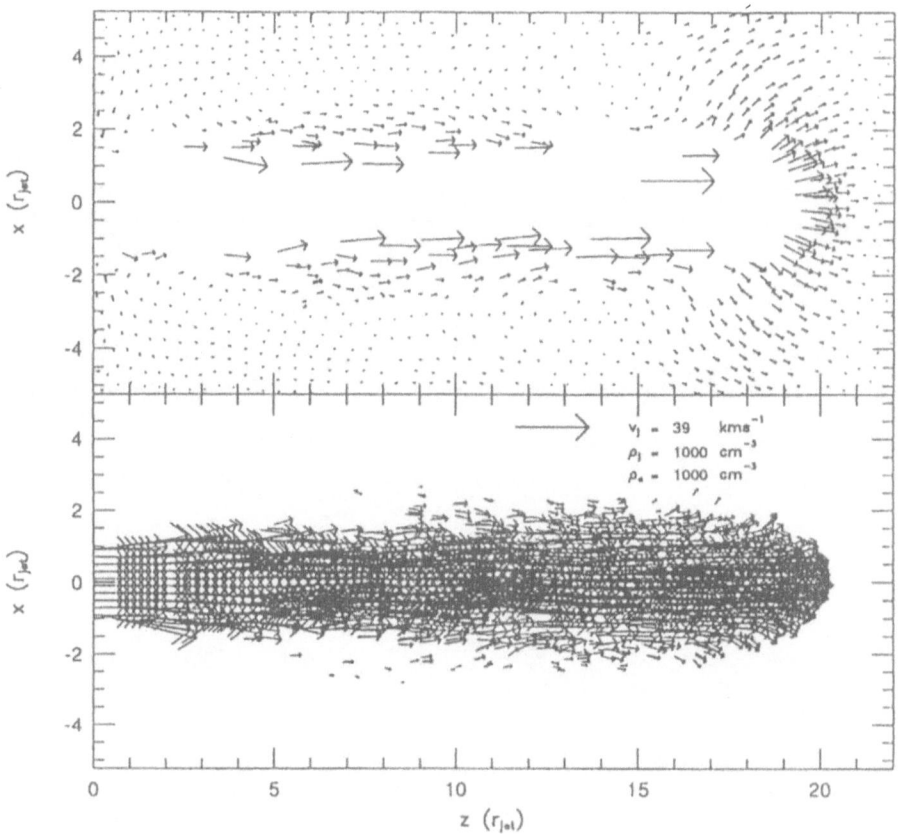

Fig. 5.   Velocity field of sph particles for a Mach three jet. In the upper panel note the turbulent entrainment of the ambient medium along the jet beam.

primarily through the working surface and not through turbulent entrainment along the sides of the jet. A wider outflow can only be obtained with a wandering jet.

References

Chernin, L. M., Masson, C. R., Gouveia Dalpino, E., and Benz, W. 1993, *Ap.J.*, submitted.

De Young, D. 1986, *Ap.J.*, **307**, 62.

Fukui, J. 1989, In *Low Mass Star Formation and Pre-Main Sequence Objects*, Ed. B. Reipurth, ESO, Garching.

Masson, C. R., and Chernin, L. M. 1993, *Ap.J.*, in press (MC).

Mundt, R., Brugel, E., and Bührke, E. 1987, *Ap.J.*, **319**, 275 (MBB).

Richer, J., Hills, R. E., and Padman, R. 1992, *MNRAS*, **254**, 525.

Shu, F. H., Lizano, S., Ruden, S. P. and Najita, J. 1988, *Ap.J.*, **328**, L19.

Jets

S. A. E. G. Falle
Department of Applied Mathematical Studies,
The University, Leeds LS2 9JT, U.K.

Abstract. Despite its title, this paper is mainly concerned with some of the latest theoretical work on the Herbig-Haro (HH) objects associated with stellar jets. Recent measurements of the proper motions of HH objects show that even those near the source are moving at about the same speed as the jet fluid. This means that they cannot be due to steady shocks in the jet and the only obvious explanation is that the source is varying.

1 Introduction

The title of this paper suggests that I am going to discuss all kinds of jets from honey flowing off a spoon to extragalactic radio jets. Although I am obviously not going to do this, it is worth briefly exploring the similarities between terrestrial and astrophysical jets. We know that jets from young stars are supersonic and composed of ordinary cosmic gas and we suspect that extragalactic jets are as well. It therefore makes sense to look at supersonic laboratory jets since they are mostly under our control and we can make detailed measurements on them.

One of the things we learn from laboratory jets is that global Kelvin-Helmholtz instabilities are of no importance in high Reynolds' number jets. Although the jet boundary is certainly Kelvin-Helmholtz unstable, this instability manifests itself as a turbulent boundary layer and it appears to be impossible to excite global modes if the Reynolds' number exceeds 4000 (Crighton, 1981). We can therefore safely ignore the vast literature on such global instabilities (e.g. Hardee, 1979; Ray, 1981; Birkinshaw, 1984; Hardee and Norman, 1990), unless we believe that astrophysical fluids are orders of magnitude more viscous than treacle. In any case Raga (1993) has pointed out that the knot spacings in stellar jets can only be consistent with those predicted by global Kelvin-Helmholtz instabilities if the temperature in the surroundings is about 10^5 K, which seems highly unlikely.

Apart from that, the most obvious feature of supersonic terrestrial jets is the regularly spaced series of steady shocks which arise when there is an imbalance between the initial jet pressure and that in the surroundings. Since both extragalactic radio jets and stellar jets also show more or less evenly spaced knots of radio or optical emission, it seemed natural to suppose that they were due to the same mechanism. However, we shall see that this is not true in general, which is a pity since we could have deduced quite a lot about the nature of the jets from the structure of such steady shocks.

It now seems clear that the reason why the shocks are not steady is that the sources of both stellar jets and extragalactic jets vary. This ought not to surprise us since we have known for some time that the optical and radio luminosity of these

Astrophysics and Space Science **216**: 119–125, 1994.
© 1994 *Kluwer Academic Publishers.*

120 S. A. E. G. Falle

sources varies considerably. However, one might have hoped that these variations were confined to short timescales and that the sources could be regarded as steady if averaged over the dynamical timescale of the jets.

2 Steady Jets

It is easy to show that, if the source is steady and there are no other disturbances, then the jet will be steady from the source out to a distance L such that $L \ll v_j t$, where v_j is the jet speed and t is the age of the source (Falle, 1991). In many cases this would imply that the jet is steady along a significant fraction of its length. Since both stellar and extragalactic jets often have regularly spaced knots of emission near the source, several authors have computed the flow in such steady jets (Sanders, 1983; Falle and Wilson, 1985; Wilson and Falle, 1985; Wilson, 1986, 1987; Falle, Innes and Wilson, 1987; Raga, Binette and Cantó, 1990).

However, the spacing l_k between the knots in a steady jet is related to the jet Mach number M_j and radius r_j by

$$l_k \simeq 2M_j r_j.$$

For extragalactic jets there is no independent way of determining the Mach number and it is thus a free parameter which can be chosen to give the observed knot spacing. This cannot be done for stellar jets since the observations tell us both the jet speed and its temperature so that the Mach number is known. Unfortunately this Mach number leads to a distance between the HH objects which is about four times the observed one. It is possible to get something like the observed knot spacing by playing around with the cooling length and allowing the external pressure to be non-uniform, but this is so contrived that it makes much more sense to abandon the whole idea of steady shocks.

In fact we have to do this anyway. Recent VLA observations of the M87 jet show that the inner knots move at relativistic speeds, so they can obviously not be due to steady shocks. It also seems that whenever one measures the proper motions of the HH objects near the source of a stellar jet, they move at about the speed of the jet fluid (Reipurth, 1989; Eislöffel, Mundt and Ray, 1989; Eislöffel and Mundt, 1992; Reipurth, Heathcote and Vrba, 1992; Eislöffel, 1992).

As if this were not enough, there are other observations of stellar jets which are in conflict with steady shock models. In many cases the fluid velocity increases significantly along the jet (Reipurth, 1989). For a steady jet Bernoulli's theorem gives

$$\frac{1}{2}v^2 + \frac{c^2}{(\gamma - 1)} = \text{ const.}$$

These jets typically have Mach numbers of twenty or more, so $v \gg c$, which implies that the jet velocity must be nearly constant. The final difficulty is that the observed velocity dispersions in the knots are of order 50 km s^{-1}, but theoretical models predict that it should only be 20–30 km s^{-1}.

3 Variable Sources

Since there does not seem to be any way in which fluctuations in the surroundings can produce a regularly spaced series of knots, we are forced to consider variable sources. This was originally suggested by Rees (1978) for extragalactic jets and such flows have been computed for adiabatic jets by Wilson (1983, 1984). More recently Raga *et al.* (1990), Raga and Kofman (1992) and Kofman and Raga (1992) have looked at simple one dimensional models of stellar jets from unsteady sourcess. They exploited the fact that the high Mach numbers in stellar jets mean that pressure forces are negligible, so the equation of motion reduces to

$$\frac{\partial v}{\partial t} + v\frac{\partial v}{\partial x} = 0$$

if the flow is one dimensional. This is the inviscid Burgers' equation and the solution is

$$v(x,t) = v_s(\tau) = \frac{x}{(t-\tau)},$$

where $v_s(t)$ is the velocity at the source and $\tau(x,t)$ is the time at which a fluid element which is at distance x at time t was ejected from the source.

If $v_s(t)$ is not a monotonic function of t, then it is easy to show that at large distances from the source the solution consists of regions in which $v \propto x$ separated by discontinuities (shocks) i.e. it is a saw-tooth function. This is extremely promising since this is exactly what the the observed jet velocity looks like in the HH46/47 system (Reipurth, 1989). Kofman and Raga (1992) were also able to show that the emission line intensities predicted by this model agree quite well with the observed ones.

4 Numerical Calculations

Although these simple one dimensional calculations suggest that a variable source can produce the observed properties of the HH objects, there are a number of details that need to be considered. The first question that arises is the equation of motion of the knots. For Burgers' equation the speed, s, of a discontinuity is given by

$$s = \frac{1}{2}(v_u + v_d)$$

where v_u and v_d are the speeds on the up and downstream sides of the discontinuity.

However, the discontinuity really consists of two gas dynamic shocks and, in the frame of the discontinuity, the gas flows in through these shocks and out the side. Ram pressure balance then gives

$$s = \frac{v_d + \beta v_u}{1 + \beta}$$

where

$$\beta = \sqrt{\frac{\rho_u}{\rho_d}}$$

and ρ_u, ρ_d are the densities up- and down-stream of the knot.

We can see from this that Burgers' equation is appropriate provided the densities are the same on either side of the knot. Raga and Kofman (1992) show that this will be true at large times provided that the velocity at the source varies periodically. Recently, Hartigan and Raymond (1993) have carried out one dimensional gas dynamical calculations with a realistic cooling function and have found that the shock structures are reasonably well described by the above analysis.

The only remaining difficulty is that the flow is will only be one dimensional if the jet is much less dense than the gas through which it propagates and this does not appear to be the case. In most systems the working surface moves at about half the jet speed, which suggests that the density in the undisturbed ambient medium is roughly the same as that in the jet. It therefore seems unlikely that the jet is surrounded by much denser gas except very near the source.

We therefore have to consider two dimensional effects and although we can make some simple estimates about what will happen, the only way to be sure is to compute such flows. This has already been done by a number of people (Stone and Norman, 1993; Falle and Raga, 1993; Biro and Raga, 1993; Ojouveia dal Pino and Benz, 1993), but these calculations are very demanding since the cooling length tends to be much smaller than the jet radius. Since one typically needs about ten mesh points in a cooling length, it is essentially impossible to compute such flows accurately on a uniform grid. The problem is even worse if the knots are allowed to move across the grid at the jet speed. If v_j is the jet speed and Δv_j is the velocity jump at the shocks in the knot, then for any shock capturing scheme, the width of the shock will be increased by a factor $v_j/\Delta v_j$ over what it would be for a shock propagating into gas which is stationary on the grid. Since $v_j/\Delta v_j$ is about four, this means that the shocks will be at least ten cells wide.

The upshot of all this is that the we have to do two things if we want to make a reasonable job of computing these flows. We can get rid of the wide shocks by ensuring that the knot is moving slowly relative to the computational grid and we can achieve the necessary resolution by using an adaptive grid. Falle and Raga (1993) have done this for the case of a symmetric knot in which the densities are the same on both sides of the knot and the external density is so small that the knot moves subsonically with respect to the gas surrounding the jet. The problem then reduces to that of a symmetric collision between two gas streams and their calculations show that, apart from the stand-off shocks which constitute the knot itself, there are shocks in the gas which is splattered sideways. Unfortunately the density in this gas is so low that it is unlikely that the emission due to these additional shocks can be detected.

Falle and Raga used an approximate equation of state which assumed equilibrium ionisation of hydrogen. The effect of this is that the gas stays approximately isothermal with a temperature of about 1.3 10^4 K if the cooling length is small compared to the jet radius. In that case the stand-off shocks are nearly perpendicular to the jet and the distance, d_s, between them is approximately give by

$$d_s = \frac{\sqrt{e}c_0 r_j}{\Delta v_j}.$$

Here c_0 is the sound speed at the equilibrium temperature and Δv_j is the velocity jump at the stand-off shock. The gas is ejected sideways with velocity $\sim c_0$ and since this is much less than v_j, the jet stays well collimated.

This is the simplest possible case in which the ambient medium merely exerts a uniform pressure and the jet is uniform on either side of the knot. In fact if the surroundings are so hot that the external pressure is constant, then it will drive shocks into the jet between the knots. This is because the flow between the knots is being stretched by the velocity gradient and so the pressure in the jet must eventually become small compared to the external one. These shocks are driven into the jet with a velocity of order v_j/M_j if the jet is initially pressure matched, which means that they will reach the axis after the knot has travelled a distance $r_j M_j$. Although these shocks have a negligible effect on the jet velocity if the Mach number is high, they are strong shocks and so the density behind them will be large if the jet gas can cool. The result is that the jet is not uniform on either side of the knot and the knot structure will changes significantly over a distance of about $r_j M_j$ i.e. over the same distance as the spacing between steady shocks.

Now suppose that the jet is surrounded by cold gas whose sound speed is much less than the speed of the knot. The gas that is ejected sideways will now produce a bow shock in the external gas upstream of the knot. The pressure behind this bow shock is roughly $\rho_e v_k^2$ where ρ_e is the external density and v_k is the speed of the knot relative to the surroundings. This pressure drives a shock with velocity $v_k\sqrt{\rho_e/\rho_j}$ into the jet upstream of the knot and this reaches the jet axis after the knot has travelled a distance $r_j\sqrt{\rho_j/\rho_e}$. This does not affect the jet velocity if the external density is small, but it makes the density non-uniform in the jet and so tends to disrupt the knot.

The figure shows the results of an adiabatic calculation in which two equal jets with unit radius collide at $z = 5$. The two jets have velocity ∓ 5, the external gas has velocity -20 and its density is 10^{-3} that of the jet. At the time shown, the knot has travelled 60 jet radii and the upstream shock has gone more than halfway to the axis. The density change induced by this shock in the jet has made the knot asymmetric and in fact it disrupts once the shock reaches the axis. It is therefore clear that such knots would look very different if the external density is of the same order as that in the jet.

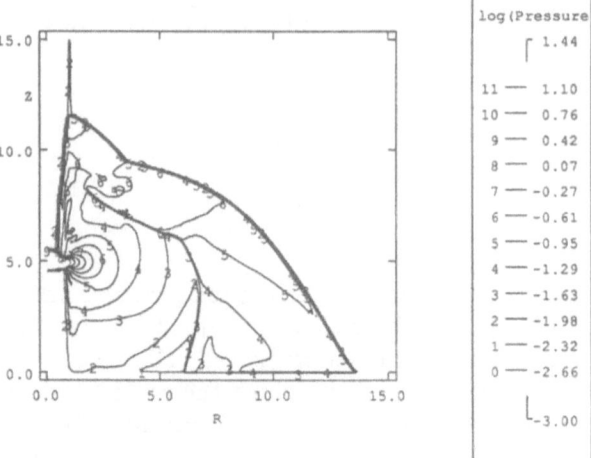

Figure 1a) Logarithmic pressure contours for the entire domain.

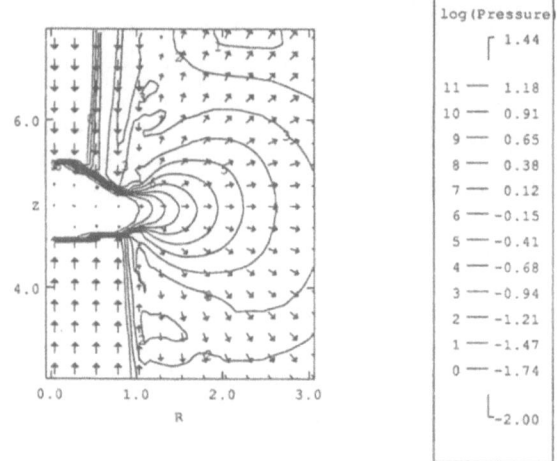

Figure 1b) Logarithmic pressure contours and velocity vectors for knot region.

For this density ratio, it would be impossible to detect the shocks in the surrounding gas, but they would be visible if the external density were higher. This is rather nice because these external shocks look very much like the mini-bowshocks observed in the HH34 jet by Reipurth and Heathcote (1992). We ought to be able to use such observations to determine the density ratio, but in order to do this we will need to improve the numerical calculations by including radiative cooling and non-equilibrium ionisation. This work is now in hand, but it requires extremely high resolution and can only be done with an adaptive grid.

5   Conclusions

In this paper I have tried to give some idea of current theoretical work on variable astrophysical jets. Although I have concentrated on stellar jets, it seems likely that the sources of extragalactic jets also vary and that the regions of enhanced radio emission in these jets are also formed by the same mechanism. It is in fact much easier to compute the flow in extragalactic jets since they are supposed to be adiabatic, but radio observations tell us so little about the flow that it is very hard to compare hydrodynamic calculations with the radio maps.

References

Biro, S. and Raga, A. C.: 1994. These proceedings.
Birkinshaw, M.: 1984, *Mon. Not. Roy. Astr. Soc.*208,887.
Crighton, D. G.: 1981, *J. Fluid Mech.*106,261.
Eislöffel, J.: 1992 PhD Thesis, Univ. Heidelberg.
Eislöffel, J. and Mundt, R.: 1992, *Astron. Astrophys.*263,292.
Eislöffel, J., Mundt, R. and Ray, T. P.: 1989, *Astron. Ges. Abstr.*, Ser. 3, 35.
Falle, S. A. E. G.: 1991, *Mon. Not. Roy. Astr. Soc.* 250,581.
Falle, S. A. E. G. and Raga, A. C.: 1993, *Mon. Not. Roy. Astr. Soc.* **261**,573.
Falle, S. A. E. G., Innes, D. E. and Wilson, M. J.: 1987, *Mon. Not. Roy. Astr. Soc.* **225**, 741.
Falle, S. A. E. G. and Wilson, M. J.: 1985, *Mon. Not. Roy. Astr. Soc.* **216**,79.
Hardee, P. E.: 1979, *Astrophys. J.* **234**,47.
Hardee, P. E. and Norman, M. L.: 1990, *Astrophys. J.* **365**, 134.
Hartigan, P. and Raymond, J.: 1993, *Astrophys. J.* **409**,705.
Kofman, L. and Raga, A. C.: 1992, *Astrophys. J.*390, 359.
Ojouveia dal Pino, E. and Benz, W.: 1993, *Astrophys. J.* **410**, 686.
Raga, A. C.: 1994, in T. Montmerle (ed.), Proceedings of the Moriond Conference, in press.
Raga, A. C., Binette, L. and Cantó, J.: 1990, *Astrophys. J.* **360**, 612.
Raga, A. C., Cantó, J., Binette, L. and Calvet, N.: 1990, *Astrophys. J.* **364**, 601.
Raga, A. C. and Kofman, L/: 1992, *Astrophys. J.* **386**, 222.
Ray, T. P.: 1981, *Mon. Not. Roy. Astr. Soc.* **196**, 195.
Rees, M. J.: 1978, *Mon. Not. Roy. Astr. Soc.* **184**, 61p.
Reipurth, B.: 1989, *Low Mass Star Formation and Pre- Main Sequence Objects*, ESO Conf. & Workshop Proc. No. 33, p247.
Reipurth, B. and Heathcote, S.: 1992, *Astron. Astrophys.* **257**, 693.
Reipurth, B., Heathcote, S. and Vrba, F.: 1992, *Astron. Astrophys.* **256**, 225.
Sanders, R. H.: 1983, *Astrophys. J.* **266**, 73.
Stone, J. M. and Norman, M. L.: 1993, *Astrophys. J.* **413**, 210.
Wilson, M. J.: 1983, PhD Thesis, Univ. Cambridge.
Wilson, M. J.: 1984, *Mon. Not. Roy. Astr. Soc.* **209**, 923.
Wilson, M. J.: 1986, *Mon. Not. Roy. Astr. Soc.* **224**, 155.
Wilson, M. J.: 1987, *Mon. Not. Roy. Astr. Soc.* **226**, 447.
Wilson, M. J. and Falle, S. A. E. G.: 1985, *Mon. Not. Roy. Astr. Soc.* **216**,971.

A Simulation of a Jet with the Hiccups

S. Biro
Astronomy Dept., The University, Manchester M13 9PL, U.K.

and

A.C. Raga
Mathematics Dept., UMIST, P.O. Box 88, Manchester, M60 1QD, U.K.

Abstract. We present results of a numerical simulation of a stellar jet from a source with supersonic variations in the outflow velocity. The simulation is compared with both analytical predictions and observations.

1 Introduction

Observations of Stellar Jets (strings of Herbig-Haro objects) show evidence for the existence of time-dependent phenomena. Proper motions, multiple bow shocks, e.g. HH111 (Reipurth et. al., 1992) and velocity gradients along the jets (Reipurth, 1991) have been detected. These phenomena have been modeled in terms of a jet from a source with supersonic velocity variations (Raga et. al., 1990), (Kofman & Raga, 1992), which produce pairs of shocks (or "internal working surfaces") along the body of the jet. In the present work we show the results of a simulation of such a jet.

2 Results

We have simulated a cylindrically symmetrical jet of compressible, inviscid, non-adiabatic gas of cosmic abundance with periodic variations in the magnitude of the injection velocity. Our numerical code is based on the Flux-Vector-Splitting scheme (van Leer, 1982). The parameters used are similar to those observed in actual HH jets. These are: $T_j = 10^4$ K (initial jet temperature), $n_j = 10$ cm^{-3} (jet gas density), $n_j/n_{ext} = 10$ (jet-to-environment density ratio), $V_j = 100$ km s^{-1} (jet velocity), $d_j = 4 \times 10^{16}$ cm (jet diameter), $\Delta V = 50$ km s^{-1} (half amplitude of velocity variation), $\Delta \tau = 1.2 \times 10^{10}$s ~ 380 yrs (period of velocity variation).

In Figure 1 we show a map of the pressure structure at the end of the simulation ($t \sim 1900$ yrs). The contours are logarithmic and regions with "piled up" contours indicate shocks. The head presents a complex shock structure consisting of a bow shock and a Mach disk. Along the body of the jet we see "internal working surfaces" consisting of two shocks with an arched shape.

The evolution of the size, shape and emission of the knots in the jet has been studied. The knots grow as they move away from the source and material enters them through both shock surfaces. They develop a bow shape as excess material flows out sideways and is swept backwards. Emission in the $H\alpha$ and [SII] 6717+31

Astrophysics and Space Science **216**: 127–128, 1994.
© 1994 *Kluwer Academic Publishers.*

lines is calculated and we find that the peak $H\alpha$ is stronger than [SII] by a factor of ≥ 2. The emission as a function of position at the final time has also been measured for both lines. The $H\alpha$ and [S II] peaks are offset, an effect which might be detected in observations of stellar jets (Reipurth & Heathcote, 1991). These results are reported further detail in (Biro & Raga, 1993).

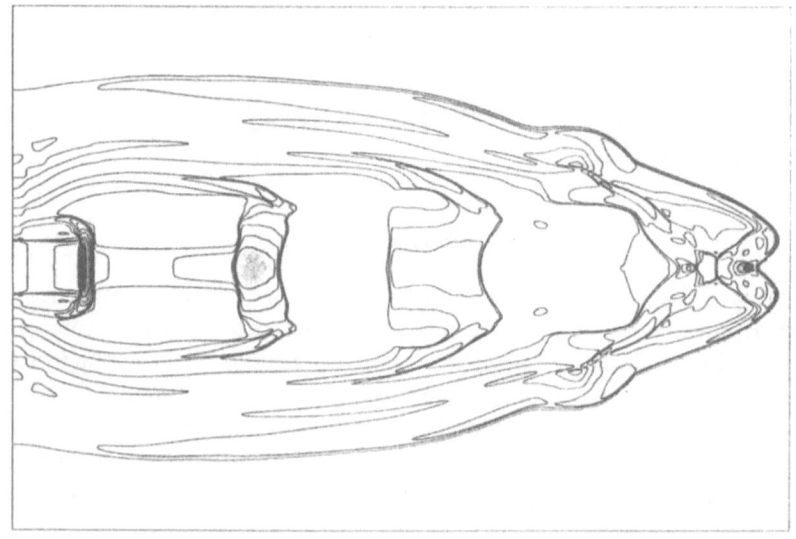

Fig. 1. Map of the pressure of the jet at time t $\sim$ 1900 yrs. Greyscales (with black indicating the highest and white the lowest pressures) and contours at logarithmic intervals (two successive contours corresponding to a pressure ratio of 2) are combined. The internal working surfaces form and develop along the body of the jet.

References

Biro, S. &Raga, A.C., 1993, *Ap.J.*, **submitted**
Kofman, L., & Raga, A.C., 1992, *Ap.J.*, **390**, 359
Raga, A.C., Cantó, J., Binette, L., & Calvet, N., 1990, *Ap.J.*, **364**, 601
Reipurth, B., & Heathcote, S., 1991, *Astron. Astroph.*, **257**, 693
Reipurth, B., Raga, A.C., & Heathcote, S., 1992, *Ap.J.*, **392**, 145
Reipurth, B., 1992, *The Physics of Star Formation and Early Stellar Evolution*, , p497
van Leer, B., 1992, *Lect.Notes in Phys.*, **170**, 507

Interactions between molecular outflows and optical jets

Rachael Padman and John S. Richer*
Mullard Radio Astronomy Laboratory
Cavendish Laboratory
Madingley Rd.
Cambridge CB3 0HE.

Abstract. Young stars produce both molecular outflows and, at a later evolutionary stage, well-collimated optical jets. The simplest explanation is that the molecular outflows are driven by *obscured* optical jets, rather than directly, by a disk wind for example, but the optical jets appear to have too small a momentum flux. Recent statistical studies however show that the molecular flows must be quasi-stationary, which means that the dynamical lifetime is a gross underestimate of the true age. As a consequence much less thrust is required. We present recent observations of RNO 43, which has well-defined optical and molecular outflows lying close to the plane of the sky. Excellent agreement with the observations is obtained with a simple kinematic model for the molecular material, which supposes that it lies in a parabolic shell around the optical jet with the highest velocities at the working surface. Together with our modelling of the NGC2024 outflow, this is very strong evidence that molecular outflows are produced by prompt entrainment of molecular material in a neutral or weakly-ionized jet.

Key words: Star formation, Molecular outflows, Jets, YSO's

1 Introduction

Are molecular outflows driven by the so-called 'optical jets' from young stars? Many of us would like to believe that they are, on purely aesthetic grounds. Attempts to unify these phenomena have, so far however, mostly fallen at the first hurdle: the inferred momentum flux in optical jets is an order of magnitude or more below that apparently required to drive the molecular outflows. Since optical jets and molecular outflows are rarely observed in the same sources (only HH 111, HH 46-47 and RNO 43 spring to mind in this connection), there is a depressing lack of evidence in support of such a unification.

Of course, even if we could show that molecular outflows were powered in this way, this of itself explains neither how the optical jets produce the observed molecular outflows, nor how optical jets are in their turn powered and collimated. Nonetheless, it seems to us that an essential first step to understanding these fascinating, and undoubtedly *related*, phenomena, is to establish what, if any, causal relationship there is between them. In this paper we attempt to show that the aesthetic judgement — that molecular outflows are driven directly by optical jets — is a reasonable one. We adduce three arguments in support of this claim. First, we use statistical studies to argue for outflow lifetimes much greater than the dynamical timescales of these sources. It follows that the required momentum injection rate is correspondingly reduced. Second, we use the particular example

---

* Current address D-Cubed Ltd, 68 Castle st, Cambridge CB3 0AJ, U.K.

Astrophysics and Space Science **216**: 129–134, 1994.
© 1994 *Kluwer Academic Publishers.*

of the unipolar outflow in NGC 2024 to show that it is possible to construct a self-consistent model in which a "typical" (but obscured) optical jet drives the molecular flow in a momentum-conserving interaction. Third, we present new observational data on the source RNO 43, in which we observe the interaction directly. The molecular gas near the head of the optical jet lies in an expanding shell, with the highest velocities coincident with the exact head of the jet. We leave it to others to explain the detailed physics of this observed interaction.

2   Observational evidence

2.1   STATISTICAL STUDIES

Two studies in particular bear directly on this problem. Parker, Padman and Scott (1991) searched for CO outflows in a sample of Lynds' class VI dark clouds selected for their association with deeply embedded IRAS point sources. Of these, 8 or 9 out of the 12 sources mapped were found to have outflow activity. On the assumption that these sources are nearly all proto-T Tauri stars, for which the stellar birthline occurs at about 2×10^5 yr, Parker *et al.* deduce a typical outflow lifetime of $\sim 1.5 \times 10^5$ yr. This is nearly an order of magnitude greater than the usual estimates, which come from the dynamical timescale R/V, where R is the greatest observed extent of an outflow lobe and V is the highest observed velocity. Parker *et al.* therefore conclude that the dynamical timescale is not a good estimator for the age, and that the molecular flows are in general *quasi-stationary* — the flow pattern evolves much more slowly than the motions of material in the flow.

Fukui *et al.* (1993) arrive at a similar estimate of outflow lifetime of 10^5 yr using the Nagoya survey of L 1641, and instead relating the frequency of outflows to the number of (optically identified) T-Tauri stars in the region, along with an estimate of the stellar ages. The dynamical timescales of outflows in this sample are somewhat larger than in ours ($\sim 50\,000$ yr), probably because the sources were observed with a larger beam and typically had somewhat lower linewidths.

The momentum supply rate required to power a molecular outflow is usually calculated by dividing the observed momentum ($\int_0^\infty Mv\,dv$) by the dynamical lifetime ($\tau_d = R/V$, as above). Because outflows (and their driving sources) cover a range in luminosity of 10^4 or more, it is hard to quote typical values. Taking L 483 as an example of an embedded source with a *low* luminosity more or less comparable with those of the visible sources powering optical jets, the momentum supply rate, or thrust, deduced thus is of order 3–8×10^{24} kg m s^{-2}, and this is reduced by a factor of ~ 10 if we use instead the statistical lifetime derived above.

2.2   NGC 2024 AS A JET-DRIVEN OUTFLOW.

Dynamical models of extragalactic jets usually involve a *light* supersonic jet propagating in a denser external medium. For such a jet the usual two-shock structure is formed at the head, and the speed of advance of this *bowshock* is determined by the conservation of momentum and the need to accelerate the swept-up material.

That is,

$$v_{\mathrm{bs}} = \frac{v_{\mathrm{j}}}{1 + \sqrt{\rho_{\mathrm{amb}}/\rho_{\mathrm{j}}}}, \qquad (1)$$

where ρ_{amb} and ρ_{j} are the ambient and jet densities respectively, and v_{j} is the jet velocity. The age of these sources is thus many times greater than the dynamical timescale, as we observe for stellar outflows (see previous section). It seems natural therefore to analyse stellar outflows in the same terms. In particular, we hypothesize that the observed molecular outflows are driven by underlying well-collimated high-velocity flows, which, if not obscured, would be identified as optical jets.

There is strong circumstantial evidence that the unipolar molecular outflow in NGC 2024 is of this kind (Richer, Hills and Padman, 1992). The collimation of the molecular flow increases with increasing velocity offset from the ambient cloud, and we observe shocked $H_2 S(1)$ emission from the (very abrupt) end of the molecular outflow. We can thus produce a simple dynamical model in which the high-velocity neutral jet sweeps up ambient (molecular) material to produce the observed outflow. Unfortunately we do not have any direct observations of the neutral jet which would allow us to determine the Mach number, density or pressure, but these can be constrained to "reasonable" values using a few simple assumptions.

Because of the high sensitivity of the CO observations, we have good estimates for the total mass, momentum and kinetic energy in the flow (subject to uncertainty in the angle of inclination, however). The jet velocity is assumed to be double the highest observed velocity in the molecular material, and the jet temperature is taken to be $\sim 8000\,\mathrm{K}$, above which the cooling rate rises very steeply. We assume that the jet is pressure-confined near its base, and has a radius equal to that of the highest-velocity CO. The *ambient* pressure can be estimated using observations of column-density tracers such as $C^{18}O$ and CS, together with an idea of the source size and temperature, and in conjunction with the jet temperature gives a pretty good handle on the jet number density. Then by equating the ram pressures of the jet and the ambient cloud material, as in (1) above, we can derive the position of the bow-shock as a function of time after the jet turns on.

Fig.1 shows the results for two models; one in which the jet is somehow confined for its whole length, so that the jet pressure stays constant, and one in which the jet is confined only to the edge of the dense core, beyond which it expands freely. Even for this very luminous and high-velocity CO outflow we deduce ages of 60 000 and 40 000 years for the two models respectively, which can be compared to the dynamical timescale of $\tau_{\mathrm{d}} \simeq 6000\,\mathrm{yr}$ (if the highest CO flow velocity is taken as $100\,\mathrm{km\,s^{-1}}$). We conclude that it is easily possible to produce self-consistent models for this source in which an optical jet of reasonable parameters provides the momentum necessary to power the observed molecular outflow, and we suggest that this may be true for other similar sources also.

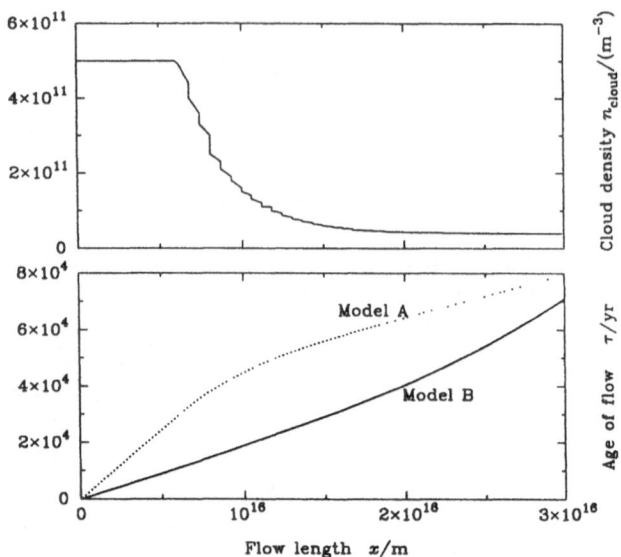

Fig. 1. Simple model of the propagation rate of a jet through the NGC 2024 cloud core. The upper panel shows the assumed average core density as a function of distance down the jet. and the lower panel shows the time after the jet turns on that the working surface passes that point.

2.3   INTERACTIONS IN RNO 43

Despite the success of the model for the NGC 2024 outflow, it still depends on the existence of an *unseen* optical jet. Our case would be greatly strengthened if we could observe *directly* an interaction between a jet and outflow. We noted earlier that there are very few sources in which these are both observed, but one such is RNO 43, for which we have recently obtained JCMT CO $J = 3 \rightarrow 2$ data. The jet in RNO 43, lies close to the plane of the sky; whilst this makes it hard to observe motions along the jet it does mean that the source presents a particularly simple aspect, which is more amenable to modelling than the usual case.

There is indeed a very close correlation between the Hα images (Mundt, 1988) and the CO maps. The CO is very bright close to the optical "blobs", and the channel maps (Fig.2a) show that the highest velocities are coincident with the end of the jet. The ambient velocity channel (at $v_{lsr} = 10\,\mathrm{km\,s^{-1}}$) is suggestive of a limb-brightened parabolic shell with its vertex at the end of the Hα jet. We have devised a simple *kinematic* model for the molecular gas which reproduces the observed channel maps extremely well (Fig.2b). We assume that the molecular material is impulsively accelerated in the radial direction (away from the jet axis) at the head of the jet, and thereafter decelerates slowly, perhaps as it sweeps up the surrounding material (for ram-pressure deceleration in a uniform density medium,

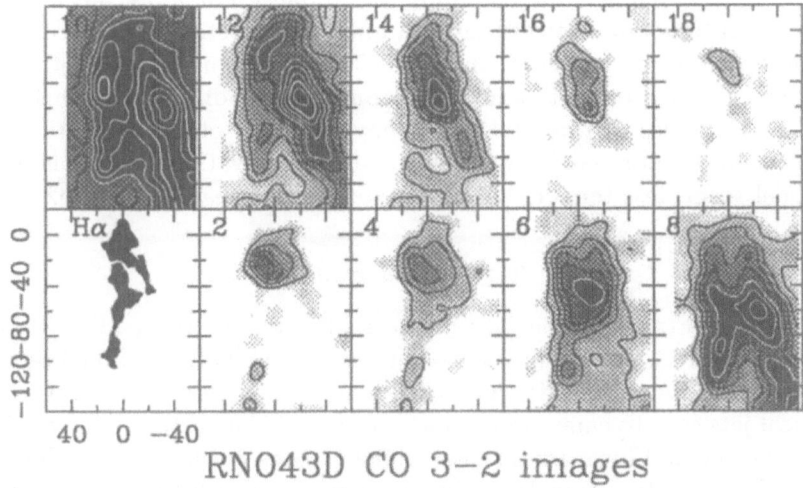

RNO43D CO 3−2 images

...and kinematic bow−shock model

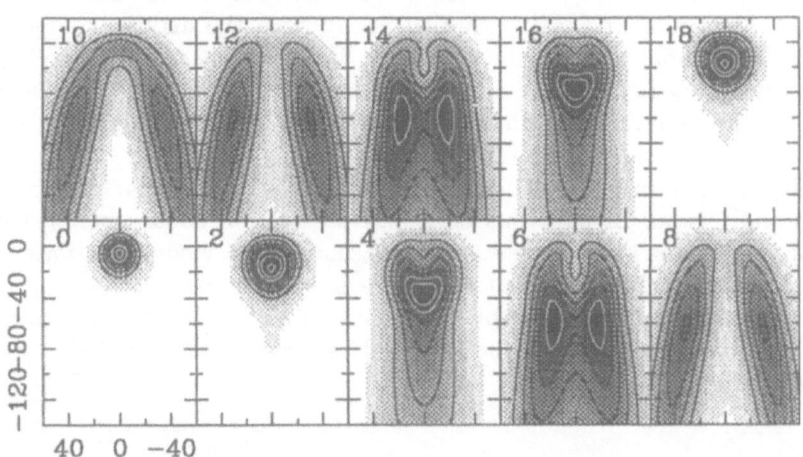

Fig. 2.   CO $J = 3 \rightarrow 2$ channel maps of RNO 43 observed with a 15 arcsec beam. The centre velocity for individual maps is indicated at the top left of each panel. (Top): JCMT data (bottom left panel shows the Hα emission from Mundt, 1988). (Bottom): Mock channel maps for the model presented in the text.

this leads naturally to a parabolic shape for the shell). We further assume that the shell has a density profile which is gaussian about the mean position. The column density of gas in each radial velocity interval can then be very simply computed, and convolution with the telescope beam gives a set of mock channel maps.

The model presented here includes only the simplest dynamics, but nevertheless constrains the interaction between jet and ambient material. The acceleration of ambient material seems to occur, in RNO 43 at least, primarily at the head of the jet ('prompt' entrainment) and not through some form of Kelvin-Helmholtz instability at the sides of the jet (which may well be isolated from the ambient medium by a hot cocoon — although we have no observational evidence for this). Such a model seems less tenable for *decelerated* outflows such as L 1262, however.

3   Conclusions

The three arguments presented here make a strong case that, at least in some sources, the appearance of the molecular outflow can be best explained by a model involving a direct interaction with an underlying, mostly hidden, neutral jet similar to the optical jets seen to emanate from less obscured young stars. The assumption that the molecular outflows are quasi-stationary eases constraints on the momentum flux, but seems to imply that the outflows are driven by *light* jets — *i.e.*, jets which are underdense with respect to the surrounding medium. This is not in conflict with the optical observations (which give minimum jet densities somewhat larger than the ambient gas). For example, in the case of NGC 2024 we see that there is a strong ambient density gradient away from the driving source, so that a collimated supersonic jet which is initially 'light' can become 'heavy', with no change in density, as it propagates into the lower density medium. As the density falls, so does the extinction, so that optically-visible jets are indeed likely to have densities at least comparable to their surroundings, as observed.

Optical jets have in general been observed only from low-mass stars, mostly in Taurus and Orion. In part this is clearly a selection effect, due to the small number of very young early-type stars in our immediate vicinity. It may however also reflect the faster evolution of the more massive stars, so the jets in these objects are shrouded from view for their entire lifetime. Observations of many more sources like NGC 2024 will be required before we can establish a generally applicable model.

References

Fukui, Y., Iwata, T., Mizuno, A., Bally, J. and Lane, A. P.: 1993, "Molecular outflows", in *Protostars and Planets III*.

Mundt, R.: 1988, "Flows and jets from young stars." in *Formation and evolution of low mass stars*, A.K.Dupree and M. T. V. T. Lago (eds), Kluwer, Dordrecht, Holland , 257.

Parker, N. D., Padman, R. and Scott, P. F.: 1991, "Outflows in dark clouds: their role in protostellar evolution." *Mon. Not. Roy. Astr. Soc.*, **252**, 442.

Richer, J. S., Hills, R. E. and Padman, R.: 1992, "A fast outflow in Orion B." *Mon. Not. Roy. Astr. Soc.*, **254**, 525.

Proper Motion Measurements in the HH 46/47 Outflow

Jochen Eislöffel
Max-Planck-Institut für Astronomie, Heidelberg, Germany and Dublin Institute for Advanced Studies, Dublin, Ireland

Abstract. Proper motion measurements have been carried out for the HH 46/47 outflow system. The results of these measurements and some implications for the physics of the outflow and its modelling are discussed.

Key words: Herbig-Haro objects, proper motions

HH 46/47 is a well studied bipolar jet system emanating from an isolated Bok globule in the Gum Nebula. It consists of the bright, blue-shifted HH 46/47 jet, which shows a wealth of internal structure (see, e.g., Reipurth & Heathcote 1991) and ends in the bright knot HH 47A, which is thought to be its bow shock (e.g., Meaburn & Dyson 1987). However, a second, fainter, but much more extended bow shock HH 47D is found ahead of HH 47A. On the opposite side of the jet source, seen only in the IR (Graham & Elias 1983), a faint counter jet points towards the redshifted bow shock HH 47C (Meaburn & Dyson 1987).

A detailed study has been carried out of the proper motions of the knots in the HH 46/47 jet and counterjet, as well as of the condensations in the associated bow shocks HH 47C and HH 47D. In the jet most tangential velocities are in the range $100 - 300 \, \mathrm{km \, s^{-1}}$ (see Fig. 1), while in the other parts of the system the measured values are mainly somewhat lower at $70 - 170 \, \mathrm{km \, s^{-1}}$. Given reasonable assumptions about the pattern speed at the apex of the presumed bow shock HH 47A the orientation of the HH 46/47 outflow is determined to be $34°.1 \pm 2°.8$ (Eislöffel 1992, Eislöffel & Mundt 1993). This allows us to correct the observed radial and tangential velocities in the jet for projection effects, enabling the local flow speed of the jet and the knot pattern speed to be derived. A typical flow speed of about $290 \, \mathrm{km \, s^{-1}}$ is found in the jet. The ratio ζ between the pattern speed of the knots and the flow speed of the particles in the jet displays a bimodal distribution with the two values $\zeta = 0.62 \pm 0.05$ and $\zeta = 0.95 \pm 0.06$. This bimodal distribution, if real, is difficult to understand with current jet models. Ambient material may be entrained into the jet along parts of the jet channel and seen as a separate flow component.

We coadded images taken under excellent seeing conditions into a single high-signal/noise image and deconvolved the latter using the Richardson-Lucy algorithm. The deconvolved image has a seeing of $0''.47$ (FWHM) and contains a wealth of detail. The jet clearly shows several kinks and along most of its length consists of two well-separated bright rims. Such a limb-brightening effect has been seen in only a few other jets so far.

Proper motions were also measured for condensations in the arc-shaped HH objects HH 47D and HH 47C, which are assumed to be the bow shocks of the jet

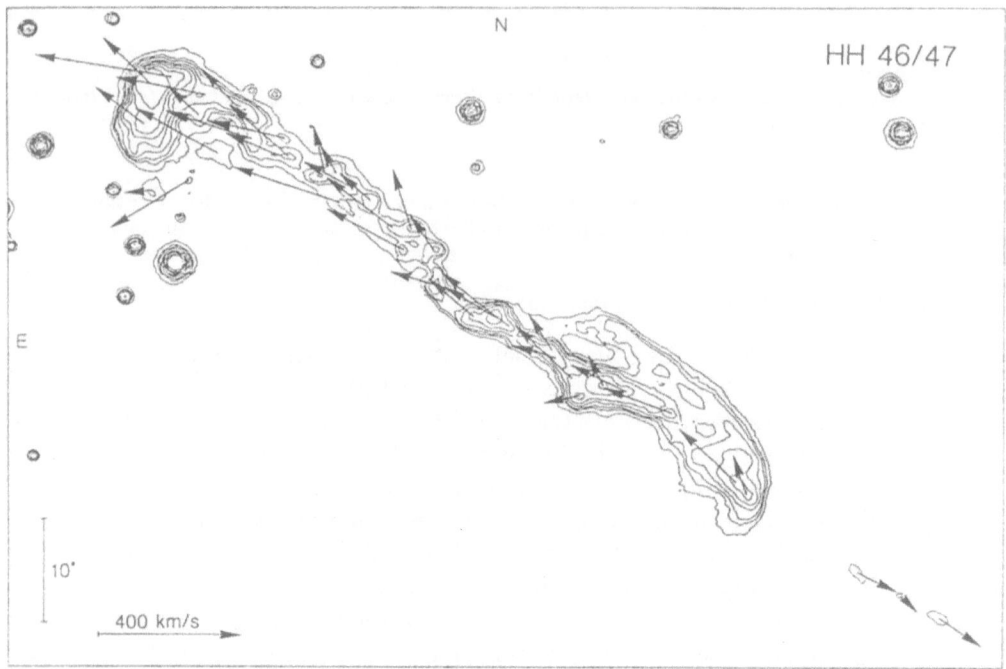

Fig. 1.   Proper motions of the knots in the HH 46/47 jet and counter jet. The bright knot at
the end of the jet in the upper left corner is HH 47A.

(from an earlier outflow episode) and the counterjet, respectively. We derived the
internal pattern of motion of these condensations, which is in good agreement with
predictions from simple bow shock models.

Acknowledgements

The author wishes to thank Drs. Reinhard Mundt and Tom Ray for many discussions on various aspects of this work. He acknowledges partial support from
the Human Capital and Mobility Programme of the European Community under
grant ERBCHBGCT920205.

References

Eislöffel, J. 1992, Ph.D. Thesis, University of Heidelberg
Eislöffel, J. Mundt, R. 1993, A&A, submitted
Graham, J.A., Elias, J.H. 1983, ApJ 272, 615
Measburn, J., Dyson, J., 1987, MNRAS 225, 863
Reipurth, B. Heathcote, S. 1991, A&A 246, 511

The Serpens Radio Jet: Evidence of Precession of Nutation

S. Curiel and J.M. Moran
Harvard-Smithsonian Center for Astrophysics

and

L.F. Rodríguez and J. Cantó
Instituto de Astronomía, UNAM

Abstract. Previous VLA observations of the triple radio continuum source in Serpens showed that it has very unusual and extraordinary characteristics. While this source is associated with a star forming region, its outer components exhibit a combination of thermal and nonthermal spectra and large proper motions, Furthermore, the NW lobe has knotty and extended emission connecting the central source with the bright outer knot. Here, we present results of new VLA radio continuum high-angular resolution observations of this Radio Jet. Combining these observations with those obtained previously, we find that: a) one of the knots along the main body of the radio jet (knot G) exhibits proper motions similar to those observed in the outer NW and SE components, and the time variable knot A; b) the outer knots are moving away from the central source in slightly different directions; and c) the orientation of the central source seems to change with time. These results are consistent with a central precessing source that undergoes periodic ejection of material.

1 Results

Combining our 3.6 and 6 cm results with our previous observations (see Rodríguez *et al.* 1989; and Curiel *et al.* 1993), we find that the four brightest knots (NW, Y, Z and G) in this radio jet have similar tangential velocities of about 0.12″ per year but are moving away from the central source in slightly different directions (see Fig. 1). The outer NW, Y and Z knots have dynamical ages between 58 and 70 years, while knot G has a dynamical age of ∼ 14 years, indicating that each knot was ejected at a different time. Furthermore, the variation of ∼18° in the direction of the proper motions seems to be significant since the statistical error is of only ∼2° for each knot. These results indicate that the ejection of material by the central source is produced during eruptive (and probably periodic) events, and that these knots were ejected at different times and different directions, suggesting that the energy source may be precessing or nutating.

At both, 3.6 and 6 cm wavelengths the central source appears elongated, having similar deconvolved sizes and position angles. However, when we compare these sizes and position angles with those obtained at 6 cm in 1984 by Rodríguez *et al.* (1989) and at 3.6 cm in 1990 by Curiel *et al.* (1993), we find that the position angle has changed about 10° between 1984 (at 6 cm) and 1990 (at 3.6 cm), and about 3° between 1990 (at 3.6 cm) and 1993 (at 6 cm). If we take into account these three epochs, the major axis of the central source seems to be rotating towards the E, and with an angular speed of about 1.5° per year. This result is also consistent

Astrophysics and Space Science **216**: 137–138, 1994.

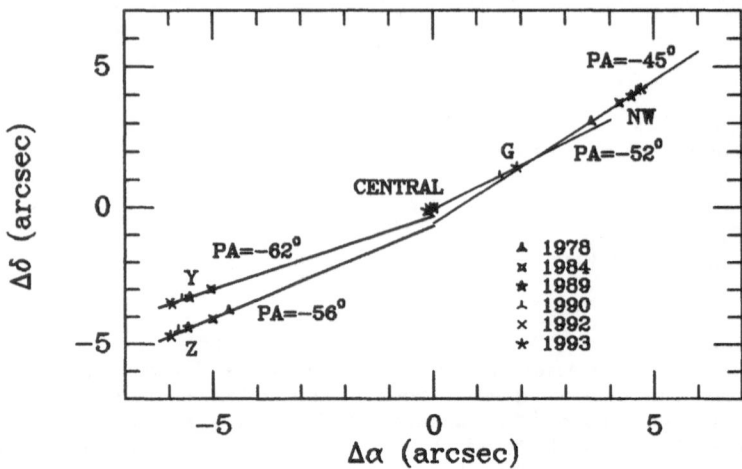

Fig. 1.   Graphical representation of the proper motions of the four brightest knots (NW, Y, Z, and G) in this radio jet. All the positions obtained for these knots between 1978 and 1993 were included. The angular displacement of the knots is given with respect to the position of the central source in the 1984 epoch. The position angles of these proper motions were obtained by making a least-squares fit to the position of a given knot at all the observed epochs, and are shown in this figure as solid lines.

with a central source that precesses or nutates.

If we trace back in time the position of knot G for the 1984 epoch (using its proper motions), we find that: a) the direction of the proper motions of this knot nearly coincides with the orientation of the major axis of the central source in 1984, and b) the predicted position of knot G for the 1984 epoch lies within the central source. These two results suggest that the elongation of the central source in 1984 was the result of insufficient angular resolution to separate this source and knot G. In such a case, one could speculate that the elongated morphologies observed in the central source in these observations could also be the result of recent ejection of small condensations.

Finally, the knotty structure observed in the NW lobe and the proper motions obtained for the brightest NW, Y, Z and G knots, suggest that the central source has been going through periodical ejection of material during the ∼70 year dynamical age of the system. Furthermore, the different orientation of the proper motions in these knots and the change in position angle of the central source major axis at different epochs, indicate that these knots were ejected at different times and in different directions by a precessing or nutating energy source.

References

Curiel, S., Rodríguez, L.F., Moran, J.M., and Cantó, J. 1993, *ApJ*, (in press).
Rodríguez, L.F., *et al.* 1989, *ApJL*, Vol. **346**, L85.

Fragmentation and Heating of Streamers in Orion

J. J. Wiseman and P. T. P. Ho
Harvard-Smithsonian Center for Astrophysics

Abstract. Long filamentary structures composed of chains of denser clumps fan out from the high mass star-forming KL core region in OMC-1. We present a high resolution VLA NH₃ study of the structure, kinematics, and temperature of the region over a large, multi-field scale. The region appears to contain multiple superimposed cloud components. Clumps along the filaments may be the result of instabilities and fragmentation; some show fast velocity gradients. Sheaths around the clumps along the filaments are heated. This is possibly the result of external radiation or of interaction with high velocity outflows from young stellar objects in OMC-1.

1 Observations of Filamentary Structure

Higher resolution studies of molecular clouds are revealing that filamentary structures, often containing dense fragments and based in high-mass active cores of star formation (e.g. Ophiuchus), are a common pattern. Recent improvements in receiver sensitivity have allowed studies of fainter regions surrounding dense molecular cloud cores, and such studies in Orion Molecular Cloud 1 (OMC-1) have revealed a striking pattern of clumpy filaments fanning out from the dense, active Orion-KL core region (Martin-Pintado *et al.*, 1990; Murata *et al.*, 1990; Wiseman and Ho, 1993). To clarify the structure and kinematics of the region, we have used the VLA to observe with high (0.3 km s^{-1}) velocity resolution and high (8″) angular resolution the NH$_3$ (1,1) and (2,2) rotation-inversion lines over 10 adjacent fields covering a 3′ × 5′ region to the north of the KL core. The images were cleaned, arranged in a linear mosaic, and the intensity integrated across the line channels to produce the maps of structure and kinematics presented here. The resulting mosaic is presented in **Figure 1**; long streamers branch 0.5 pc to the north and northwest of the KL region at the bottom of the image. The contours of **Figure 2** represent the integrated intensity of line emission, and show the clumpy makeup of the streamers. The streamers are chains of denser cores; the regular spacing of the cores suggest that they are possibly the result of instability and fragmentation. The individual cores are also of interest. A localized outflow has been discovered around the core at α 05 32 49, δ -05 22 30 (Wilson and Mauersberger, 1991), strong evidence that these fragments may be sites of low mass star formation.

2 Kinematics

The greyscale shades of **Figure 1** display the first moment integral of the intensity over the (1,1) line channels, and represent the velocity field of the region. The darker shades show that the main ridge filamentary material is at about 10 km s^{-1}. The northwestern filaments are at about 8 km s^{-1} (lighter shades) and appear

Astrophysics and Space Science **216**: 139–142, 1994.

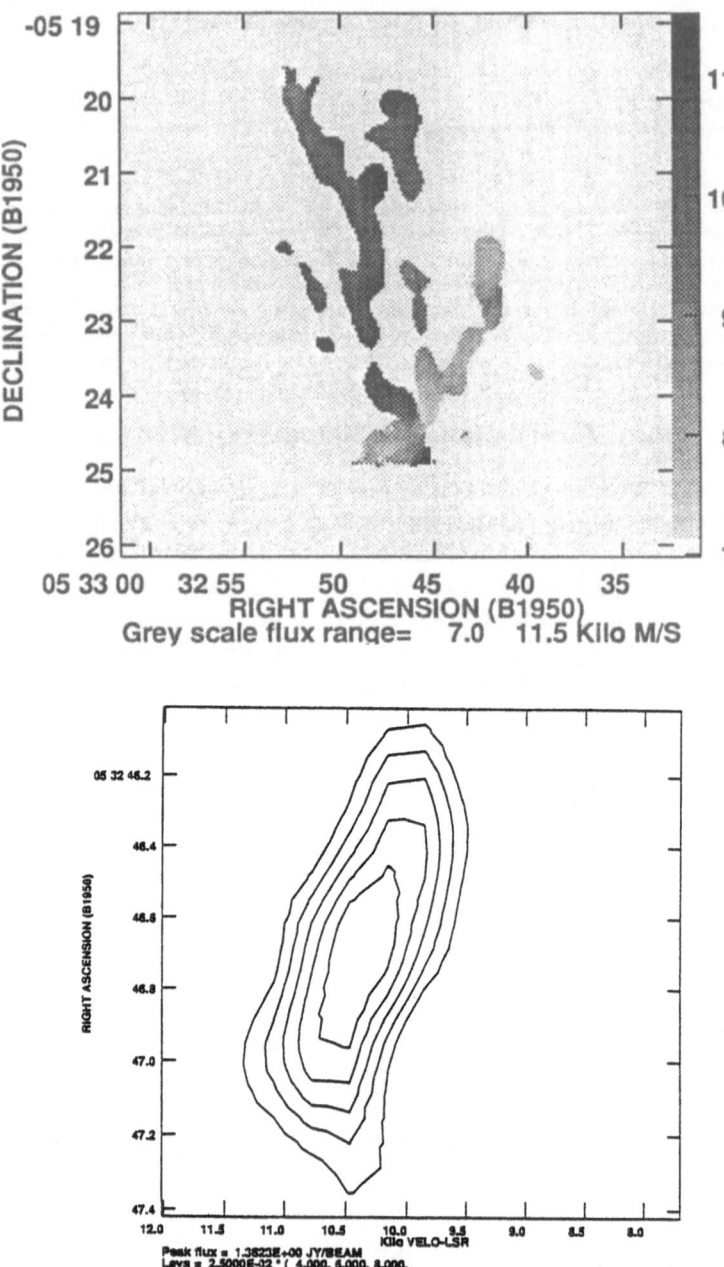

Fig. 1.   The filaments in OMC-1, mapped with 10 NH₃ (1,1) VLA fields. The grayscale shades (top section) are the first moment integrated intensity and represent the velocity field of the material; darker material has a higher mean LSR velocity. Note the lower velocity of the overlapping western filaments. The bottom section shows a position-velocity plot of a northern clump (see text). The smooth, fast velocity gradient seen here is 30 km s^{-1} pc^{-1}, and may indicate core rotation.

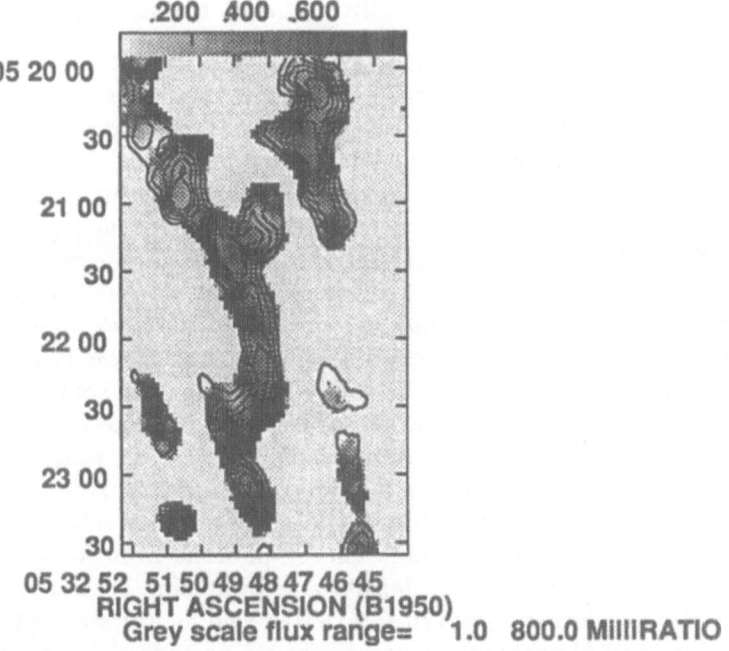

Fig. 2.   Close-up view of the inner filaments. The grayscale shades are the ratio of the integrated NH₃ (2,2) and (1,1) intensities, and represent the temperature of the region. Contours are the integrated (1,1) intensity, at levels of 60, 120, 180, 240 and 300 mjy/8″ beam. Note the edge heating toward the outflow source IRc2 to the south of the image, and the apparent interclump heating along the filaments.

to lie superimposed with the 10 km s⁻¹ material in the active KL region in the southern part of the image, suggesting the possibility of the energetic interaction of two clouds (see Womack *et al.*, 1991; Wang *et al.*, 1993). It is also apparent from **Figure 1** that there are fast velocity gradients across several of the cores. Closer spectral inspection have shown that some of these apparent gradients are actually the result of superimposed material of different velocities, again suggesting the presence of multiple cloud components in the region. But some cores show fast gradients with no apparent superposition. The bottom plot of **Figure 1** shows a position-velocity cut taken across the velocity gradient of the core at α 05 32 47, δ -05 20 00. The gradient is smooth and fast (30 km s⁻¹ pc⁻¹), and confirms previous evidence for core rotation in the region (Harris *et al.*, 1983). Fragmenting filaments may produce unstable cores which collapse, spin up, and if they can lose their angular momentum, become sites for low mass star formation.

3   Heating of Clumps and Filaments

The OMC-1 filaments lie in a region of high velocity outflows emanating from young stellar objects in the Orion-KL core region, such as IRc2. Lobes of high velocity CO, shocked H_2, and an array of Herbig-Haro objects moving away from IRc2 with velocities of hundreds of kilometers per second (Taylor, 1984; Lane, 1989) are all evidence that high velocity material is injecting energy into the surrounding environment. **Figure 2** is a temperature map of the region, created from the ratio of the NH_3 (2,2) and (1,1) intensities. The darker greyscale shades are the regions where this ratio is higher and represent warmer regions. Two striking features are apparent here: Near the bottom of the image, facing IRc2 to the south, the core edges show strong heating effects. Shocks from large-scale outflows may have penetrated and heated the core edges facing the source. Sheaths along the interface of dense cores within less dense flows are likely sites of turbulence and prolonged heating (Hartquist and Dyson, 1993). The other striking feature is the apparent inter-clump heating all along the ridge. This feature may also be the result of interaction with outflowing gas, or alternatively this could be the result of radiative heating, with high energy photons from foreground O and B stars penetrating regions of lower density and producing heating gradients. Continuing studies of heating effects in OMC-1 as well as of its complex kinematics, fragmentation, and filamentary structure should provide rich information on the complexities of interaction between a star-forming core and its environment.

References

Harris, A., Townes, C. H., Matsakis, D. N., and Palmer, P., 1983, *Astrophys. J. (Lett.)*, **265**, L63.

Hartquist, T.W., and Dyson, J.E., 1993, *Quart. J. Roy. Astr. Soc.*, **34**, 57.

Lane, A.P., 1989, in *Low Mass Star Formation and Pre-Main Sequence Objects*, ESO Conference and Workshop Proceedings No. 33, ed. Bo Reipurth, p. 331.

Martín-Pintado, J., Rodríguez-Franco, A., and Bachiller, R., 1990, *Astrophys. J. (Lett.)*, **357**, L49.

Murata, Y., Kawabe, R., Ishiguro, M., Morita, K., Kasuga, T., Takano, T., Hasegawa, T., 1990, *Astrophys. J.*, **359**, 125.

Taylor, K. N. R., Storey, J. W. V., Sandell, G., Williams, P. M., and Zealey, W. J., 1984, *Nature*, **311**, 236.

Wang, T.Y., Wouterloot, J.G.A., an Wilson, T.L., 1993, *Astron. Astrophys.*, **277**, 205.

Wilson, T. L., and Mauersberger, T., 1991, *Astron. Astrophys.*, **244**, L33.

Wiseman, J.J., and Ho, P.T.P., 1993, *in preparation.*

Womack M., Ziurys L. M., Wyckoff S., 1991, *Astrophys. J.*, **370**, L99.

Highly Supersonic Molecular Flows
in Wind-Clump Boundary Layers

M. T. Malone and J. E. Dyson
Department of Physics and Astronomy,
The University of Manchester,
Manchester M13 9PL, England

and

T. W. Hartquist
Max Planck Institute for Extraterrestrial Physics,
D-85740 Garching, Germany

Abstract.
   The gradual acceleration, through viscous coupling to a wind, of molecular hydrogen in turbulent boundary layers around obstacle clumps is proposed to be responsible for the widths of the emission features with full widths at zero intensity of greater than 100 km s^{-1}.

1   Introduction

Molecular hydrogen emission profiles with widths of about 100 km s^{-1} and more have been detected towards many sources (e.g. Geballe 1990) including the protoplanetary nebula CRL 618 (Burton & Geballe 1986) and the Orion Becklin-Neugebauer/Kleinmann-Low Nebula (e.g. Geballe *et al* 1986). Shocks preceding expanding wind-blown bubbles are often responsible for the acceleration of interstellar and circumstellar matter. However, for standard choices of the ambient medium's magnetosonic speed, the collisional dissociation of molecules is efficient in a magnetically moderated shock propagating at a speed greater than about 50 km s^{-1} into a dense medium (Draine, Roberge & Dalgarno 1983; McKee, Chernoff & Hollenbach 1984). Because of this speed limit for non-dissociating shocks, the presence of the broad H$_2$ profiles has been considered problematic. In the case of the Orion BN/KL region, Chernoff, Hollenbach & McKee (1982) speculated that the broad H$_2$ profile wings actually originate in a molecular reformation region in an initially molecular stellar wind that has passed through an inner dissociating shock in a wind-blown bubble. Spatially coincident emissions from SO and from SO$^+$, if found, might indicate the existence of a dissociating shock (Dalgarno 1993). Smith, Brand & Moorehouse (1991) have noted that the inner shock through which the stellar wind flows would not be dissociating if the magnetosonic speed in the preshock wind is sufficiently high.

   Hartquist & Dyson (1987) pointed out that H$_2$ can reach high speeds without being dissociated if it is in the shell of a wind-blown bubble that is expanding into an ambient medium with a density decreasing more rapidly than the wind diverges. For example, for a steady spherically symmetric wind and spherically symmetric ambient medium, the shell will accelerate if the ambient medium's density drops

Astrophysics and Space Science **216**: 143–150, 1994.
© 1994 *Kluwer Academic Publishers.*

more rapidly than the inverse square of the distance from the wind's source. The key elements of this picture are that the observed swept-up H_2 passed through a sonic point and started to move supersonically relative to the ambient medium when the shell was still moving at speeds well below those of dissociating shocks and that the subsequent acceleration occurred so gradually that the gas never heated up sufficiently for it to have made a later supersonic to subsonic transition or for the molecules in it to be dissociated.

In this paper we develop a somewhat different picture for the acceleration of H_2 to high speeds, but the same key elements are present in it. We investigate the structures of flows in boundary layers between fast winds and embedded clumps. Acceleration of gas that has become supersonic at temperatures very low compared to those required for dissociation to take place is continuous as it flows along the boundary layer. In section 2 we present our basic two dimensional model of a steady flow in a wind-clump boundary layer. In section 3 numerical results are given for the flow patterns in such a boundary layer while section 4 contains a synthetic H_2 line emission profile that would originate in a number of clumps distributed around a source of wind. In section 5 we discuss advantages of the boundary layer model for the origin of broad lines.

2   The Basic Model of Flow in a Boundary Layer

We assume that the wind has driven a converging shock through the clump, that the clump has cooled radiatively, and that the magnetic field pressure greatly exceeds the thermal pressure. We assume further that the wind is highly supersonic; as a consequence, the magnetic field is nearly parallel to the boundary layer. The variable x measures the distance along the boundary layer and z is the distance outside the inner edge of the boundary layer. The boundary layer is assumed to be very thin, so that $\partial P/\partial z = 0$ where P is the total pressure.

The continuity equation for the flow of ablated cloud material and mixed-in wind material in the boundary layer is

$$\frac{\partial}{\partial x}(\rho v_x) + \frac{\partial}{\partial z}(\rho v_z) = S \tag{1}$$

where ρ is the density, $\underline{v}$ is the fluid velocity, the subscripts x and z indicate the vector component being considered, and S is the rate per unit volume at which mass is being added from the fast wind to the bundary layer flow. The force equation is

$$\frac{\partial}{\partial x}(\rho v_x \underline{v}) + \frac{\partial}{\partial z}(\rho v_z \underline{v}) + \hat{x}\frac{\partial P}{\partial x} = \hat{x}F_x \tag{2}$$

where $\hat{x}F_x$ is the force per unit volume exerted on the flow in the boundary layer through its coupling to the fast wind. Coupling can be due to direct mixing of wind material into the boundary layer flow or through the propagation and dissipation of waves generated by the response of the wind to its encounter with the clump.

To solve equations (1) and (2) we will follow the flow along each of a number of trajectories starting at $x = x_i$, $z = 0$ where x_i has a different value for each trajectory. Along a trajectory z will be considered to be a function of x. Thus we have

$$z = \int_{x_i}^{x} \frac{v_z}{v_x} \, dx \tag{3a}$$

$$\frac{dv_x}{dx} = \frac{1}{\rho v_x} \left(F_x - S v_x - \frac{dP}{dx} \right) \tag{3b}$$

$$\frac{dv_z}{dx} = -\frac{v_z S}{\rho v_x} \tag{3c}$$

where we have used (1) and (2) and written partial derivatives with respect to z as $(v_x/v_z)(d/dx - \partial/\partial x)$.

S and F_x are, in principle, functions of x and z, but in this first study we will take S and F_x to be functions, which we specify, of x only. We also specify P as a function of x. In fact,

$$P = P_T + P_M \tag{4a}$$

where P_T is the thermal pressure and P_M is the magnetic pressure. Since the magnetic field is nearly parallel to the layer, we take

$$\frac{dP}{dx} = \frac{dP_T}{dx}. \tag{4b}$$

In later work we will consider the thermal balance in a boundary layer more thoroughly, but in this first study we assume that the gas in the boundary layer is isothermal with a temperature

$$T = T_0. \tag{5}$$

Thus,

$$\rho = \frac{\mu P_T}{k_B T_0} \tag{6}$$

where k_B is Boltzmann's constant and μ is the mean mass per particle in the boundary layer. In this work we will specify

$$P_T = \beta P = \frac{\beta \alpha^2 \rho_w v_w^2}{\alpha^2 + 4x} \tag{7a}$$

$$S = \frac{\alpha \rho_w v_w}{(\alpha^2 + 4x)^{\frac{1}{2}} L} \left(\frac{d}{\delta} \right) \tag{7b}$$

$$F_x = \frac{2\alpha x^{\frac{1}{2}} \rho_w v_w^2}{(\alpha^2 + 4x) L} \left(\frac{d}{\delta} \right). \tag{7c}$$

Here δ is the boundary layer thickness and d^2 the cross-sectional area presented by the clump to the wind. The ratio (d/δ) will be taken to be ~ 10 (Hartquist and Dyson 1988). α is a constant and has the dimensions of the square root of a length. β is also taken to be a constant. ρ_w and v_w are the distant upstream wind density and speed. L is the total length of the boundary layer. For very small α the expression for P corresponds to that expected for the pressure on the post-shock side of a very long thin parabolic bow shock. Other forms for P, S and F_x can be selected, and we hope that at some point sufficiently high spatial and high spectral resolution data can be obtained to allow us to infer *detailed* information about the functional forms of P, S and F_x appropriate for a particular source.

3   A Boundary Layer in the Becklin-Neugebauer/Kleinmann-Low Nebula

The flow in a model boundary layer is determined by the choices made for the values of T_0, α, β, L, ρ_w and v_w. In this section we present model results for choices of those values that are appropriate for the BN-KL Nebula where H_2 features with profiles having widths somewhat in excess of 100 km s^{-1} are formed. The choices for T_0, β, L and v_w are 2000 K, 0.1, 10^{-2} pc and 200 km s^{-1}. α was selected to be $(4L/9)^{1/2}$ to give a value of 0.1 for the ratio $P(x = L)/P(x = 0)$. ρ_w was set to 1.0×10^{-20} gm cm^{-3} which is the density of a 200 km s^{-1} wind from a star at a distance of 0.05 pc which loses mass at a rate of $10^{-3}(\Omega/4\pi)\,M_\odot$ yr^{-1}, where Ω is the solid angle into which the star is ejecting mass; this mass loss rate is compatible with that suggested by Chernoff *et al* (1982) and with Levreault (1988) who assumes $8.8 \times 10^{-4} M_\odot$yr^{-1} for the mass-loss rate of IRc2. Figure 1 shows v_x and z as functions of x for ten trajectories. The maximum flow speed reached is 70 km s^{-1} which approaches the halfwidth at zero intensity of the BN-KL H_2 emission features.

For the specified parameters the magnetic field strength varies from about 10 mG at $x = 0$ to 3 mG at $x = L$, while the H_2 number density varies from 9×10^5 cm^{-3} to 9×10^4 cm^{-3} over the same interval if the ratio of the He and H_2 number densities is 1/8. The spatial variation of the H_2 number density is proportional to that of P_T given by (7a) since we have assumed the flow to be isothermal.

We would like to use observations of a boundary layer to diagnose the x, z dependence of the heating rate per unit volume, H, in it just as we would like to infer more about the positional dependences of P, S, and F_x from observations. However, it is reasonable to assume that

$$H \approx \frac{\rho_w v_w^3}{L} \approx 3 \times 10^{-15} \, \text{erg cm}^{-3}\,\text{s}^{-1} \left(\frac{\rho_w}{1 \times 10^{-20}\,\text{gm cm}^{-3}} \right)$$
$$\left(\frac{v_w}{200\,\text{km s}^{-1}} \right)^3 \left(\frac{L}{0.01\,\text{pc}} \right)^{-1}. \tag{8}$$

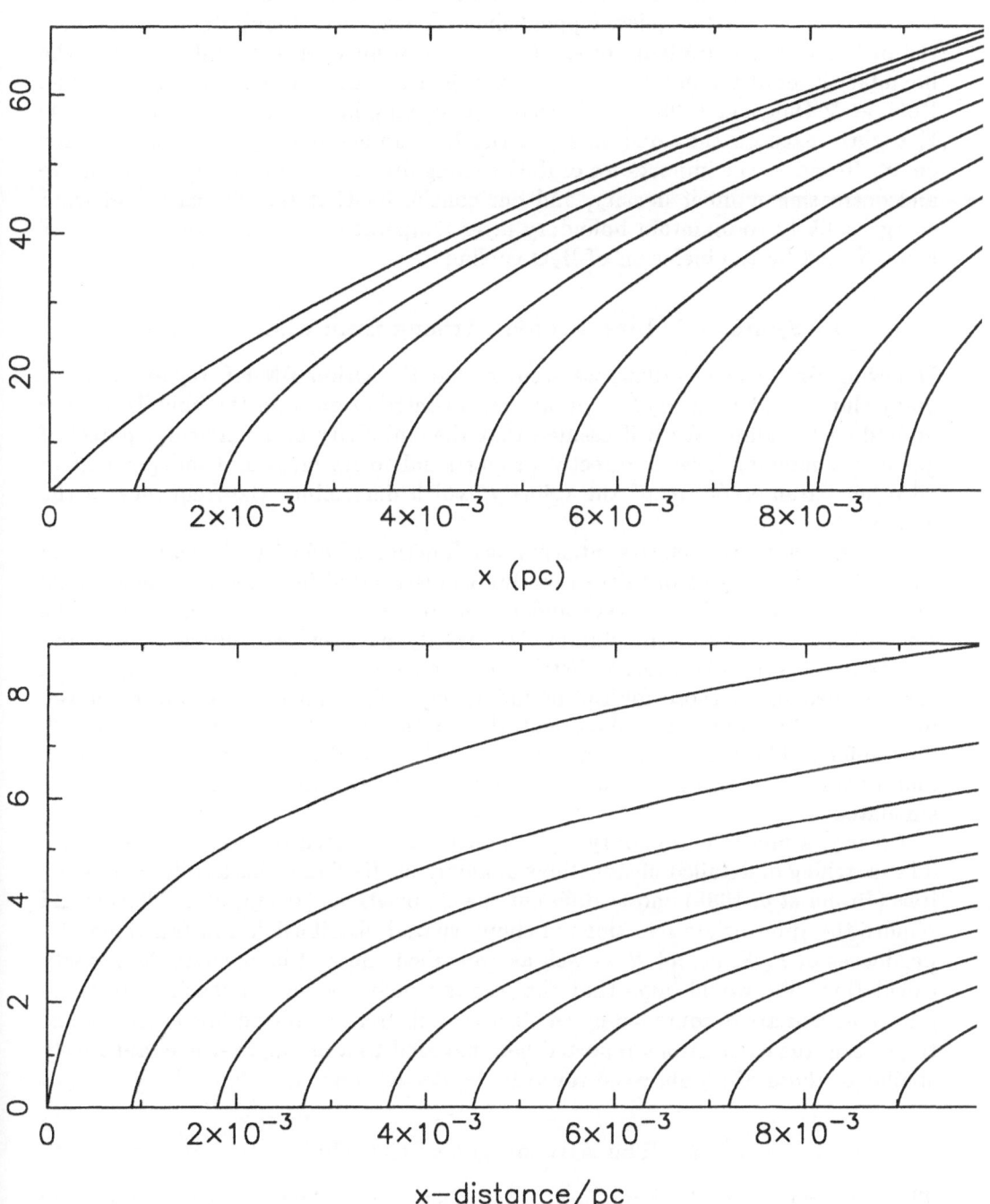

Fig. 1.   Flow in a BN/KL Boundary Layer. The model parameters are given in the text. a) The velocity component parallel to the boundary layer. b) The height of a trajectory above the base of the boundary layer.

The cooling rate per H_2 molecule in thermal equilibrium at $T = 2000$ K is 5×10^{-20} erg s^{-1} (Hartquist, Oppenheimer & Dalgarno 1980). Thus if H_2 were the only important coolant, dissipation of the wind energy would maintain the boundary layer at temperatures of a couple thousand degrees. The results of Draine, Roberge & Dalgarno (1983) imply that H_2O cooling in postshock gas at $T \approx 2000$ K slightly exceeds that due to H_2 if the H_2 number density is as high as 10^6 cm^{-3}. However, the importance of H_2O cooling drops with increasing temperature and decreasing number density, and our conclusion that the dissipation of wind energy is likely to maintain boundary layer temperatures of thousands of degrees is unaffected by the inclusion of H_2O cooling.

4   Symmetric Line Profiles Arising from Many Clumps

Following Brand *et al* (1989) we assume that the Orion BN-KL region contains many clumps. Then an H_2 line profile has a contribution from the boundary layer around each clump. We will assume that the emissivity of a particular parcel of gas in a boundary layer is directly proportional to its mass and independent of all other parameters except the velocity, which determines the frequency of the emission.

Figure 2 shows the relative intensity as a function of velocity of a feature formed by the boundary layers of thirty two clumps distributed in a single plane at equal angular intervals. The observer and the source of a symmetric wind are in the same plane, and one of the clumps lies directly between the source of the wind and the observer. The density distribution and the flow in each boundary layers are assumed to be those for the boundary layer for which results are displayed in Figure 1, but, obviously, the orientation, with respect to the observer, of each flow differs. Thermal broadening of 3 km s^{-1} in each parcel of gas is assumed, and a Gaussian instrumental broadening with a halfwidth of 16 km s^{-1} has been simulated.

Clearly, a multiple boundary layer model can be fitted to Orion BN-KL data. The matching of detailed observations of numerous H_2 line ratios at different velocities (Brand *et al* 1988) and at different spatial positions (Brand *et al* 1989) would require the appropriate selections of clump spatial distribution and functional dependences of P, S, F and H as well as a detailed thermal balance and emsisivity calculation. We would hope that the results would not be too sensitive to those selections and are encouraged by the fact that the reasonable ad hoc choices made to perform the calculations reported here have led to a profile that is qualitatively similar to those of H_2 observed towards the BN-KL region.

5   The Advantages of the Model

The acceleration in the boundary layer is sustained along its entire length by viscous coupling between the fast wind and the clump gas. The viscous coupling

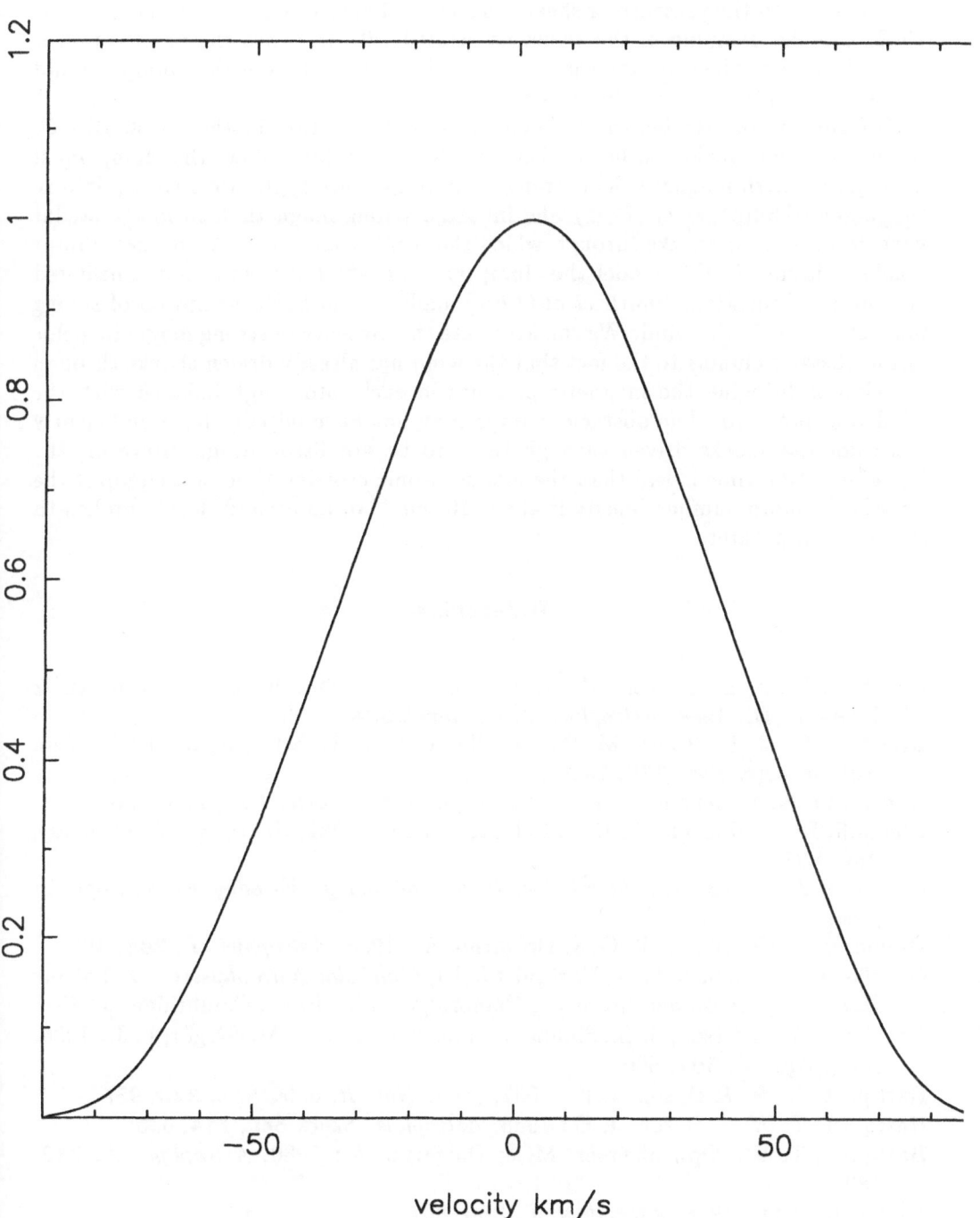

Fig. 2.   A line profile produced from the contribution of many boundary layers distributed evenly around a source of a spherically symmetric wind.

arises due to the presence of shear induced turbulence (e.g. Hartquist & Dyson 1988) and the damping of the waves associated with it. That the acceleration is gradual and sustained means that the molecules ablated from the clump are not dissociated while reaching high speeds.

For speeds comparable to the highest ones that obtain in the Orion BN-KL H_2 emission line region to be reached in a boundary layer flow, the clump must have quite a high magnetic field strength. High field strengths are also required in the model of Smith *et al* (1991) who invoke a strong magnetic field in the stellar wind to make bow shocks through which the wind passes as it encounters clumps nondissociating; in this model the clump gas and ablated gas are not considered as sources of emission. Smith *et al* (1991) made the ad hoc assumption of strong magnetic fields in the wind. We can attribute the presence of strong magnetic fields in the obstacle clumps to the fact that the wind has already driven shocks through the clumps bringing the magnetic pressure in each into rough balance with the wind ram pressure. The obstacle clumps may not have initially been sufficiently dense for the shocks driven through them to be nondissociating. However, the H_2 reformation time is less than the magnetosonic crossing time in a clump if the postshock clump number density is about 10^7 cm^{-3} or above and the clump length is 0.01 pc or greater.

References

Brand, P. W. J. L., Burton, M. G., Geballe, T. R., Moorhouse, A., Bird, M. & Wade, R. D.: 1988, *Astrophys. J. (Letters)*, **334**, L103

Brand, P. W. J. L., Toner, M. P., Geballe, T. R. & Webster, A. S.: 1989, *Mon. Not. R. astr. Soc.*, **237**, 1009

Burton, M. G. & Geballe, T. R.: 1986, *Mon. Not. R. astr. Soc.*, **223**, 13p

Chernoff, D. F., Hollenbach, D. J. & McKee, C. F.: 1982, *Astrophys. J. (Letters)*, **259**, L97

Dalgarno, A.: 1993, *The Royal Society of Chemistry: Faraday Proceedings*, in press

Draine, B. T., Roberge, W. G. & Dalgarno, A.: 1983, *Astrophys. J.*, **264**, 485

Geballe, T. R.: 1990, in T. W. Hartquist (ed.), *Molecular Astrophysics— A Volume Honouring Alexander Dalgarno*, Cambridge Univ. Press, Cambridge, p.345

Geballe, T. R., Persson, S. E., Simon, T., Lonsdale, C. J. & McGregor, P. J.: 1986, *Astrophys. J.*, **302**, 500

Hartquist, T. W. & Dyson, J. E.: 1987, *Mon. Not. R. astr. Soc.*, **228**, 957

Hartquist, T. W. & Dyson, J. E.: 1988, *Astrophys. Space Sci.*, **144**, 615

Hartquist, T. W., Oppenheimer, M. & Dalgarno, A.: 1980, *Astrophys. J.*, **236**, 182

Levreault, R.M.: 1988, *Astrophys. J.*, **330**, 897

McKee, C. F., Chernoff, D. F. & Hollenbach, D. J.: 1984, in M. F. Kessler & J. P. Phillips (eds.), *Galactic and Extragalactic Infrared Spectroscopy*, D. Reidel Publ. Co., Dordrecht, p.103

Smith, M. D., Brand, P. W. J. L. & Moorehouse, A.: 1991, *Mon. Not. R. astr. Soc.*, **248**, 730

High Density Tracers in Outflow Regions: NH₃ vs. CS

R. López, O. Morata, I. Sepúlveda, R. Estalella, G. Anglada, J. Pastor
Departament d'Astronomia i Meteorologia (U. Barcelona). Av. Diagonal 647, E-08028 Barcelona (Spain)

and

P. Planesas
Centro Astronómico de Yebes (I.G.N.) Apdo 148, E-19080 Guadalajara (Spain)

Abstract. We present $CS(J = 1 \rightarrow 0)$ observations of the high density gas in a sample of eleven star forming regions with molecular or optical outflows. The sources of this sample cover a wide range of physical sizes. All these sources had been previously mapped in NH₃ (1,1) with similar angular resolution. In all the sources of this sample CS emission was detected, indicating a global correlation between the emissions traced by the CS and the NH₃ molecules. However, the detailed characteristics of these two emissions (e.g., the extent and the location of the emission peak) show, in general, significant differences in the sources which are well resolved by the beam. As a general trend, the emission traced by the NH₃ molecule appears as compact clumps which engulf an outflow activity center. In contrast, the emission traced by the CS molecule, usually more extended, appears as a background which connects different outflow activity centers associated with each NH₃ clump.

1 Introduction

Molecular and optical outflows are ubiquitous phenomena in star-forming regions that are generally associated with the early phases of stellar evolution. Since its discovery, the focusing mechanism and the identification of the powering source of the outflow have received considerable attention. Mapping of the high-density gas has been used to get valuable information on these questions. Several high-density tracer molecules have been used to study the morphology, kinematics and physical parameters of the high-density gas. In particular, the NH₃ and CS molecules have significant emission for densities $\gtrsim 10^4$ cm^{-3}. Nevertheless, despite their critical densities being similar, the detailed characteristics of the NH₃ and CS emissions often show significant differences. In order to compare in detail the CS and NH₃ emissions, we carried out a program of CS $(J = 1 \rightarrow 0)$ mapping of outflow regions in a sample of sources that had been previously mapped in NH₃ (1,1) with similar angular resolution. We present in this contribution the results of a CS survey toward eleven star-forming regions with molecular and/or optical outflows: L1524, AFGL 5142, AFGL 5157, HH 38/43, NGC 2068, AFGL 6366S, L43, RNO 109, HHL 73, L1251A and L1251B. The sources of this sample covers a wide range of distances (from 140 to 1800 pc). Because of this, the same angular resolution corresponds to a wide range of linear sizes for the mapped regions. Results obtained from a more limited source sample were reported in Pastor *et al.* (1991). The CS $(J = 1 \rightarrow 0)$ observations were carried out with the 14 m telescope of the Centro Astron"mico de Yebes (resolution, 1.'9). The NH₃ observations had

Astrophysics and Space Science **216**: 151–152, 1994.

been carried out with the 37 m Haystack telescope (resolution, 1.'4). In the two cases, the same grid of observed positions was used.

2   General Results

The main results are the following:

1. In all the sources of the sample, CS emission was detected, indicating a global correlation between the emission traced by the CS and the NH_3 molecules.

2. In the sources that are well resolved by the beam, the CS emission is, in general, more extended than the NH_3 emission.

3. Because of the range of distances covered by the source sample, the physical sizes of the mapped regions are very different. The sizes of the CS condensations range from ~ 0.3 pc to ~ 2 pc. Depending on the distance to the source, the following cases can be distinguished: the farthest sources (d $\gtrsim 1500$ pc) are only partially resolved by the beam, thus it is not possible to resolve structures within the CS emitting region. In these sources, the characteristic sizes of the CS condensations are $\gtrsim 1.5$ pc. For intermediate-distance sources, it has been possible to resolve the extended CS emission in several subcondensations. The sizes of the CS condensations in the nearby sources (up to 200 pc) range from ~ 0.3 pc to ~ 0.5 pc, i.e., similar to the sizes of the smaller CS subcondensations detected in the intermediate-distance sources.

4. In general, both CS and NH_3 emissions increase toward the outflow activity centers. However, the detailed characteristics of these two emissions usually show significant differences in the sources that are well resolved by the beam, e.g.:

Morphology: As a general trend, the emission traced by the NH_3 molecule appears as compact clumps which engulf an outflow activity center. In contrast, the CS emission appears as a background which connects different outflow activity centers associated with each NH_3 clump. The regions HH38/43, RNO 109 and HHL 73 are good examples of this. The CS emission usually correlates well with the visual extinction as seen in the Palomar Sky Survey plates.

Location of the CS and NH_3 emission peaks: In general, the positions of the CS and NH_3 peaks are displaced by $\gtrsim 0.1$ pc (similar to the scale of abundance variations of several molecular species reported in the literature).

5. From the velocity maps, we derived for most of the sources, velocity gradients of the order of $\sim 0.5 - 1$ km s^{-1} pc^{-1} .

References

Pastor, J., Estalella, R., López, R., Anglada, G., Planesas, P. and Buj., J.: 1991, 'A CS study of star-forming regions previously mapped in ammonia', *Astron. Astrophys.* **252**, 320.

Modelling The Constancy of X

S. D. Taylor and D. A. Williams
Department of Mathematics
UMIST, PO Box 88
Manchester
M60 1QD

and

T. W. Hartquist
Max Planck Institut für Extraterrestrische Physik
D-8046 Garching bei München
Germany

Abstract.
Observations of the CO J=1–0 line are commonly used as a tracer for molecular material in clouds. The ratio of the H_2 column density to the integrated intensity of this line, X, is often taken to be constant, despite theoretical and observational uncertainty. We have tried to identify how this ratio depends on cloud parameters, testing a simple theoretical argument suggesting its invariance with respect to density. The apparent constancy can be understood if clouds are clumpy on scales of $A_V \approx 1$–2 mag.

Key words: Chemistry, Molecular Hydrogen, Cloud Models

1 Introduction

Theoretical models supporting a constant ratio for X have been based either on virialization of observed cloud ensembles, or on a constant CO/H_2 abundance ratio with CO excited by H_2 molecules. Observed small scale structure brings these assumptions into question. Calibrations of X use a variety of methods at all wavelengths to estimate $N(H_2)$, but currently give values for X in the range 1–4 $(10^{20}$ mol cm^{-2}(K kms$^{-1})^{-1})$

The aim of this work is to establish whether there is a plausible parameter space over which X is constant for individual clouds. The motivation is a recognition of the consequences of the external UV field and cosmic ray flux controlling the cloud temperature T_k. The heating rate per unit vol. for both mechanisms is proportional to the number density n and independent of temperature. The balance with heating then implies that the cooling rate is $\propto n$ also, if temperature does not affect the density. Since $N(H_2) \propto n$ it follows that X should be almost independent of density, provided the cooling in the 1–0 line of CO is always a similar fraction of the total cooling.

2 Cloud Model

We have modelled here plane-parallel clouds with cooling emission dictated by a turbulent linewidth ΔV_T, and in addition relax two assumptions that are often

Astrophysics and Space Science **216**: 153–154, 1994.

made, namely constant CO/H_2 ratio and constant T_k. No pressure structure is imposed. The molecules CO and H_2 are taken to self-shield, and a full time-dependent chemistry is used. Cooling occurs through the species C, C^+, O and CO.

3   Results

For a given ΔV_T, radiation field enhancement factor χ, cosmic ray ionization rate ζ and carbon depletion f_c X varies only slowly with A_V provided the optical depth is high enough that CO has been formed in significant amounts ($A_V \approx 1$). The line then becomes saturated, but the fractional abundance of CO is increasing still, so driving more CO emission.

More importantly the variance of the ratio with n is weak, supporting our theoretical argument. X varies by only a factor of 2 for an order of magnitude change in density (in the range $\approx$ 1–2 for $n \approx$ 200–2000 cm^{-3}).

These quoted values reflect the cloud parameters $\Delta V_T = 1 km s^{-1}$, $\chi = 1$, $\zeta = 1.3 \times 10^{-17} s^{-1}$. If density is held constant then metallicity has a significant effect on the results only if carbon is severely depleted so that the CO abundance peaks before saturation. Likewise χ appears only to shift the depth into the cloud at which the line saturates. However this saturation point, and hence the numerical value of X, is sensitive to ΔV_T although the trends outlined above still occur.

Gas-Grain Interaction in the Low Mass Star-Forming Region B335

J. M. C. Rawlings
Department of Mathematics, UMIST, PO Box 88, Manchester, M60 1QD

and

N. J. Evans II and S. Zhou
Department of Astronomy, The University of Texas at Austin, Austin, Texas, USA

Abstract. We have modelled chemical abundances and line profiles in the Bok globule B335. The chemical characteristics of this star-forming core are largely determined by gas-grain interaction in the inflowing material. By comparing high resolution observational data with our model it should be possible to determine the evolutionary status of B335 and to establish the role that surface chemistry plays in protostellar clouds.

Key words: stars, chemistry, dust

1 Introduction and background

Nearby, low-mass dense cores have simple geometry and kinematics with little turbulence, making them ideal laboratories for studying the effects of gas-grain interaction in star-forming regions. Recently, Zhou et al. (1993) observed H_2CO and CS in the Bok globule, B335, and found strong evidence for the inside-out (Shu, 1977) type of collapse.

To model B335 we have extended earlier models that were specific to the dark core L1498 (Rawlings et al. 1992). These have the advantage of predicting both column densities and line profiles for the species and transitions of interest. The essential chemical characteristics of infall regions arise from differential depletion effects (on to the surface of grains). Thus, some molecular species (eg. N_2H^+, HCO^+) demonstrate enhancements as collapse and depletion ensues. This becomes apparent in line profile broadening. Observed line profiles (of NH_3 in particular) in L1498 and other low mass cores indicate that, unlike the cloud complex from which they came, the cores are regions of exceptional quiescence. However, the chemical abundances in these cores are somewhat larger than can be explained by simple, static, gas-phase chemistries.

Our physical model assumes that low mass stars form in shock regulated cloud complexes that are both dynamically active and chemically rich. From this material isothermal pressure-balanced protostellar collapse cores form that are in critical dynamical equilibrium. At some later stage the hydrostatic cores undergo collapse. This process is initiated at the centre of the cloud and propagates outwards through the protostellar envelope.

Astrophysics and Space Science **216**: 155–157, 1994.
© 1994 *Kluwer Academic Publishers.*

2   The model of B335

B335 is a low mass core that shows strong evidence for inflow associated with protostellar collapse. It is apparently at a more advanced stage of evolution than L1498 in that it has an embedded protostar and is almost certainly undergoing collapse. B335 is probably the best example to date of the inside-out type of collapse first described by Shu (1977) and is an ideal test object for the chemical models of Rawlings et al. (1992). Parameters for the B335 core are given in table I. The density and temperature distributions, as well as the velocity field, are known for B335 so that theoretical abundances and line profiles can be accurately related to the observations.

TABLE I
Basic parameters for the B335 core

Temperature (Isothermal region)	13K
Turbulent velocity	0.12 kms^{-1}
Protostellar mass	0.42M$_\odot$
Effective sound speed	0.23 kms^{-1}
Collapse age	1.5×10^5 years
Infall radius	0.036 pc (30")

In order to make it applicable to B335, the model of L1498 has been modified and updated with a full revision of the sulphur chemistry. A one point global chemistry is modelled in order to establish the chemical initial conditions prior to protostellar collapse. Once collapse has started a 1.5D multi-point time-dependent model is used.

2.1   CONSTRAINTS AND FREE PARAMETERS

The main chemical constraints are (from Zhou et al. 1990,1993; Menten et al. 1984): $X(CS) \approx 3.6\times10^{-9}(\pm 30\%)$, $X(H_2CO) \approx 2.8\times10^{-9}$, $X(NH_3) \lesssim 2\times10^{-9}$ in the high density centrally condensed region of the core (from observations of the (2,2) line). The main free parameters in the model are: the sulphur depletion from the gas phase, the length of time that the core spends in the hydrostatic pressure-balanced state, the age of the collapse, and the extinction (A_v) of the region in which the core is embedded.

3   Results and conclusions

It is found that there are several combinations of the free parameters which satisfy the chemical and physical constraints. We have so far only considered two such combinations; one where $A_v=5.5$ and the collapse starts immediately after the formation of the hydrostatic core, another where $A_v=7.0$ and there is a pause of some 1/2 million years before collapse starts. Both models predict the same abundances for CS, H_2CO and NH_3 (consistent with the observations), but the larger depletion in the second model results in species like N_2H^+, HCO^+, CN and H_2CO having broadened line profiles. By observing appropriate lines (eg. $H^{13}CO^+$ J=1-0, HCO^+ J=3-2,1-0 and N_2H^+ J=1-0) it should thus be possible to constrain physical parameters such as the evolutionary status and the mass of the protostellar core, at the same time yielding information on the efficiency of the depletion and desorption processes.

The model is now being further generalised so as to include in greater detail previously simplified or neglected processes (such as the role of the external radiation field, surface reactions, and grain mantle desorption mechanisms). Our aim is to develop fully flexible and observationally/theoretically self-consistent chemical models of low mass star-forming regions.

References

Menten, K.M., Walmsley, C.M., Krugel, E. and Ungerechts, H.: 1984, *Astronomy and Astrophysics* **137**,108

Rawlings, J.M.C., Hartquist, T.W., Menten, K.M. and Williams, D.A.: 1992, *Monthly Notices of the Royal Astronomical Society* **255**,471

Shu, F.H.: 1977, *Astrophysical Journal* **214**,488

Zhou, S., Evans, N.J., Butner, H.M., Kutner, M.L., Leung, C.M., and Mundy, L.G.: 1990, *Astrophysical Journal* **363**,168

Zhou, S., Evans, N.J., Kompe, C., and Walmsley, C.M.: 1993, *Astrophysical Journal* **404**,232

The Structure and Dynamics of M17SW

M.P. Hobson
Mullard Radio Astronomy Observatory, Cavendish Laboratory, Madingley Road, Cambridge,
CB3 0HE, United Kingdom

Abstract. The M17SW molecular cloud core has been mapped at high resolution in the $C^{17}O$ $J = 3 \rightarrow 2$ transition and in 450, 600, 800, 1100 and 1300 μm continuum emission, using the JCMT. The clumpy nature of the cloud core is clearly revealed and the northern condensation has been resolved into 3 main clumps, each of which lies close to an H_2O maser, suggesting that they may contain young embedded stellar objects.

Key words: infrared: interstellar: continuum, interstellar medium: clouds, interstellar medium: individual objects: M17SW, interstellar medium: molecules, radio lines: molecular: interstellar

1 Introduction

The M17SW molecular cloud core is adjacent to the M17 H^+ region, at a distance of 2.2 kpc, and appears to be an excellent example of triggered star-formation. A nearby cluster of OB stars, recently formed from the molecular cloud complex, is surrounded by a prominent H^+ region which is heating and compressing the molecular material, possibly initiating its collapse to form the next generation of stars.

2 Observations and Results

Using the JCMT, M17SW has been mapped in the $C^{17}O$ $J = 3 \rightarrow 2$ transition at 15-arcsec resolution, and in 450, 600, 800, 1100 and 1300 μm continuum emission, giving 8-arcsec resolution for the shortest wavelength (Hobson *et al.*, 1993a,b). The large scale structure of these maps agrees well with previous observations (e.g. Stutzki and Güsten, 1990; Chini, 1990), and the clumpy nature of the cloud core is clearly revealed in both the optically thin spectral line and continuum emission. This close correlation suggests that the dust and molecular line emission are sampling the same volume of gas.

A more detailed comparison reveals a close agreement between individual peaks of emission, particularly in the prominent 'northern condensation' region. In both the line and continuum maps, this region has been resolved into 3 main clumps, each of which lies close to an H_2O maser, suggesting that these features may contain young embedded stellar objects. These clumps are labelled N–S as FIR1–FIR3 (Fig. 1), have masses in the range 300 to 450 $M_\odot$ with luminosities of 8000 to 12000 $L_\odot$, and appear to be gravitationally unstable. These values are consistent with them containing young protostellar objects of mass $\sim 10\ M_\odot$.

We find that the northern condensation coincides closely with the NH_3 clumps of Massi, Churchwell and Felli (1988), and is bounded to the north-east by the

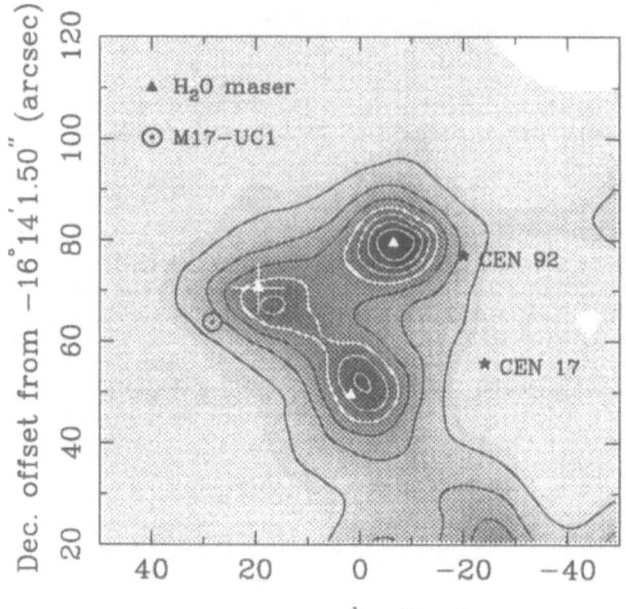

Fig. 1. The northern condensation at 800 μm, which is dominated by the components FIR1, FIR2 and FIR3. Also shown are the positions of the three known H_2O masers, the stars CEN 17 and CEN 92 (Chini et al. 1980), and the ultra-compact H^+ region M17-UC1 (Felli et al. 1984).

radio-continuum arc of Felli, Churchwell and Massi (1984), which contains the ultra-compact H^+ region M17-UC1. The overall velocity structure of the northern condensation seen in $C^{17}O$ suggests, however, that it may be the front face of a shell expanding at ~ 1.5 km s^{-1}, possibly driven by the outflows of a few unobserved low to medium mass stars. The positions of the sources FIR1–FIR3 are consistent with them lying within the shell material. The formation of the features may have been triggered by the compression of the gas from *both* sides.

References

Chini, R.: 1990, in G. D. Watt, A. S. Webster, (eds), *Submillimetre Astronomy*, Kluwer, Dordrecht, p. 19.

Chini, R., Elsässer, H. and Neckel, Th.: 1980, *Astron. Astrophys.* **91**, 186.

Felli, M., Churchwell, E. and Massi, M.: 1984, *Astron. Astrophys.* **136**, 53.

Hobson, M. P.: 1992, *Mon. Not. Roy. Astr. Soc.* **256**, 457.

Hobson, M. P., Padman, R. and Scott, P. F.: 1994, *Mon. Not. Roy. Astr. Soc.* submitted.

Hobson, M. P., Padman, R., Scott, P. F., Prestage, R. M. and Ward-Thompson, D.: 1993, *Mon. Not. Roy. Astr. Soc.* **264**, 1025.

Massi, M., Churchwell, E. and Felli, M.: 1988, *Astron. Astrophys.* **194**, 116.

Stutzki, J. and Güsten, R.: 1990, *Astrophys. J.* **356**, 513 (SG90).

The Hydrodynamics of Bipolar Explosions

H. M. Lloyd*
Chemical and Physical Sciences, Liverpool John Moores University, Byrom Street, Liverpool L3 3AF, U.K.

T. J. O'Brien
Computing and Mathematical Sciences, Liverpool John Moores University, Byrom Street, Liverpool L3 3AF, U.K.

M. F. Bode
Chemical and Physical Sciences, Liverpool John Moores University, Byrom Street, Liverpool L3 3AF, U.K.

and

F. D. Kahn
Department of Astronomy, The University, Manchester, M13 9PL, U.K.

Abstract. Bipolarity is a common feature in the outbursts of classical, recurrent and symbiotic novae. In this paper, we present the results of numerical calculations of bipolar explosions in which two parcels of gas are ejected at highly supersonic and oppositely directed velocities into a cold surrounding medium. The calculations demonstrate the formation of a dense ring of shocked material in the plane normal to the line of ejection, due to the interaction of two bowshocks. Observational candidates for this hydrodynamic phenomenon are discussed.

1 Introduction

Evidence for the occurrence of bipolar explosions in classical novae, recurrent novae and symbiotic novae is present in spectroscopic and imaging data taken during and after outburst. High-resolution radio maps made during outburst have in many cases shown the ejecta to be highly bipolar in nature, for example the recurrent nova RS Oph (Taylor *et al*, 1989), the symbiotic nova CH Cyg (Taylor, Seaquist and Mattei, 1986b) and the classical novae QU Vul (Taylor *et al*, 1986a) and V1974 Cygni (Pavelin *et al*, 1993). In the absence of resolved images, spectral lines of novae in outburst often have multiple, variable components again suggestive of non-spherical ejection, for example the recurrent novae V3890 Sgr (Gonzalez-Riestra. 1992) and V745 Sco (Cassatella *et al*, 1985) and the classical novae HR Del, FH Ser (Hutchings, 1972) and V1974 Cygni (Shore *et al*, 1993). Optical images of classical nova remnants in the late stages of development often have a prolate structure which can sometimes be resolved into polar 'blob' and equatorial ring components (e.g. HR Del—Slavin, O'Brien and Dunlop, 1993—and RR Pic—Duerbeck, 1987).

The recurrent novae RS Oph, V3890 Sgr and V745 Sco and the symbiotic nova CH Cyg all contain M-giant secondaries, and a nova outburst in one of these systems therefore occurs close to the centre of a dense red giant wind. The presence of a dense circumstellar medium at the time of outburst in such systems can be

---

* This author is supported by the SERC.

Astrophysics and Space Science **216**: 161–166, 1994.
© 1994 *Kluwer Academic Publishers.*

inferred observationally—the interaction between the supersonic ejecta and the cold surrounding medium gives rise to hot shocked gas, which in the case of the 1985 outburst of RS Oph, was observed in X-rays using EXOSAT (Mason *et al*, 1987). Non-thermal radio emission was also observed during this outburst (Taylor *et al*, 1989). In addition, emission lines in RS Oph (Dufay *et al*, 1964), V745 Sco and V3890 Sgr (Gonzalez-Riestra, 1992) are seen to narrow progressively through outburst, suggesting that the ejecta are being decelerated by a confining medium.

In this paper, we present numerical calculations of the interaction of a bipolar explosion with a cold, dense surrounding medium. The latter case is applicable to a bipolar explosion at the centre of a wind with constant velocity and mass-loss-rate. Finally, we discuss the bearing of our calculations on observations of novae in outburst.

2   Numerical Models

We have calculated numerical models of bipolar explosions using a hydrodynamic code due to Falle (1991). The code solves the Euler equations of compressible flow in two-dimensional axisymmetry, using a second-order Godunov scheme. At the beginning of a calculation, the numerical grid is filled with cold gas at rest. The outburst is initiated using a boundary condition in the plane $z = 0$ designed to produce oppositely directed heavy, supersonic jets along the axis of symmetry. These jets are maintained for a time t_j, given by $t_j = M/2\pi r_j^2 v_j \rho_j$ where M is the total mass ejected during the outburst and r_j, v_j and ρ_j are the radius, velocity and density of the jet. After a time t_j has elapsed, the boundary condition at $z = 0$ is changed to reflected symmetry about the entire plane.

In this manner, two parcels of gas, each of mass M are projected into the cold gas with oppositely directed velocities v_j along the symmetry axis. We have used this method to calculate models for wind-like and uniform density distributions of ambient material.

3   Results

We present the results of two runs, one in which the density surrounding the outburst site varies as the inverse square of radial distance, and another in which the density is uniform. The parameters for the two runs are given below in Table I. In both cases, the numerical grid consists of 300 cells in the r direction and 600 cells in the z direction, with $\Delta r = \Delta z = 4 \times 10^{-3}$. Figure 1 shows the situation after both models have been evolved for several times the duration of the 'jet', t_j. The plots show the velocity and density fields in the r, z plane.

In both models, $v_j t_j > r_j$, and hence the ejecta are initially elongated in the z direction; however it is clear from Figure 1 that at times of order several times t_j, the ejecta have become more flattened. This is due to the high pressure in the shocked material ahead of the ejecta 'squeezing' it out to larger radii. In both

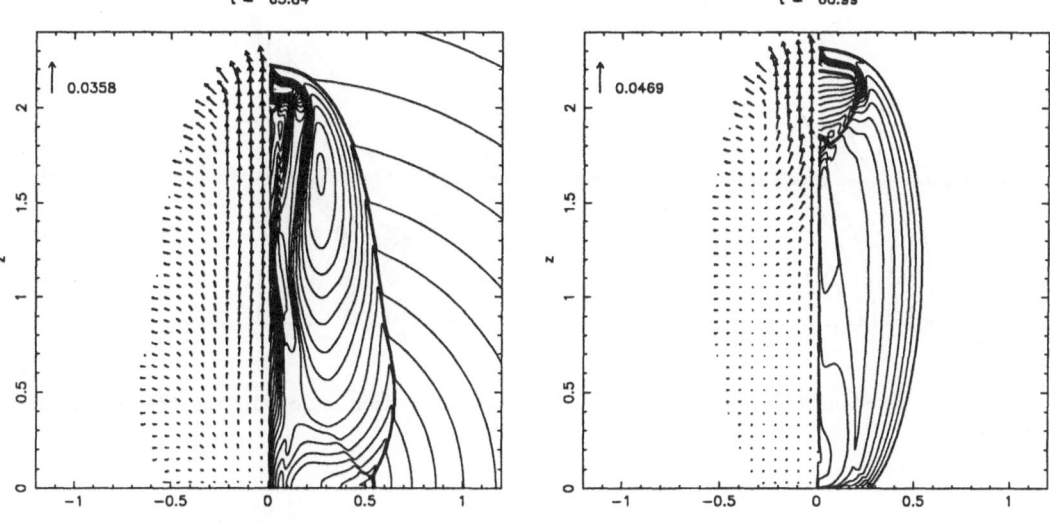

Fig. 1.   Logarithmic density contours and velocity fields (left) for the $1/r^2$ model at $t = 63.8$ and (right) the uniform density case at $t = 61.0$

TABLE I
Model parameters for the numerical calculations.   The ambient density distribution is
given by $\rho = \rho_0 r^{-\alpha}$.

	$1/r^2$ model	Uniform density model
α	2	0
ρ_0	850	10^4
v_j	3.8×10^{-2}	3.8×10^{-2}
ρ_j	1.5×10^5	10^5
r_j	1.6×10^{-2}	1.6×10^{-2}
M	50	48

cases, the bowshock driven by the ejecta does not cross the plane $z = 0$ normally, and hence this shock interacts with an identical wave coming in from the negative z direction. This is a Mach interaction, which can be seen better in Figure 2. This figure depicts contours of density for the interaction region in the $1/r^2$ case and clearly shows a three-shock pattern with a tangential discontinuity separating the two shocks leaving the line of intersection. There is therefore a dense ring of shocked material on either side of the equatorial plane, each bounded by a weak shock and a tangential discontinuity.

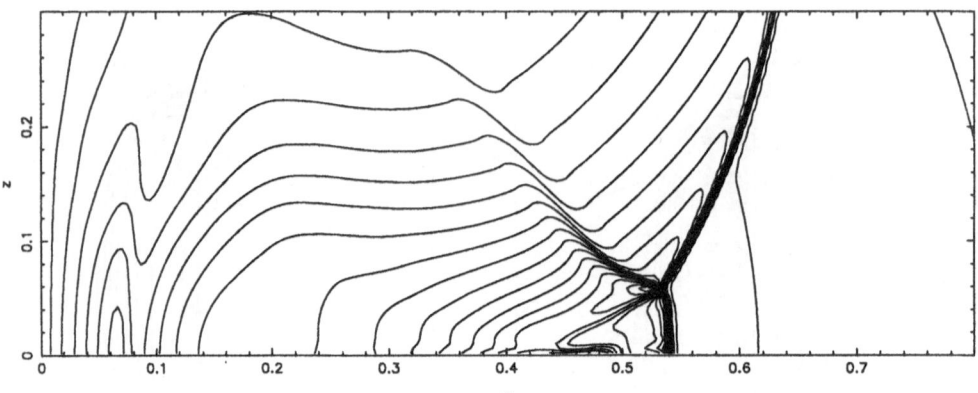

Fig. 2.   Logarithmic density contours in the Mach interaction region for the $1/r^2$ model at
$t = 63.8$.

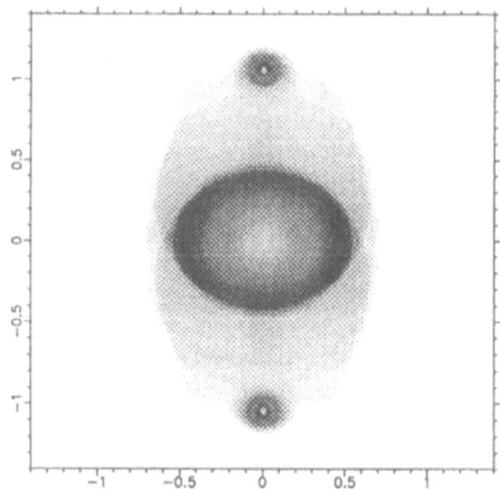

Fig. 3.   Map of ρ^2 for the $1/r^2$ case integrated along lines-of-sight at an angle of 60° to the
equator.

In order to assess the observability of such a ring, we have created synthetic
images from the dynamical solutions by assigning an emission coefficient equal to
the square of the density in each cell, and integrating along several lines-of-sight.
Figure 3 shows the result for the $1/r^2$ case at $t = 63.8$, viewed along a line of sight
at an angle of 60° to the equatorial plane. The ring is clearly seen in the image.
    The mechanism is also operating in the uniform density case, although the ring

is less pronounced due to the fact that the post-shock density is uniform. The ring is more obvious in the wind-like models because the shape of the bowshock ensures that the density in the remnant is highest at the equator; this density is further enhanced by the passage of the second shock.

4 Discussion

In this paper we have described a hydrodynamic phenomenon in which a dense ring of material is formed by the passage of oppositely directed supersonic 'bullets' through a cold medium. The effect is more pronounced in the case of an outburst at the centre of a wind. This mechanism provides an appealing model for the formation of the 'equatorial ring-polar blobs' geometry commonly observed in the ejecta of classical novae, although the progenitor star in this case would have no dense wind. It is possible that the dense wind is provided by an early, slow phase of the outburst whereas the 'bullets' are due to a later, bipolar ejection episode. However, calculations of the common-envelope phase of classical nova outbursts (Livio et al, 1990) predict a slow, asymmetric outflow at early times followed by a fast, point symmetric wind. The calculations we have performed are thus perhaps more relevant to the case of a bipolar recurrent or symbiotic nova outburst.

The radio observations of RS Oph (Taylor et al, 1989) and CH Cygni (Taylor et al, 1986b) can be understood in terms of these calculations. In both cases, the radio maps show two lobes on either side of a central component. The central component will be provided by the equatorial ring, and we need only postulate the bipolar ejection of material in order to explain all three components. Modelling of the radio emission from RS Oph (1985) using hydrodynamic models of the type described in this paper has been successful in reproducing the details of the radio map and will be described in a paper to follow.

References

Cassatella, A., Hassall, B. J. M., Harris, A. and Snijders, M. A. J.: 1985, in *Recent Results on Cataclysmic Variables*, ESA SP-236, p. 281.
Duerbeck, H. W.: 1987, ESO Messenger **50**, 8.
Dufay, J., Bloch, M., Bertaud, C. and Dufay, M.: 1964, *Ann. Astrophys.* **27**, 555.
Falle, S. A. E. G.: 1991, *Mon. Not. Roy. Astr. Soc.* **250**, 581.
Gonzalez-Riestra, R.: 1992, *Astron. Astrophs.* **265**, 71.
Hutchings, J. B.: 1972, *Mon. Not. Roy. Astr. Soc.* **158**, 177.
Livio, M., Shankar, A., Burkert, A. and Truran, J. W.: 1990, *Astrophys. J.* **356**, 250.
Mason, K. O., Córdova, F. A., Bode, M. F. and Barr, P.: 1987, in M. F. Bode (ed.), *RS Ophiuchi (1985) and the Recurrent Nova Phenomenon*, VNU Science Press, Utrecht, p. 167.
Pavelin, P. E., Davis, R. J., Morrison, L. V., Bode, M. F. and Ivison, R. J.: 1993, *Nature* **363**, 424.
Shore, S. et al.: 1993, IAU Circular no. 5614.
Slavin, A. J., O'Brien, T. J. and Dunlop, J. S.: 1993, *Annals of the Israel Physical Society*, in press.
Taylor, A. R., Seaquist, E. R., Hollis, J. M. and Pottasch, S. R.: 1986a, *Astron. Astrophys.* **183**, 38 .

Taylor, A. R., Seaquist, E. R. and Mattei, J. A.: 1986b, *Nature* **319**, 38.
Taylor, A. R., Davis, R. J., Porcas, R. W. and Bode, M. F.: 1989, *Mon. Not. Roy. Astr. Soc.*
237, 81.

Shock-heated Gas in the Outbursts of Classical Novae

T. J. O'Brien
Computing and Mathematical Sciences
Liverpool John Moores University, Byrom Street, Liverpool L3 3AF, U.K.

and

H. M. Lloyd*
Chemical and Physical Sciences
Liverpool John Moores University, Byrom Street, Liverpool L3 3AF

Abstract. We present a simple model for interactions between the slow and fast phases of mass ejection in classical nova outbursts. A situation develops in which a layer of hot gas sandwiched between two shocks progresses through the slow wind until it 'blows-out', expanding and cooling adiabatically. We calculate the X-ray emission from this model and find that the dynamical timescale is at least as important as the radiative cooling time in determining the form of the X-ray light curves. Models of this type may help to explain the origin of X-ray emission in some classical nova outbursts.

1 Introduction

Classical nova outbursts occur in close binary systems in which mass transfer occurs from a main sequence star onto a white dwarf. The outburst is due to a thermonuclear runaway (TNR) on the surface of the white dwarf which results in the ejection of $\sim 10^{-5} - 10^{-4}$ $M_\odot$ of material at up to several thousand km s^{-1} (Starrfield, 1989). Several classical nova outbursts have been detected in X-rays, a summary of the observations is provided in Table I. Thus far three models have been proposed for the origin of these X-rays: emission from the remnant of the white dwarf during the constant bolometric luminosity phase following the TNR (Ögelman, Krautter and Beuermann, 1987); shock-heating of circumstellar material due to the high ejection velocities (Brecher, Ingham and Morrison, 1977); and, most recently, Compton scattering to X-ray energies of gamma rays produced in the decay of ^{22}Na (Livio *et al.*, 1992). So far none of these models have provided a complete explanation for the observed X-rays. The first two models appear to produce X-rays which develop on timescales significantly different from those observed. In the first case the model light curve (Ögelman, 1990) is characterised by a steep rise to maximum as the remnant shrinks and increases in effective temperature, followed by an equally steep decline as it cools at constant radius, whilst the observed X-rays develop on longer timescales. ROSAT observations of two recent classical novae, Nova Her 1991 (Lloyd *et al.*, 1992) and Nova Cyg 1992 (Krautter, Ögelman and Starrfield, 1992), have shown that emission from gas shock-heated to a few keV may be quite common. It has been suggested however that the timescale for cooling of the shock-heated gas is in conflict with the observations (Ögelman

---

* This author is supported by the SERC

Astrophysics and Space Science **216**: 167–172, 1994.
© 1994 *Kluwer Academic Publishers.*

TABLE I

A summary of X-ray observations of novae in outburst, showing outburst date, speed class, instrument, time since outburst of observation and the previously suggested origin of the X-rays as the WD remnant (WDR) or shock-heated gas (SHG). RN denotes recurrent nova.

Nova	Outburst	Speed	Instrument	Epochs	Interpretation
V1500 Cyg	Aug 1975	VF	SAS-3	~ 0, ~ 22 d	Not detected[0]
			EINSTEIN	1390 d	SHG/WDR[0],[0]
NQ Vul	Oct 1976	M	Ariel V	~ 0 d	Not detected[0]
GQ Mus	Jan 1983	M	EXOSAT	$460 \rightarrow 900$ d	WDR[0]
PW Vul	Jul 1984	M	EXOSAT	8 d	Not detected[0]
				107, 313 d	WDR[0]
QU Vul	Dec 1984	F	EXOSAT	115, 307 d	WDR[0]
Nova Her	Mar 1991	VF	ROSAT	5 d	SHG[0]
				365 d	Marginal detection[0]
Nova Cyg	Feb 1992	F	ROSAT	$60 \rightarrow 258$ d	SHG[0]
RS Oph	Jan 1985	F	EXOSAT	$55 \rightarrow 93$ d	SHG[0]
(RN)				251 d	WDR[0]

et al., 1987). In particular, X-rays seem to turn off in nova outbursts on a much shorter timescale (months to a few years) than that suggested by a simple calculation of radiative cooling. The third model relies upon the Comptonization of gamma rays of energy 1.275 MeV resulting from the decay of ^{22}Na with a half-life of 2.6 years. However a detailed calculation (Livio *et al.*, 1992) shows that this mechanism can only contribute at a significant level at energies above about 7 keV.

In this paper we develop the shock-heated gas model and demonstrate that it can provide X-rays comparable in magnitude and timescale to those observed in some novae.

2   Intra-ejecta interactions

It is known from spectroscopic studies of novae in outburst that mass ejection takes place over a period of time, and that during this period the speed of ejection often increases (see Bode and Evans, 1989 and references therein). The velocity ranges observed during outburst are ~ 1000 to 4000 km s^{-1} for fast novae and ~ 500 to 2000 km s^{-1} for moderate novae. This naturally leads to a situation in which material ejected at late times overruns that previously ejected, shock-heating the gas to X-ray emitting temperatures.

In order to make quantitative calculations of the X-rays produced from such interactions we have considered a simple model in which mass-loss occurs at a constant rate $\dot{M}$ and speed u_1 for a time t_1 after outburst. After time t_1 the speed of the mass-loss increases ($\dot{M}$ remaining constant) until it reaches a speed u_2 after time t_2. We have performed numerical simulations using an Eulerian second-order Godunov method (Falle, 1991), assuming spherical symmetry. The grid is initially filled with low density (10^{-24} g cm^{-3}) cold gas. At some small radius (in this case 20 cells $\equiv 10^{12}$ cm) an inner boundary condition is set up which injects material at a rate $\dot{M}$ and with speed u_1. The grid therefore steadily becomes filled with wind material with density proportional to $1/r^2$ and bounded by a shock wave driven into the cold surrounding medium. This situation holds for a time t_1 after which the wind speed is steadily increased until time t_2. The faster wind ploughs into the slower wind setting up a double-shock structure in which a layer of hot gas is sandwiched between a forward shock driven into the slow wind and a reverse shock where the fast wind is decelerated. This shock sandwich moves outwards until it reaches the outer boundary of the slow wind at which point it 'blows out' into the surrounding low-density environment and cools adiabatically as it expands (see Fig. 1). In itself this provides an interesting hydrodynamical situation in which a variety of outcomes can result depending on the values of the parameters t_1, t_2, u_1 and u_2. For the moment, we have confined ourselves to a calculation of the X-ray light curve produced by thermal emission from the hot shock sandwich for two particular cases and neglecting the effects of radiative cooling. The two cases are: (1) a generic fast nova with $t_1 = 5$ days, $t_2 = 10$ days, $u_1 = 1000$ km s^{-1}, $u_2 = 4000$ km s^{-1}; and (2) a generic moderate nova with $t_1 = 30$ days, $t_2 = 60$ days, $u_1 = 500$ km s^{-1}, $u_2 = 2000$ km s^{-1}. In both cases the mass-loss rate $\dot{M}$ is 10^{-5} M$_\odot$ yr^{-1}. X-rays are calculated using a Raymond-Smith equilibrium ionisation code (Cox and Raymond, 1985) assuming an absorbing column of 3.0×10^{21} cm^{-2}. The light curves for the two cases are presented in Fig. 2.

3   Discussion

An important point to note from these calculations is that a shocked gas model does not necessarily imply a very long timescale for the X-ray emission. Ögelman et al. (1987) assume that the radiative cooling timescale defines the form of the light curve whereas in fact the dynamical timescale is also important and in principle can be much shorter.

The cooling time behind the forward shock may be written in terms of the pre-shock density and the shock velocity. Assuming that the shock velocity is approximately equal to $u_2 - u_1$, the post-shock temperature is given by $T = (3/16)(\bar{m}/k)u_1^2(f - 1)^2$ where $\bar{m}$ is the mean particle mass, k is Boltzmann's constant and $f = u_2/u_1$. For temperatures above about 5×10^7 K (corresponding to $u_1 > 640$ km s^{-1}, assuming $f = 4$) the cooling is dominated by bremsstrahlung, and the cooling time is given by

$$\tau_c = 0.54 \frac{u_8^4 (f - 1)^3 t_d^2}{\dot{M}_{-5}} \text{ days} \tag{1}$$

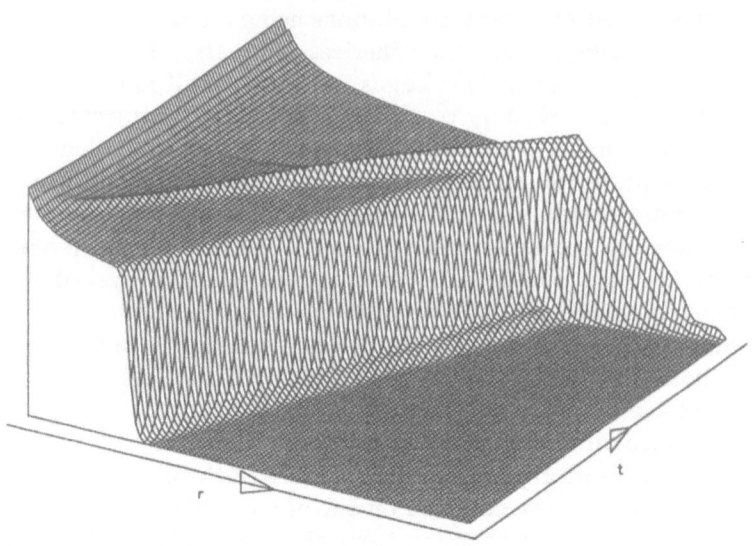

Fig. 1.   Hidden-line plot showing the logarithm of density on the vertical axis against radius
(r) and time (t). The plot shows the development of the slow wind, the shock interaction region
moving through the slow wind, and finally the blow-out and adiabatic expansion.

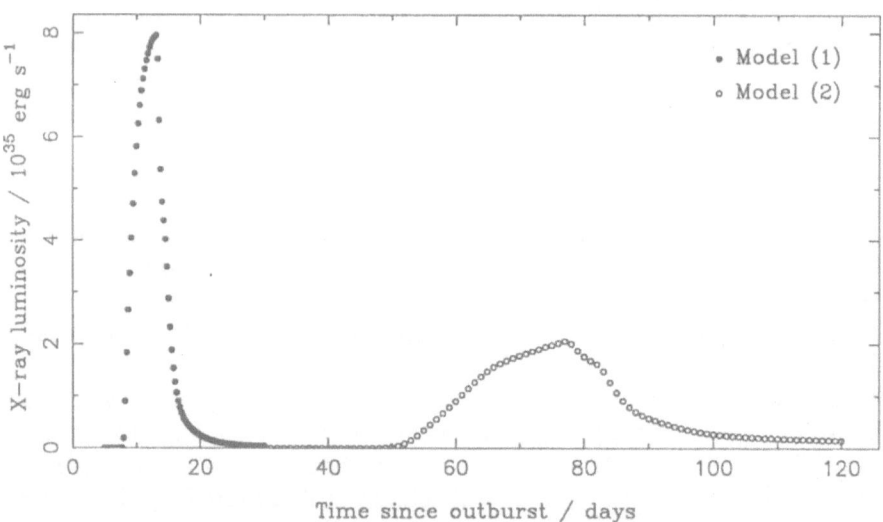

Fig. 2.   X-ray light curves calculated from the hydrodynamical models discussed in the text. The
X-ray luminosity is calculated in the pass-band $0.01 - 2.4$ keV (equivalent to that of the ROSAT
PSPC) and subject to an absorbing column of 3.0×10^{21} cm^{-2}.

where u_8 is u_1 in units of 10^8cm s^{-1}, t_d is the time since the beginning of the shock interaction in days and $\dot{M}_{-5}$ is the mass-loss-rate into the wind in $10^{-5} \, M_\odot \, \text{yr}^{-1}$. For temperatures below 5×10^7 K ($u_1 < 640 \text{km s}^{-1}$), line radiation becomes important and we may use Kahn's (1976) cooling law to show that

$$\tau_c = 0.15 \, \frac{u_8^6 \, (f-1)^5 \, t_d^2}{\dot{M}_{-5}} \text{ days.} \tag{2}$$

These two expressions are thus valid for fast and moderate novae respectively.

The time taken from the start of the interaction for the forward shock to 'blow out' of the slow wind is given by $t_b = t_1/(f-1)$, and the cooling time at this epoch can then easily be calculated from equations (1) and (2). Taking $u_1 = 1, t_1 = 5 \, \text{d}$ for the former and $u_1 = 0.5, t_1 = 30 \, \text{d}$ for the latter, with $f = 4$, $\dot{M}_{-5} = 1$ in both cases, we obtain $\tau_c = 12.2 \, t_b$ for fast novae and $\tau_c = 3.5 \, t_b$ for moderate novae. Radiative cooling is thus less important than the dynamical timescale in determining the development of X-ray light curves at late times. It should be noted however that cooling is important at early times because $\tau_c \to 0$ as $t \to 0$. In a later paper we will investigate the effect of including radiative energy losses.

It is likely that the X-rays observed much later as the nova returns to quiescence are a combination of accretion and/or continued burning although even then some hot gas may remain and contribute to the emission at a non-negligible level.

4   Concluding Remarks

We have demonstrated that a simple shocked-gas model for the origin of X-rays in classical nova outbursts can provide significant emission on reasonable timescales, the latter determined by the length of time the shock sandwich stays within the early slow wind. However a spherically symmetric adiabatic model such as the one presented here is simplistic and cannot be universally applied. For instance it is thought that classical novae pass through a common envelope phase during outburst in which the slower mass loss takes place preferentially in the equatorial plane (Shankar, Livio and Truran, 1991). In this case any subsequent fast mass loss will drive a slow shock in the equatorial plane but will blow out quite quickly in the polar directions. We have developed a 2-dimensional hydrodynamical code similar to that used here but in spherical polar $r - \theta$ coordinates and are currently using it to model this situation. It is expected that the X-ray emission will develop with different timescales in different regions of the remnant. Models of this type may also help explain the morphology of the resolved remnants of old novae such as DQ Her (Martin, 1989) and HR Del (Slavin, O'Brien and Dunlop, 1993).

References

Bode, M .F. and Evans, A.: 1989, eds. *Classical Novae*, Wiley, p. 163.
Brecher, K., Ingham, W. H. and Morrison, P.: 1977, *Astrophys. J.* **213**, 492.

Chlebowski, T. and Kaluzny, J.: 1988, *Acta Astron.* **38**, 329.
Co,x D. P. and Raymond, J. C.: 1985, *Astrophys. J.* **298**, 651.
Cruise, A. M.: 1977, *Nature* **267**, 685.
Falle, S. A. E. G.: 1991, *Mon. Not. Roy. Astr. Soc.* **250**, 581.
Hoffman, J. A., Lewin, W. H. G., Brecher, K., Buff, J., Clark, G.W., Joss, P.C. and Matilsky, T.:
    1976, *Nature* **261**, 208.
Kahn, F. D.: 1976, *Astron. Astrophys.* **50**, 145.
Krautter, J., Ögelman, H. and Starrfield, S.: 1992, IAU Circ. no. 5550.
Lloyd, H. M., O'Brien, T. J., Bode, M. F., Predehl, P., Schmitt, J. H. M. M., Trümper, J.,
    Watson, M. G. and Pounds, K. A.: 1992, *Nature* **356**, 222.
Livio, M., Mastichiadis, A., Ögelman, H. and Truran, J.W.: 1992, *Astrophys. J.* **394**, 217.
Martin, P. G.: 1989, in M. F. Bode and A. Evans (eds.) *Classical Novae*, Wiley, p. 73.
Mason, K. O., Cordova, F. A., Bode, M. F. and Barr, P.: 1987, in M. F. Bode (ed.), *RS Ophiuchi
    (1985) and the Recurrent Nova Phenomenon*, VNU Science Press, p. 167.
O'Brien, T. J., Lloyd, H. M. and Bode, M. F.: 1994, in preparation.
Ögelman, H.: 1990, in IAU Colloq. 122, *The Physics of Classical Novae*, eds. Cassatella & Viotti,
    Springer-Verlag.
Ögelman, H., Krautter, J. and Beuermann, K.: 1987, *Astron. Astrophys.* **177**, 110.
Shankar, A., Livio, M. and Truran, J. W.: 1991, *Astrophys. J.* **374**, 623.
Slavin, A. J., O'Brien, T. J. and Dunlop, J. S.: 1993, *Annals of the Israel Physical Society*, in
    press.
Starrfield, S.: 1989, in M. F. Bode and A. Evans (eds.) *Classical Novae*, Wiley, p. 39.

The Crab Nebula Revisited

L. Woltjer and M. P. Véron-Cetty
Observatoire de Haute-Provence, F-04870 Saint Michel l'Observatoire

Abstract. The evolution of pulsar driven supernova remnants is briefly reviewed with special reference to the Crab Nebula. Simple models account for the integral properties of the Nebula. New data on the optical synchrotron continuum show strong spectral variations over the Nebula which will require more complex models of the particle diffusion.

1 Introduction

Some 25 years ago Franz Kahn and one of us (Kahn and Woltjer 1967) studied the dynamical evolution of a bubble of relativistic particles injected into the interstellar medium. In the meantime it has become clear that pulsar driven supernova remnants are such bubbles, but that the injection is not instantaneous. Moreover matter ejected by the supernova or just before its explosion usually plays a more important role than the interstellar medium, in contrast to the situation in the more numerous and usually older shell–like supernova remnants. The evolution of the sysnchrotron radiation from such a bubble was taken by Setti and Woltjer (1972) who took into account the input of new relativistic electrons by the pulsar, but not that of a new magnetic field, and by Pacini and Salvati (1973) who assumed that half of the pulsar energy loss is transformed into energy of relativistic electrons and the remainder into magnetic field energy. While this model appears to be closest to the situation in actual pulsar driven remnants, it contains a number of *ad hoc* assumptions, in particular that the relativistic particles are injected with a power–law spectrum with time independent index and maximum energy.

The prototype of a pulsar driven supernova remnant is the Crab Nebula and we shall discuss next how well the Pacini-Salvati (PS) model accounts for its integrated properties.

2 The Crab Nebula

The distance of the Crab Nebula will remain uncertain until a parallax for its pulsar can be determined. However the pulsar dispersion measure (57 pc cm^{-3}) suggests a distance of about 2 kpc, a result compatible with the usual geometric distance determination of also about 2 kpc, with an uncertainty of ± 1 kpc. The total energy radiated by the Nebula then becomes 0.7 x 10^{38} ergs^{-1}, or about 15% of the estimated energy input by the pulsar. The mass of the filamentary shell of the Nebula is about $1 - 2M_\odot$ with an uncertain correction for cool, dense cores in the filaments. The motion of the shell is slightly accelerating due to the pressure of magnetic fields and relativistic electrons inside; the corresponding increase in its energy is a few times 10^{38} ergs^{-1}, at the expense of the energy inside the shell.

Astrophysics and Space Science **216**: 173–177, 1994.
© 1994 *Kluwer Academic Publishers.*

From the observed synchrotron radiation and the Compton radiation at TeV energies, the magnetic field may be estimated as on average 0.3 mG or somewhat less (De Jager and Harding 1992), not very far from the value expected if equipartition prevails between magnetic fields and relativistic electrons and positrons. Both the energetics of the Nebula and the observed acceleration of the shell then show that the energy in the form of relativistic protons can be at most a few times that of the electrons and positrons.

The pulsar should not only put enough energy into relativistic electrons and positrons, but it should also create the required number of a few times $10^{39}s^{-1}$. The most likely origin of these particles is in showers of electron pairs and gammas in the pulsar magnetosphere. Some particles will escape into the Nebula, others will strike the surface of the neutron star, the precise proportions depending on the structure of the magnetic field. The positrons will annihilate and generate 511 keV gammas. In fact, recent observations (Massaro et.al. 1991) have shown a flux of gammas at 440 keV, corresponding to the annihilation of 10^{39} positrons s^{-1}, in agreement with expectation and with the lower energy interpreted as a gravitational redshift for plausible neutron star models. While the correspondence with the Nebular requirement is suggestived, it should be stressed that the observations still are rather uncertain and that confirmation is needed, especially in view of some reports of gammas at around 560 keV.

If electrons and positrons are equally abundant, it follows that objects like the Crab Nebula cannot be relevant to the origin of galactic cosmic rays which have few, if any, primary positrons and about 100 times more energy in baryons than in electrons. Recent results about the abundances of cobalt isotopes appear to confirm that supernovae are not the main accelerators of cosmic rays (Leske et.al. 1992).

The Pacini–Salvati model accounts for several properties of the Crab Nebula. First of all the value of the magnetic field strength is predicted to be about 10^{-3} Gauss, perhaps somewhat larger than, but not far from, the observed value.

Secondly, the fragmentary data on the time variation of the Nebula are well reproduced by the model. At 8 GHz the observed decline is $0.167 \pm 0.015\%$ per year (Aller and Reynolds 1985). In the optical, comparison of O'Dell's (1962) data and ours yields an uncertain decline of $0.1 \pm 0.2\%$ per year. The predicted values from the PS model are 0.18 and 0.11% respectively (Véron–Cetty and Woltjer 1991).

The continuous spectrum of the Nebula consists of a number of components:

1. The Radio Spectrum up to about 10^{12} Hz is a power law with $\alpha = 0.30$.

2. Between 10^{12} Hz and 2×10^{13} Hz there is an IR bump probably caused by dust, heated by the synchrotron radiation from the Nebula.

3. Between 2×10^{13} Hz and 10^{15} Hz the spectrum may be represented by a power law with $\alpha = 0.73$, the precise value depending upon uncertain corrections for

interstellar reddening.

4. Between 2 and 50 keV the spectrum is well fitted by a power law with $\alpha = 1.14$.

5. Around 100 keV the spectrum further steepens to $\alpha = 1.4$ which ultimately steepens to $\alpha = 1.7 - 1.9$ around 100 MeV (Nolan et.al. 1993).

6. Finally around 10^{12} eV emission has been detected, which is believed to be Compton radiation as mentioned before.

The PS model would predict a spectral break at about 10^{13} Hz where the radio spectrum with $\alpha = 0.3$ would steepen to $\alpha = 0.8$ by the synchrotron losses of the electrons, rather close to the value observed for component 3. However components 4 and 5 probably represent features of the injection spectrum of the relativistic particles.

3   Variations over the Nebula

While the main features of the integrated Nebula appear to be more or less understood, the situation becomes more complex when the Nebula is studied in detail. The radio spectrum appears to be largely the same throughout the Nebula, except for some local variations in the general area of the pulsar, (Bietenholz and Kronberg, 1992). The optical spectra, however, cover a much larger range.

To study the optical spectra we have taken CCD images with the 120 cm telescope at the Observatoire de Haute–Provence through interference filters (50 − 100Å wide) centred at 9241, 6450, 5364 and 3808 Å, these wavelengths being chosen to avoid most emission lines from the filamentary shell. In the outer parts the contribution of some weaker lines and also of the Balmer and 2-photon continua nevertheless is important. After subtraction of these components the integrated spectrum of the Nebula does not significantly deviate from a power law; within the accuracy of the respective data sets the shape of the integrated spectrum agrees well with that found by O'Dell (1962) indicating that no major spectral variations have occurred over the last 30 years.

The distribution of the optical spectral indices over the Nebula is shown in Figure 1. There is a general steepening of the spectra with distance to the pulsar, as previously found by Scargle (1969) who obtained photoelectric photometry at a number of points in the Nebula, but the detailed picture is rather complex. The flattest spectra ($\alpha = 0.57$) are found NW of the pulsar, where also the X-rays are strongest. East of the pulsar the spectra steepen very rapidly, but to the SE there is an extended region where they remain rather flat. Nearer to the edge of the Nebula the spectra are steep with $\alpha = 1.0 - 1.1$ being reached. Very near to the pulsar Scargle (1969) reported some very flat spectra. We have not been able to confirm these and believe that these may be due to contamination from light of the star 5" to the NE of the pulsar which turns out to be bluer than the nebular

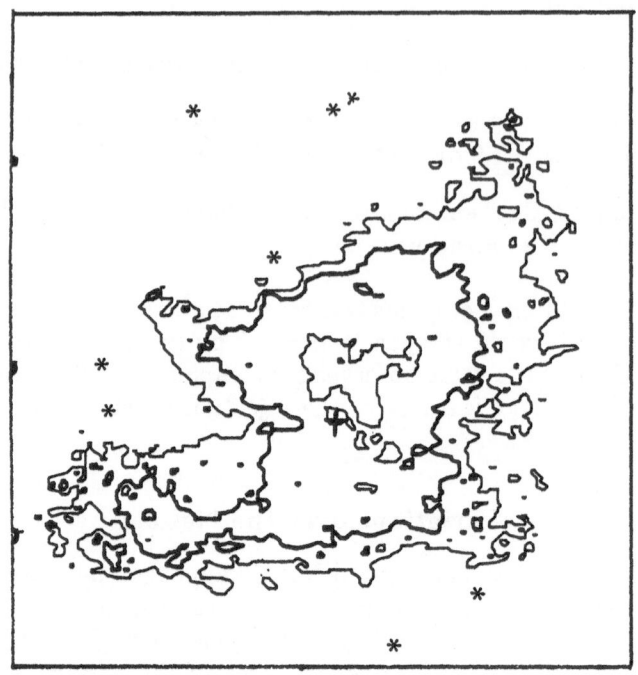

Fig. 1.

continuum. Alternatively real time variations associated with the "wisp" activity cannot be excluded.

These spectra show that the simple picture of a break where the spectral index increases by 0.5 due to synchrotron losses is too simple. Perhaps a mixture of the effects of energy losses and diffusion of relativistic electrons through the Nebula can account for these data. A detailed comparison with the high resolution X and UV data that now are becoming available should allow further progress to be made.

References

Aller, H. D. and Reynolds, S. P.: 1985, *Astrophys. J.*293, L73.
Bietenholz, M. F. and Kronberg, P. P.: 1992, *Astrophys. J.*393, 206.
De Jager, O. C. and Harding, A. K.: 1992, *Astrophys. J.*396, 161.
Kahn, F. D. and Woltjer, L.: 1967, in *Radio Astronomy and the Galactic System, IAU Symposium No.91*, H. van Woerden (ed), Academic Press, London and New York, p.117.
Leske, R. A., Milliken, B and Wiedenbeck, M. E.: 1992, *Astrophys. J.*376, L11.
Massaro, E. *et al*: 1991, *Astrophys. J.* 376, L11.
Nolan, P. L. *et al*: 1993, *Astrophys. J.* 409, 697.
O'Dell, C. R.: 1962, *Astrophys. J.* 136, 809.
Pacini, F. and Salvati, M.: 1973, *Astrophys. J.* 186, 249.
Scargle, J. D.: 1969, *Astrophys. J.* 156, 401.

Setti, G. and Woltjer, L.: 1972, *Astrophys. J.* **178**, L17.
Véron-Cetty, M. P. and Woltjer, L.: 1991, *Astron. Astrophys.* **251**, L31.
Véron-Cetty, M. P. and Woltjer, L.: 1993, *Astron. Astrophys.* **270**, 370.

PULSAR MAGNETOSPHERES: CLASSICAL AND QUASI-CLASSICAL DESCRIPTIONS

A. A. da Costa
Centro de Electrodinâmica, Instituto Superior Técnico,
1096 Lisboa Codex, Portugal

Abstract. The classical description of the motion of plasmas in pulsar magnetospheres, following the single particle approach, is unacceptable when radiative effects need to be taken into account in the equation of motion. When this happens, a quasi–classical description is required, with deterministic trajectories becoming stochastic, because quantum corrections must be considered. A new extension to the kinetic theory of plasmas, with these macroscopic radiative effects present, is developed in this paper.

Key words: pulsars, gamma-rays, stochastic chaos, fractals

1   Foreword

On the 1st October 1972 I met Professor Franz Daniel Kahn for the first time. It was a magic moment. There was the man who was going to lead me to my goal: a research career. Four years later, under his supervision, I was awarded my PhD. By then a long lasting friendship had developed.

Now, and after all these years of friendship, which made this paper possible, I just wish that we remain close in his well deserved, long and happy retirement.

2   Introduction

The classical description of plasmas in pulsar magnetospheres has been done through the single particle approach. Particle trajectories obey the Lorentz-Dirac equation of motion ($\chi \sim 6.27 \times 10^{-24}$ s for electrons)

$$\dot{u}^i = \frac{e}{m} F^{ik} u_k + \chi \ddot{u}^i + \frac{\chi}{c^2} (\dot{u}^j \dot{u}_j) u^i \tag{1}$$

when radiative effects are taken into account. The three terms in the right hand side of the equation are, respectively, the Lorentz, The Schott, and the Dirac terms.

The L-DEM is a full classical description of motion, and it assumes that the radiation field acts in a continuous way, and therefore is associated with a classical description of the radiated field. However, in classical pulsar magnetospheres the spectral content of the radiated power by the individual electrons peaks at frequency in the hard γ-ray domain, when the Dirac term is of the same order of magnitude as the Lorentz term, and electrons and positrons are accelarated up to Lorentz factors $\gamma \sim 10^7$. Then the radiation field requires a quantum description, using photons, and therefore its effects are discontinuous.

Astrophysics and Space Science **216**: 179–184, 1994.
© 1994 *Kluwer Academic Publishers.*

Thus the classical L-DEM must be changed so that the new equation of motion accommodates these effects, and eventually shows new and unexpected phenomena. In this new quasi–classical (q-classical) approach particle trajectories are stochastic due to the random nature of radiation of γ-ray photons. They create random charge and current densities distributions, which generate electromagnetic fields and plasma with random characteristics.

In this paper we shall make a summary of the most distinctive features of this new q-classical single particle approach, showing how it will provide the basis for a collective treatment of magnetospheric plasma, with new phenomena.

3   The Single Particle Approach

The q-classical regime, is a quantum correction to the classical regime, and has to be applied to pulsar magnetospheres (da Costa & Kahn, 1991). In the q-classical description the motion of charged particles is driven, as usual, by the Lorentz force. However there are impulsive changes of momentum and energy due to the radiation of photons at random times with random linear momentum $\hbar \mathbf{k}$. These individual photons are radiated with non-normalized density distribution probability $P_r(\mathbf{p}, \mathbf{n}, \omega, t) = (\hbar\omega)^{-1} P_{\mathrm{Rad}}(\mathbf{p}, \mathbf{n}, \omega, t)$ where $P_{\mathrm{Rad}}(\mathbf{p}, \mathbf{n}, \omega, t)$ is the single particle power radiated spectrum. This means that the total power radiated is, apart from a constant, an average over the all possible radiated frequencies. $P_r(\mathbf{p}, \mathbf{n}, \omega, t) = f\{\omega[1 + (\hbar\omega/\gamma mc^2)]\}$ (γ–the Lorentz factor) as given by Schwinger (1954), is a semi–classical distribution, when $\hbar\omega_{\mathrm{max}} \lesssim 0.1\gamma mc^2$, where, with ρ the radius of curvature of the trajectory, $\omega_{\mathrm{max}} = \gamma^3 c/\rho$ is the peak angular frequency of $f(\omega)$. When $\hbar\omega_{\mathrm{max}} \ll \gamma mc^2$ then $P_r(\mathbf{p}, \mathbf{n}, \omega, t) = f(\omega)$, as given by Schwinger (1949), is the full classical expression. In both cases $\mathbf{n} = \mathbf{k}/k$ is inside a cone whose axis is parallel to $\mathbf{p}$, with aperture γ^{-1}. The semi–classical description might now be used because we are no longer limited to the classical case, as it happened in the L-DEM description.

The radiation of individual photons establishes a diffuse boundary between classical and q-classical domains when $\hbar\omega \sim mc^2$, and this presents a remarkable similarity with thermal plasmas, whose boundary between non-relativistic and relativistic domains is related to its temperature by $kT \sim mc^2$.

The above mentioned ω will be called $\omega_{Q-CL} \sim 10^{21}\,\mathrm{Hz}$, and therefore photons with $\omega < \omega_{Q-CL}$ will not affect the motion of the particles in this way. Moreover, particles radiate mostly at ω_{max}, with a band $[0, \kappa\omega_{\mathrm{max}}]$, such that $\kappa > 1$, where most of the radiation is concentrated, translated by the relation

$$\int_0^{\kappa\omega_{\mathrm{max}}} d\omega\, P_r(\mathbf{p}, \mathbf{n}, \omega, t) \sim \int_0^{\infty} d\omega\, P_r(\mathbf{p}, \mathbf{n}, \omega, t) \tag{2}$$

This means that photons with $\omega > \kappa\omega_{\mathrm{max}}$ occur with negligible probability, and then only when $\omega_{Q-CL} < \omega_{\mathrm{max}}$ is there a significant probability that particle motion will be affected by quantum effects due to radiation. Thus particles leave the classical domain when $\kappa\omega_{\max} \sim \omega_{Q-CL}$ and for this value of ω_{Q-CL} there is a $p\,(= p_{Q-CL})$, such that $c\delta p \sim \hbar\,\kappa\omega_{\max}$. Therefore photons with $(k/c) = \omega \lesssim \omega_{Q-CL}$, will not play any role in the motion of the particles, and particles with $p < p_{Q-CL}$ will not qualify for quantum effects, since they are unlikely to produce high energy photons.

In this new q-classical description then

$$dp = e(\mathbf{E} + \mathbf{v} \times \mathbf{B})dt - \mathbf{R_p}(t) \tag{3}$$

with

$$\mathbf{R_p}(t) = \sum_i \hbar\mathbf{k}_i \delta(t - t_i) \tag{4}$$

a stochastic variable, and t_i a random variable with non-uniform distribution in the interval $(t - (\Delta t/2), t + (\Delta t/2))$, with the number of photons N'_{ph} radiated in the same interval Δt given by the probability density

$$p_t(x = N'_{ph}) = \exp\left[-\int_{t-\frac{\Delta t}{2}}^{t+\frac{\Delta t}{2}} dt_1\,\lambda(\mathbf{p}, t_1)\right] \frac{\left[\int_{t-\frac{\Delta t}{2}}^{t+\frac{\Delta t}{2}} dt_1\,\lambda(\mathbf{p}, t_1)\right]^{N'_{ph}}}{N'_{ph}!} \tag{5}$$

as this is a generalized Poisson process. $\lambda(\mathbf{p}, t)$ is the probability of radiating one photon per unit time which is equal to the average number of photons radiated per unit time (dN_{ph}/dt) such that

$$\frac{dN_{ph}}{dt} = \lambda(\mathbf{p}, t) = \int_{\substack{all\,\omega \\ all\,\mathbf{n}}} d\omega\,d\mathbf{n}\,\mathrm{P_r}(\mathbf{p}, \mathbf{n}, \omega, t) \tag{6}$$

The interval of time τ between two successive radiated photons is also a random variable with distribution

$$f_t(\tau) = \lambda(\mathbf{p}, t + \tau)\exp\left[-\int_t^{t+\tau} dt_1\,\lambda(\mathbf{p}, t_1)\right] \tag{7}$$

and the probability that a particle, after radiating at time t, will radiate later than $t + \tau$ is

$$P\{X > \tau\} = 1 - F_t(\tau) = \exp\left[-\int_t^{t+\tau} dt_1\,\lambda(\mathbf{p}, t_1)\right] \tag{8}$$

The general principles so far outlined have an application in pulsar magnetospheres. The goal is to check the difference between this and the L–D description, and in particular whether some of the predictions made in da Costa & Kahn (1991) are real. At the same time we are looking for suggestions how to deal with the collective approach.

The calculations were done using four of magnetic field lines close to speed of light cylinder of the cylindrical model described in da Costa & Kahn (1982) and da Costa (1983). We used a Monte–Carlo Method to generate the random pathlengths and frequencies of the processes. If P_{Lor} denotes the power acquired by the particle, then we observe that during the computations P_{Rad}/P_{Lor} starts below 10^{-2}, and reaches a value close to unity. Thus we tried to monitor what happen when $P_{Rad}/P_{Lor} > 0.75$, the L–DEM indicates that the energy should grow very slowly.

The q-classical calculation gives completely different results from the classical L–DEM. This is due to the different nature of the calculations. In the L–DEM calculation, P_{Rad}/P_{Lor} is the most significant quantity; but in the q-classical calculation, what really matters is the ratio between the acquired energy between the radiation of two successive photons, and the energy of the next radiated photon.

The results allow the conclusion that two different particles, will follow different trajectories, but they are very close to each other. We believe, supported by the Law of Large Numbers, that this will happen for all particles, and therefore we shall get a bundle of trajectories. The equivalent temperature of a thermal plasma with the same spread of energies exibited by this bundle of trajectories would be about 10^{14}K, which shows that the radiative process is largely dominant, and it is not affected by any others. When the trajectories are very far from E>cB, ($C_* \gtrsim .80$), then there is an influence of Larmor precession in the trajectories but not as described in da Costa & Kahn (1991), as the strength of the rotating electromagnetic wave will almost fully dominate the process (Kirk, 1980).

The calculation clearly show that Schwinger's (1954) semi–classical description is not needed, as the energies reach an upper limit ($\gamma \sim 10^7$) smaller, but very far from the classical limit ($\hbar\omega_{max} = \gamma mc^2$, $\gamma \sim 10^9$, for the Crab Pulsar). The same applies to the superluminal effects described by Ginzburg(1989), when $\gamma B \gtrsim 10^{11}$T.

The difference in the particle trajectories is such that after certain stage of the calculations, with some of them it is impossible to continue, due to numerical instabilities. This, so far, suggests a transition to stochastic chaos. At the same time due to the random nature of the trajectories, we might assume that a random media will be created. These conclusions will be reassessed in the context of collective phenomena.

4   The Collective Phenomena

The kinetic equation for a radiative plasma in the indicated conditions is

$$\frac{\partial f}{\partial t} + \mathbf{v} \cdot \nabla f + \mathbf{F} \cdot \nabla_p f = \int d\mathbf{p_1}\, W(\mathbf{r}, \mathbf{p}|\mathbf{p_1}, t)\, f(\mathbf{r}, \mathbf{p_1}, t) - \lambda(\mathbf{p}, t)\, f(\mathbf{r}, \mathbf{p}, t) \quad (9)$$

In the above integral the first term gives the density of particles per unit time which enter $(\mathbf{r}, \mathbf{p}, t)$ at fixed position $\mathbf{r}$ and time t. At the same time the second term gives the density of particles per unit time which are available for radiation, and therefore leave $(\mathbf{r}, \mathbf{p}, t)$ but keeping fixed position and time.

$W(\mathbf{r}, \mathbf{p}_\alpha \,|\, \mathbf{p}_\beta, t)$ is the probability per unit time that a particle will jump from $\mathbf{p}_\beta \rightarrow \mathbf{p}_\alpha$ at a given position $\mathbf{r}$ at time t (Gardiner, 1990). $W(\mathbf{r}, \mathbf{p}_\alpha \,|\, \mathbf{p}_\beta, t) = 0$ above a certain thereshold $\mathbf{p}_\alpha^{thr}$ as no transitions $\mathbf{p}_\beta \rightarrow \mathbf{p}_\alpha$, $(p_\alpha > p_\alpha^{thr})$ are allowed.

Then,

$$W(\mathbf{r}, \mathbf{p}_\alpha \,|\, \mathbf{p}_\beta, t) = \begin{cases} P_r(\mathbf{p}_\beta, \mathbf{n}, \omega, t) & \text{for allowed } \mathbf{p}_\beta \rightarrow \mathbf{p}_\alpha \\ 0 & \text{otherwise} \end{cases} \tag{10}$$

As seen before only particles which radiate photons with $\hbar\omega > mc^2$ will change their trajectories. Then we may define

$$\lambda'(\mathbf{p}, t) = [1 - P(\omega < \omega_{Q-CL} | \mathbf{p})]\lambda(\mathbf{p}, t) \tag{11}$$

and

$$\begin{aligned}
\mathcal{N}_c(\mathbf{r}, \mathbf{p}, t) &= \int_{c|\mathbf{p}_1 - \mathbf{p}| > \hbar\omega_{Q-CL}} d\mathbf{p}_1 \, W(\mathbf{r}, \mathbf{p}|\mathbf{p}_1, t)\, f(\mathbf{r}, \mathbf{p}_1, t) \\
&= \int_{\omega_{Q-CL}}^{\infty} dn\, d\omega\, P_r\left(\mathbf{p} + \frac{\hbar\omega}{c}\mathbf{n}, \mathbf{n}, \omega, t\right) f\left(\mathbf{r}, \mathbf{p} + \frac{\hbar\omega}{c}\mathbf{n}, t\right) \cdot \\
&\qquad \frac{\partial[\mathbf{p} = (\hbar\omega/c)\mathbf{n}]}{\partial(\omega, \mathbf{n})}
\end{aligned} \tag{12}$$

and write

$$\frac{\partial f}{\partial t} + \mathbf{v} \cdot \nabla f + \mathbf{F} \cdot \nabla_p f = \mathcal{N}_c(\mathbf{r}, \mathbf{p}, t) - \lambda'(\mathbf{p}, t)\, f(\mathbf{r}, \mathbf{p}, t) \tag{13}$$

This equation has the following integral form

$$\begin{aligned}
f(\mathbf{r}, \mathbf{p}, t) = &\int dA_s\, g(A_s) \int_0^t dt_s\, J[\mathbf{r}_1(t, A_s, t_s), \mathbf{r}_s] \\
&\exp\left\{-\int_{t_s}^t d\tau\, \lambda'[\mathbf{p}_1(\tau, A_s, t_s), \tau]\right\} \cdot \\
&\delta[\mathbf{r} - \mathbf{r}_1(t, A_s, t_s)]\, \delta[\mathbf{p} - \mathbf{p}_1(t, A_s, t_s)] + \\
&\int_{t_{in}}^t d\eta\, \mathcal{N}_c[\mathbf{r}_2(\mathbf{r}, \mathbf{p}, \eta), \mathbf{p}_2(\mathbf{r}, \mathbf{p}, \eta), \eta] \cdot \\
&J[\mathbf{r}_2(\mathbf{r}, \mathbf{p}, t), \mathbf{r}]\, \exp\left\{-\int_\eta^t d\chi\, \lambda'[\mathbf{p}_2(\mathbf{r}, \mathbf{p}, \chi), \chi]\right\}
\end{aligned} \tag{14}$$

where

$$\mathbf{r}_1(t, A_s, t_s) = \mathbf{r}_s(A_s) + \int_{t_s}^t d\tau\, \mathbf{v}_1(t - t_s, A_s, t_s) \tag{15}$$

$$\mathbf{p}_1(t, A_s, t_s) = \int_{t_s}^t d\tau\, \mathbf{F}[\mathbf{r}_1(\tau - t_s, A_s, t_s), \mathbf{p}_1(\tau - t_s, A_s, t_s), \tau] \tag{16}$$

with (*), defining quantities with origin in the source region, and

$$r_2(r, p, \tau) = \int_{t_{in}}^{\tau} d\eta\, v_1[r_2(r, p, \eta), \eta] \tag{17}$$

$$p_2(r, p, \tau) = \int_{t_{in}}^{\tau} d\eta\, F[r_2(r, p, \eta), p_2(r, p, \eta), \eta] \tag{18}$$

with t_{in} defined such as $p_2(r, p, t_{in}) = 0$.

These expressions show a clear relationship with the single particle approach, in a continuously recursive way. They must be considered as describing an average behaviour of the magnetospheric plasma. However the single particle approach, due to the randomness of the phenomena, does not fully support these expressions, because it is very difficult or even impossible to establish average smooth trajectories. Nevertheless, their recursive nature combined with the stochasticity of the particle trajectories allow us to foresee the magnetospheric plasma as a random media with fractal characteristics. This might explain the difference among the individual pulses, and the "noise" observed in some pulsars (Rickett *et al.*, 1975).

Acknowledgements

This work has been conducted in strict collaboration with Professor F. D. Kahn, in particular during my stay at the Department of Astronomy, Manchester University, in the academic year of 1991/92, as Honorary Research Associate, during a sabbatical leave. It has been funded since January 1993 by Junta Nacional de Investigação Científica e Tecnológica, Portugal, under contract STRDE/C/PRO/996/92.

We thank Instituto Superior Técnico for granting the sabbatical leave.

References

da Costa, A.A.: 1983, 'Pulsar Electrodynamics: Cylindrical Model and Radio and Gamma-Ray Radiation', *Mon. Not. R. astr. Soc.*,**204**, 1125-1144

da Costa, A.A. and Kahn, F.D.: 1982, 'High Energy Electrons in Pulsar Magnetospheres', *Mon. Not. R. astr. Soc.*,**199**, 211-217

da Costa, A.A. and Kahn, F.D.: 1991, 'Relativistic Electrons in Pulsar Magnetospheres: Quantum Corrections to the Classical Regime', *Mon. Not. R. astr. Soc.*,**251**, 681–686

Gardiner, C.W.: 1990, *Handbook of Stochastic Methods*, Springer-Verlag, Berlin

Ginzburg, V.L.:1989, *Applications of Electrodynamics in Theoretical Physics and Astrophysics*, Gordon Breach Science Publishers, London

Kirk, J.G.: 1980, 'Coherent Curvature Radiation', *Astron. Astrophys.*, **82**, 262-264

Rickett, B.J., Hankins, T.H. and Cordes, J.M.: 1975, 'The Radio Spectrum of Micropulses from Pulsar PSR0950+08', *Astrophys. J.*, **201**, 425-430

Schwinger, J.: 1949, 'On the Classical Radiation of Accelarated Electrons', *Phys. Rev.*,**75**, 1912-1925

Schwinger, J.: 1954, 'The quantum Correction in the Radiation by Energetic Accelarated Electrons', *Proc. Nat. Acad. Sci*, **40**,132–136

The Global Structure of the Interstellar Medium

T. W. Hartquist
Max-Planck-Institut für extraterrestrische Physik
D-85740 Garching, Germany

Abstract. In the context of a review of work on the global structure of the interstellar medium, supernova remnant evolution, flows in multiphase media, cosmic ray moderation of flows, theories of the Galactic halo gas, and the nature of the local superbubble are considered. Speculations about the nature of a one parameter fully self-consistent model of the interstellar medium-supernova-radiation and cosmic ray background system are offered.

1 Introduction

I first met Franz Kahn in 1981. He is not only a scientist of breadth and depth; he is also a gentleman.

A thorough knowledge of the global structure of the interstellar medium (ISM) and of the processes that determine that structure is necessary in order to understand the reasons that the Galactic thickness and star formation rates are those that obtain. It is also required for theories of the properties of starbursts and winds in other galaxies and of line forming regions in active galactic nuclei dominated by power sources smaller than a light year. Whatever self-regulation existed in the formation of galaxies and objects, such as global clusters, in them can only be addressed with confidence after that thorough knowledge of the ISM has been attained.

The global structure of the ISM remains a topic of much dispute. Some scientists (e.g. Slavin and Cox, 1992) favour a model in which the filling factor of hot ($T \approx 10^6 \, K$) plasma is about 0.1 while others (e.g. McKee, 1993) advocate a model in which hot plasma fills most of the volume of the Galaxy's disk.

The filling factor of hot plasma and other properties of the ISM depend greatly on the nature of supernova remnant (SNR) evolution. Hence, after reviewing in section 2 the input and output parameters in two and three phase models of the ISM I concentrate in sections 3 through 5 on the evolution of SNRs. Section 3 is a summary of key numerical hydrodynamic simulations of SNR development performed in roughly the last 20 years; in some of these simulations the effects of a magnetic field were approximated. A SNR's structure and evolution are influenced tremendously by the multiphase structure of the ambient ISM. Section 4 contains a review of our knowledge of the influence of embedded clumps on the flows of more tenuous plasmas. Section 4 also includes a discussion of proposed studies that may help show whether mixing of plasmas in different phases occurs in magnetized turbulent boundary layers and the extent to which such mixing, if it takes place, enhances the radiative cooling rate in a SNR. Section 5 concerns the possible influence of cosmic rays on remnants.

Astrophysics and Space Science **216**: 185–200, 1994.

In section 6 I describe observations of and theories of the gas in the Galactic halo; a fountain driven by large associations of supernovae may account for the observations, but the relative importance of thermal pressure and cosmic ray pressure in driving a fountain is unclear.

Section 7 is about the local bubble of hot gas that surrounds the Earth and the uncertainty of how it was formed; in that section I stress the importance of the distinction between local bubble, halo, and what might be called 'transition region' emissions.

Section 8 contains speculations concerning what the one free parameter in an ISM model should be and a mention that the problem of why the Galactic disk is the thickness that it is requires attention.

2   Two Phase and Three Phase Models

In the standard steady state two phase model of the ISM (Field, Goldsmith, and Habing, 1969) the specification of the radiation and cosmic ray backgrounds, the grain photoelectric properties, and the average ISM number density determines the interstellar pressure as well as the filling factors, the fractional ionizations, and the temperatures of the cold cloud and warm ($\approx 10^4$ K) phases of the ISM. Bottcher et al. (1970) argued that bursts of radiation from the birth and evolution of the brightest stars and supernovae are sufficiently frequent to prevent a two phase steady state from obtaining.

Cox and Smith (1974) considered the mechanical effects of supernovae on the ISM and argued that supernovae are frequent enough that their remnants overlap, create a 'tunnel' system, and occupy around ten percent of the Galactic disk's volume. If SNRs do overlap and have a substantial filling factor then the pressure of the ISM is determined by the mechanical energy input rather than the radiative energy input as envisaged by the inventors of the two phase model.

McKee and Ostriker (1977) also proposed a three phase model of the ISM. The input parameters in their model include all of those listed above in the paragraph on two phase models. In addition, the shape of the spectrum of cloud masses and the supernova rate per unit volume, S, were specified. For a particular ISM pressure (an output parameter of the model), the shape of the cloud mass spectrum, the radiation and cosmic ray backgrounds, and the grain properties determine the ratios of the cloud and warm phase filling factors as well as the cloud and warm phase temperatures and ionization structures. The supernova rate per unit volume and the nature of SNR evolution (which depends on the average ISM number density, the mass spectrum of the clouds, the interstellar pressure, and the filling factors of all three phases of plasma) determine the hot phase filling factor and temperature distribution as well as the ISM pressure distribution.

If supernovae are uncorrelated, the filling factor of the hot phase is given roughly by (Cox and Smith, 1974; McKee and Ostriker, 1977; McKee, 1990)

$$f_h = [1 - \exp(-Q)](1 - f_c - f_w) \tag{1}$$

where f_c and f_w are the cloud and warm phase filling factors and

$$Q = S \int_0^\infty V_{SNR}(t')dt' \tag{2}$$

$V_{SNR}(t)$ is the volume of a supernova remnant of age t. Hence, the global nature of the ISM is sensitive to the behaviour of SNRs.

3   Numerical Simulations of Supernova Remnants

One necessary condition for the disk ISM to be a closed system requires that SNRs radiate in the disk and lose very little energy to the halo. Once a SNR enters its radiative era its cooling time generally becomes comparable to or less than the recombination times of a number of ions important for the cooling. Thus, much of the hot phase plasma is not in ionization equilibrium and radiative cooling rates for equilibrium plasmas are inappropriate for the study of some important aspects of SNR evolution. In addition, many of the OVI, NV, CIV, and SIV ultraviolet absorption features observed to diagnose the hot phase gas probably arise in regions in which ionization equilibrium does not obtain. Similarly, extreme ultraviolet and far ultraviolet, as well as some of the lowest energy soft X-ray, emissions used to study the ISM most likely originate in nonequilibrium plasmas. The importance of nonequilibrium ionization should be borne in mind as one reads this section.

Chevalier (1974) performed important early numerical simulations of SNR evolution for constant density ambient media. As are all models described in this section Chevalier's are one-dimensional, spherically symmetric, time-dependent hydrodynamic models. Chevalier did include, in an approximate manner, the effects of a magnetic field on SNR shell development. (I will explicitly state that such effects were included in those simulations in which they were.) For each simulation Chevalier chose initial conditions which led to Sedov flow during the adiabatic era. Radiative losses for ionization equilibrium were included, and remnant evolution into the radiative era was followed.

McKee and Ostriker (1977) and Cowie, McKee, and Ostriker (1981) considered the effects of cloud evaporation as well as the sweeping up of intercloud plasma on remnant evolution. I will return in section 4 to a fuller discussion of the effects of mass entrainment on SNRs. Radiative losses and the nonequilibrium ionization were not treated rigorously in either of these important papers.

Falle (1981) used a cooling law appropriate for an equilibrium plasma when he showed that catastrophic cooling and multishock structure develop near the leading edge of a SNR entering its radiative era as it expands into a uniform ambient medium.

Cioffi, McKee, and Bertschinger (1988) assumed a type of initial conditions that differ from those assumed by Chevalier. They used cooling rates for equilibrium plasmas and showed that because a SNR enters its radiative era on a timescale less than that required for a small amplitude sound wave to cross a remnant, the Sedov solution does not ever describe the SNR flow structure. However, the outer radius of a SNR propagating into an ambient medium does increase as the SNR age to the two fifths power as in Sedov flow. The work of Cioffi *et al.* shows clearly that even in the radiative era the internal pressure of a SNR affects its expansion rate; that is, there is no purely momentum driven SNR shell.

In 1992 Slavin and Cox presented the first results for coupled hydrodynamic nonequilibrium ionization models of SNRs. A uniform ambient medium with a magnetic field of $5\,\mu$G and number density of $0.2\,\mathrm{cm}^{-3}$ was assumed to be the site of a 5×10^{50} erg supernova. Slavin's and Cox's paper gives the first SNR model data necessary for some sort of detailed comparison with many observational ISM data; their extensive presentation of figures for model spectroscopic results shows that they were fully aware of this fact.

On the basis of their calculations Slavin and Cox (1992) estimated f_h to be about 0.1. McKee (1990) attributed this conclusion to their choice of supernova energy, which he considers to be too small by a factor of two, and to their value of the magnetic field strength, which he considers too high. McKee (1990) favours a hot phase filling factor exceeding 0.5. The large value of the magnetic field adopted by Slavin and Cox (1992) is consistent with the view expressed by Boulares and Cox (1990) that the magnetic field supports much of the weight of the ISM against collapse on the disk. McKee's (1990, 1993) assessments of observational work on gas kinematics and magnetic field strength measurements have led him to conclude that much of the ISM is supported kinematically. In any case, Slavin and Cox have shown only that for their assumed supernova and magnetic parameters the filling factor of warm gas will remain large if it is initially large; they did not demonstrate that the filling factor of hot gas will become small even if it is initially large.

4 The Effects of Clumpiness on Remnant Evolution

Hartquist and Dyson (1993) have recently reviewed the subjects of the boundary layers between hot flowing tenuous astrophysical plasmas and embedded clumps, the intermediate scale structures formed by tenuous plasma - embedded clump interactions, and the effects of those interactions on the global structures of diffuse astrophysical sources.

In their model of the ISM McKee and Ostriker (1977) included the effects of cloud evaporation on SNR evolution. They assumed that the evaporation is driven by conductive heat transfer and that the rate that mass is evaporated into the tenuous plasma is a constant times $T_\infty^{5/2}$, where T_∞ is the temperature in the diffuse plasma well away from the clouds. They argued that if a SNR's mass increases predominately by evaporation at this rate, rather than by the sweeping up of ambient tenuous plasma, the SNR radius increases as $t^{3/5}$ (instead of $t^{2/5}$ as for a Sedov flow), where t is the SNR age. They estimated that due in part to the nonequilibrium ionization structure of evaporated gas the effective radiative cooling rate coefficient is about an order of magnitude greater than that that obtains in equilibrium plasmas.

McKee and Ostriker assumed that a typical SNR in the disk ISM evaporates the same amount of mass as it sweeps up. This was a key, though totally ad hoc, assumption in their treatment of mass exchange between different ISM phases. If the ISM is a closed system, for it to be in some sort of statistical steady state the average rate of mass evaporation per unit volume has to be counterbalanced by an equal average rate of condensation of hot gas to form cloud and warm phase gas. McKee and Ostriker assumed that cloud gas is replenished by the formation and fragmentation of SNR shells during the radiative eras of the SNRs. However, numerical calculations by Cowie, McKee, and Ostriker (1981), who adopted a radiative cooling rate coefficient equal to the equilibrium one, showed that for a $T_\infty^{5/2}$ dependance of the evaporation rate per unit volume the radiative losses are much more distributed throughout a SNR than if mass is added almost solely by the sweeping up of diffuse plasma. Thus, there is some question about whether cloud gas is replenished by shell formation in SNRs. Perhaps, as a SNR cools the replenishment takes place by condensation, possibly induced by conduction (c.f. McKee and Cowie, 1977; Borkowski, Balbus and Fristrom, 1990), onto the larger clouds. Then, the larger clouds may grow sufficiently to become gravitationally unstable and produce stars.

Clearly, mass pickup by tenuous hot plasma from embedded clouds has tremendous importance for SNR evolution and the global structure of the ISM. However, the relevance of conduction in the establishment of evaporation rates is less clear. The tenuous plasma flows around clouds, and hydrodynamic effects other than conduction also induce cloud evaporation.

Two dimensional hydrodynamic studies of the response of a cloud, initially confined by the pressure of a warmer more tenuous medium, to the passage of a shock were conducted by Woodward (1976) and Nittmann, Falle, and Gaskell (1982). The results suggested that those parts of a cloud that do not become bound by their own gravity are dispersed in a time of the order of the sound crossing time in the cloud. Klein, McKee, and Colella (1990) have drawn similar conclusions after following the dispersal for a longer time than in previous studies.

Hartquist and Dyson (1988) argued that interstellar cloud–hot flowing plasma interfaces are turbulent boundary layers with thicknesses of several to about ten percent of the cloud sizes. Hartquist et al. (1986) pointed out that even if the flow around the cloud is subsonic the Bernoulli effect and momentum transfer to a gravitationally unbound cloud will result in its eventual expansion and dispersal; they suggested that in very subsonic flows the rate of mass entrainment by the tenuous plasma (time averaged over the period necessary to entirely evaporate the cloud) is proportional to the four thirds power of the upstream Mach number of the tenuous plasma as measured in the cloud's frame.

The importance of hydrodynamic effects in the evaporation of embedded obstacles is strikingly illustrated by the beautiful images of the "cometary tails" of clumps being ablated by a fast wind in the Helix Nebula (Meaburn *et al.*, 1992). The length to width ratio of the order of 10 for each clump suggests that each is being impinged by shocked subsonic wind and evaporating subsonically (Dyson, Hartquist and Biro, 1993).

The mass evaporation rate is only one of the important parameters governing the influence of evaporation on SNR evolution. The degree to which the ionization structure deviates from equilibrium in the evaporated material and whether lowly ionized ablated gas mixes with the hotter, more tenuous plasma are also significant since the radiative cooling properties of a SNR are affected by the ionization structure of the gas. Possibly, because the boundary layers are magnetized mixing doesn't occur, but the process of turbulent driven ambipolar diffusion or the extremely poorly understood mechanism of magnetic reconnection (c.f. the discussion of reconnection in a section 6) may result in mixing. Perhaps, ablated gas remains in the boundary layer long enough as it flows along a cloud surface that heating is slow enough that ionization equilibrium obtains.

Diagnostic studies, rather than first principles theoretical attacks, are required to establish the properties of boundary layers and the consequences of evaporation through them for the global structures of diffuse astrophysical sources. The reader can refer to the paper by Hartquist and Dyson (1993) for a more complete review of the possibilities of diagnosing boundary layer properties. Spatial resolution comparable to that of the boundary layer thickness is desirable. For instance, the lack of spatial resolution makes the demonstration that OVI absorption arises in conduction fronts (McKee and Ostriker, 1977) rather than the interiors of radiative SNR and superbubbles (Innes and Hartquist, 1984; Slavin and Cox, 1992) difficult. Probably, observations of the OVI line profiles imply that at least some of the features are not formed in evaporating flows driven by conduction (e.g. Hartquist and Dyson, 1993). In any case, OVI data do not provide a good means of learning about boundary layers. Observations of various molecular emissions at the interfaces between dark cores and winds in low mass star forming regions are probably the most promising means of determining the extent of the dissipation layer and whether mixing occurs, at least in such a situation (Charnley *et al.*, 1990). The dark core TMC-1 is an obvious candidate for an observational program (Williams and Hartquist, 1992). Work on that problem is currently being extended in several directions by J.E. Dyson, M. Malone, L.A.M. Nejad, S.D. Taylor, D.A. Williams, and myself. However, at the moment our knowledge of cloud–hot plasma interactions is so minimal that we have no idea how great their effect on SNR evolution is. Possibly, evaporation results in Q, given in eqn (2), being rather small.

5   The Effects of Cosmic Rays on Remnant Evolution

The cosmic ray, thermal, magnetic, and kinematic contributions to the pressure of the ISM are comparable (e.g. Boulares and Cox, 1990; McKee, 1990, 1993). Cosmic rays may affect SNR evolution substantially. For them to be of importance in a SNR about to overlap with another SNR or to merge with the ISM the timescale for the cosmic rays that supply most of the pressure to diffuse out of the SNR must be greater than the SNR's age. The bulk of the cosmic ray pressure is due to cosmic rays with kinetic energies of the order of 1 GeV. Analysis of cosmic ray composition and the distribution of radio synchrotron emission in the Galaxy leads to the conclusion that the spatial diffusion coefficient, κ, of 1 GeV cosmic rays is typically about $10^{29}\,\mathrm{cm}^2\,\mathrm{s}^{-1}$ (e.g. Dogiel, 1991). For this value of κ the diffusion timescale across a SNR having a radius, R, of 100 pc would be about $R^2/\kappa \approx 3 \times 10^4$ years, which is very short relative to the age of a SNR about to merge with the ISM. A necessary condition for cosmic rays to continue to have a significant influence on a SNR throughout its expansion is that $\kappa \leq 10^{27}\,\mathrm{cm}^2\,\mathrm{s}^{-1}$ for GeV cosmic rays.

Maybe the composition and radio data provide a measure of κ in that part of the ISM volume that contains only SNRs that have overlapped and come into some sort of equilibrium with one another. Possibly until overlap occurs $\kappa \ll 10^{29}\,\mathrm{cm}^2\mathrm{s}^{-1}$. In another paper in this volume Hartquist and Morfill (1994) make further comments on the value of κ in the typical ISM; those comments are consistent with the possibilities raised in the previous two sentences.

The only semiempirical evaluation of κ in a specific SNR was made in a beautiful paper by Boulares and Cox (1988). They argued that the Balmer line emission towards an X-ray emitting shocked region in the Cygnus Loop arises in a cosmic ray heated shock precursor and implies $\kappa \leq 2 \times 10^{25}\,\mathrm{cm}^2\,\mathrm{s}^{-1}$. The Cygnus loop is only about 10^4 years old, is in its adiabatic stage of evaluation, and has a radius of only 20 pc. It is difficult to extrapolate κ to later times.

An additional requirement for cosmic rays to have a large effect on a SNR is that at least somewhere between 10^{-4} and 10^{-3} of the suprathermal protons in the vicinity of the shock front must be injected as low energy cosmic rays to be accelerated (e.g. Bell, 1987). Since young SNRs are strong radio sources it seems likely that this level of injection is achieved. A numerical study of the process in the vicinity of the Earth's bow shock has been made by Scholer, Trattner, and Kucharek (1993) using a hybrid electron fluid–heavy particle code including magnetic and electric forces.

The roles that relatively low energy cosmic rays can play in the ISM are often forgotten. Hartquist and Morfill (1983, 1994) have suggested that cosmic ray protons with energies of only a few MeV provide much of the ionization in cold diffuse molecular clouds. On the basis of their analysis of the roughly 0.2 to 10 MeV bremsstrahlung gamma-ray background Skibo and Ramaty (1993) have argued that cosmic ray electrons with energies comparable to those of the gamma rays they considered induce ionization at a rate of $10^{-15}\,\mathrm{s}^{-1}$. They noted that this ionization rate is sufficient to maintain the so-called Reynolds Layer which contributes at heights of up to about 1 kpc above the Galactic plane to the dispersion of pulsar signals (Reynolds, 1989).

6   The Galactic Halo Gas

Because they do not suffer radiative losses, cosmic rays might be expected to be particularly important for the levitation of gas in the Galactic halo, far from the supernovae in the disk. Chevalier and Fransson (1984), Hartquist, Pettini, and Tallant (1984) and Hartquist and Morfill (1986) considered observations of SiIV and CIV absorption made against early-type stars in the halo at distances of up to several kiloparsec above the disk (Pettini and West, 1982; Savage, 1990; Sembach and Savage, 1992) within the context of models of cosmic ray supported halo gas. Pettini and West (1982) had concluded that NV absorption had not been detected, that the CIV to SiIV equivalent width ratio varied little and implied halo gas temperatures below 1×10^5 K, and that the HWHM of the features are typically $25\,\mathrm{km\,s}^{-1}$ and the average speed of the gas relative to the disk is about $10\,\mathrm{km\,s}^{-1}$. On the basis of those observational conclusions the gas appeared to be too cold to be thermally supported and too slow to be kinematically supported to the heights of several kiloparsec to which it extends. Hartquist and Morfill (1986) inferred that the CIV and SiIV absorption regions must have a large volume filling factor since a configuration of $T \leq 10^5$ K gas having a small filling factor embedded in a hotter less dense medium would be Rayleigh-Taylor unstable resulting in the CIV and SiIV line forming regions falling to the disk at speeds higher than those observed. Whether at 8×10^4 K and collisionally ionized or at $T \approx 10^4$ K and ionized by the absorption of photons of extragalactic origin the CIV and SiIV absorbing gas would have a number density of about $10^{-3}\,\mathrm{cm}^{-3}$ if its filling factor were nearly unity (Chevalier and Fransson, 1984; Hartquist et al., 1984).

Subsequent observations (e.g. Savage, 1990; Sembach and Savage, 1992) revealed the existence of NV absorption features formed in the halo gas and that the low speed CIV and SiIV are predominant only in the Southern hemisphere towards which the Pettini and West (1982) observations were made. The HWHM of CIV and SiIV features detected towards Northern hemisphere halo stars is up to about $55\,\mathrm{km\ s}^{-1}$ and the gas in which they originate is moving with an average velocity of about -40 km s^{-1} (Savage, 1990). These Northern hemisphere data raised again the possibility that much of the halo gas is in a Galactic fountain (Shapiro and Field, 1976; Bregman, 1980; Kahn, 1981) driven by disk supernovae and possibly the source of high-latitude, high-velocity and intermediate-velocity clouds (Bregman, 1980). Specifically, they removed the requirements that most of the halo volume be filled with $T \leq 10^5$ K gas.

Benjamin and Shapiro (1992) have shown that gas in a fountain cooling radiatively in a steady flow from an initial state of ionization equilibrium at a temperature of about 10^6 K would give rise to relative abundances of N^{4+}, C^{3+}, and Si^{3+} consistent with the observed halo ultraviolet absorption features. The calculations included the effects due to the radiation from the cooling gas on its nonequilibrium ionization structure. Benjamin and Shapiro (1992) found that the absorption data and the CIV $\lambda 1550$ emission data (Martin and Bowyer, 1990) are matched in cooling fountains with a range of parameters. McKee (1993) has argued that the Benjamin and Shapiro models are compatible with measured upper bounds to OVI $\lambda\lambda$ 1032, 1037 emission (Edelstein and Bowyer, 1993) and do not produce too many soft X-rays if the initial temperature is equal to or lower than 1.0×10^6 K.

Raymond (1992) performed calculations similar to those of Benjamin and Shapiro and found similar agreement between model results and observational data. His choice of the initial temperature distribution was based on an analogy between the heating of halo gas and microflaring in the solar corona. Magnetic field reconnection must occur for such halo microflaring to occur. It is also important in models of the Galactic dynamo that are based on the assumption that dissipation occurs in the Galactic halo (Kahn, 1990; Parker, 1992) since the diffusivity in the disk is not strong enough to give rise to sufficiently rapid dissipation of random field fluctuations to leave the ordered large scale Galactic magnetic field. However, the theory of reconnection is poorly understood, and numerical simulations have shown no Petschek-type rapid reconnection in high magnetic Reynolds number environments unless localized regions of higher resistivity were included at the outset of a simulation (e.g. Scholer, 1991). Thus, Galactic halo microflaring and the role of halo field reconnection in the Galactic dynamo cannot be treated with any theoretical rigour.

Ignoring cosmic rays, McKee (1993) has argued that all of a supernova's energy will be radiated in the disk unless it is part of a correlated burst of at least about 800 supernovae. Thus, a thermally driven fountain must be powered by superbubbles rather than a set of individual SNRs. From diffuse optical and radio emission data he has estimated that the distribution of stellar associations with respect to the number, N_{*h}, of supernova progenitors produced within the lifetime of the association to be

$$N_a \left(N_{*h} \right) = 5.5 \left(\frac{6200}{N_{*h}} - 1 \right) \tag{3}$$

with 6200 being the maximum that N_{*h} attains in the Galaxy. From this distribution for N_a he concluded that about ten percent of the disk surface supplies a superbubble driven fountain. This conclusion coupled with the Benjamin and Shapiro results impies that a fountain with a mass injection rate of about 2–3 $M_\odot \, y^{-1}$ into each hemisphere supplies the halo gas.

However, the possible relevance of cosmic ray levitation should be given more consideration. For the moment we ignore the data for CIV emission which may, in fact, originate in the lower halo or transition region energized by that vast majority of superbubbles which, according to McKee (1993), do not break into the halo. We suppose that the CIV, SiIV, and NV absorbing gas fills a substantial fraction of the volume in the halo; then the number density, n, typically is around $10^{-3}\,\mathrm{cm}^{-3}$. Though the models envisaged by Chevalier and Fransson (1984) and by Hartquist and Morfill (1986) were of static cosmic ray supported halo gas, dynamical models of cosmic ray supported halo gas in which velocity gradients are substantial over a few kiloparsec can be constructed (e.g. Breitschwert, McKenzie and Völk, 1987). Thermal instability in such cosmic ray driven flows has not been investigated, but it may play a role giving rise to cloud formation and fountain-like return of mass to the disk. Thus, a cosmic ray driven fountain consistent with the dynamics implied by the absorption data is imaginable. Furthermore, a natural mechanism for heating gas to temperatures required for the collisional ionization necessary to produce C^{3+}, Si^{3+}, and N^{4+} operates in a cosmic ray driven fountain with $n \approx 10^{-3}\,\mathrm{cm}^{-3}$. As discussed by Hartquist and Morfill (1986), in a static cosmic ray supported gas the heating rate per unit volume of gas due to the dissipation of high freAquency waves resonant with the cosmic rays is $L_w = \rho g v_A$ where ρ, g, and v_A are the gas density, gravitational field strength, and Alfvén wave speed. L_w is somewhat higher in a slowly accelerated flow. For typical values of g and the ISM magnetic field strength L_w in a static halo is sufficient to balance radiative losses in $10^5\,\mathrm{K}$ gas in ionization equilibrium if n is not more than $10^{-3}\,\mathrm{cm}^{-3}$. Since the equilibrium cooling rate coefficient is a flat function of temperature at temperatures around $10^5\,\mathrm{K}$, SiIV absorbing gas and NV absorbing gas need differ only slightly in number density.

Thus, the absorption data do not rule out the possibility that halo gas is supported by cosmic rays.

We now assume that the CIV emission does arise in the halo. Then the CIV spectral features must originate in gas that is considerably denser and has a much smaller volume filling factor ($\approx 10^{-2}\,\mathrm{cm}^{-2}$ and 0.1 roughly) than discussed above. A dense hot fountain could possibly be formed by the loss of energy by cosmic rays at the top of the transition layer, which is, perhaps, coincident with the Reynolds layer. Assume, as Slavin and Cox (1992) would argue, that warm ($\approx 10^4\,\mathrm{K}$) gas rather than hot ($\approx 10^6\,K$) gas fills most of the volume of the disk. Assume further that most of the volume containing the transition layer is also warm gas and that the ionization rate per neutral hydrogen is independent of position. Then as the transition layer ends at high altitude, the neutral density drops as the square of the number density. This corresponds to a rapid drop, with density, in the ion-neutral damping rate of Alfvén waves on which cosmic rays resonantly scatter. Possibly κ is typically very large in the transition region and in the disk and only at the top of the transition region and above is cosmic ray streaming successful in inducing sufficient wave activity for κ to become large enough to limit cosmic ray streaming. If such a picture were correct cosmic rays could stream along magnetic field lines at very high speeds from a supernova remnant as soon as it slowed to an expansion speed of about $100\,\mathrm{km\,s^{-1}}$ (below which the leading shock ceases to have an ionization precursor in which the cosmic rays might be but possibly aren't trapped) or earlier. Strong coupling to gas would occur as the density decreased at the top of the transition layer, and the cosmic ray pressure possibly would drive a shock into the warm medium. Compression to a number density of $10^{-2}\,cm^{-3}$ and acceleration to a speed of $100\,km\,s^{-1}$ over a distance of $100\,pc$ would give rise to sufficiently rapid dissipation of cosmic ray energy that radiative cooling during the acceleration would be marginally overcome. Such a picture needs to be studied quantitatively, but possibly fewer than 800 correlated supernovae may be required to produce a supply region for a cosmic ray driven fountain.

The dissipation of Alfvén waves with wavelengths of many tens of parsecs may provide another source of heat in the halo. In fully ionized gas, a circularly polarized Alfvén wave generally dissipates most rapidly by decaying into a backwardly propagating Alfvén wave and a sound wave. Hartquist (1983) has pointed out that if the filling factor of fully ionized gas is large and the temperature is above about $10^5\,K$, the conductive damping of sound waves is sufficiently rapid to inhibit the three-wave mechanism for dissipating Alfvén waves. Hence, if the hot phase has a large filling factor long wavelength waves may propagate into the halo and provide a heat source for hot gas whenever the temperature drops below about $10^5\,K$. The dissipation of these low frequency Alfvén waves can maintain the temperature of hot phase halo gas only if the number density is around $10^{-3}\,cm^{-3}$ or less.

7   The Local Bubble

As described in the previous section if observed far ultraviolet emissions or soft X-ray emissions are assumed to originate in the halo gas they provide constraints on the models of halo gas. However, the Earth is located in a bubble or superbubble that is a source of emission as well. Most likely a clear understanding of our local environment is a prerequisite to the construction of a reliable model of the global behaviour of the ISM.

Cox and Anderson (1982) attempted to explain the B-band and C-band soft X-ray emissions by studying the emissivity of SNRs undergoing Sedov evolution. They concluded that these emissions could be produced by a SNR that formed 10^5 years ago by the explosion of a supernova ejecting 5×10^{50} ergs into an ambient medium with a number density of $0.004\,cm^{-3}$. Such a SNR would produce very little M-band X-ray emission, and the number density of 0^{5+} in it would be about five times higher than observed.

Innes and Hartquist (1984) tried to explain the origin of the M-band emission as well as the lower energy X-ray emissions. Kahn (1976) had earlier developed a simple analytic means of calculating the temperature distribution in a very old isobaric radiative SNR. From Kahn's treatment one sees clearly that the emission spectrum of such a pressume bound SNR hardens with time if thermal conduction is inefficient. The spectrum observed at the center of an old SNR depends in this model on only two free parameters. $\chi \equiv E_o^{1/3} P^{17/9}/\rho_o^{2/9}$ and $\beta \equiv Pt$, where E_o, P, ρ_o, and t are the initial supernova energy, the ambient medium's pressure and mass density, and the SNR age. Innes and Hartquist (1984) found that the local bubble emits roughly the typically observed B-band and C-band emissions and around half the M-band emission if $\chi = (2\,k_B)^{17/9}1.8 \times 10^{31}$ and $\beta = (2\,k_B)4.4 \times 10^{18}$ where c.g.s. units are used and k_B is Boltzmann's constant. One combination of parameters that gives the above values of χ and β is $P/2\,k_B = 2 \times 10^4\,\mathrm{cm}^{-3}\,\mathrm{K}$, $t = 7 \times 10^6\,\mathrm{y}$, $\rho_o = 5 \times 10^{-26}\,\mathrm{g\,cm}^{-3}$, and $E_o = 1.5 \times 10^{52}\,\mathrm{erg}$; the radius of the bubble specified by these parameters is 200 pc, and its average O^{5+} number density is about a factor of two below the typical interstellar average. (For fixed χ and β the average number density of a species varies as P and the bubble radius scales as P^{-2}.) The high model energy density would imply that the local bubble was formed by tens of correlated supernovae. The model pressure is higher than other observational data, described below, seem to allow, a problem that might be partly resolved through the performance of a more thorough calculation using more accurate cooling and emission rates or the adoption of greater abundances of heavy species.

The work of Innes and Hartquist (1984) influenced Cox and Reynolds (1987) who, within a more general discussion, also examined the possibility that the Earth is surrounded by a radiative superbubble. Cox and Reynolds focussed on the B-band and C-band X-ray data, did not try to explain the presence of the M-band background, and argued that the combination of thermal conduction of heat outwards and radiative losses in the outer regions being rapid results in an old bubble being nearly isothermal with a sharp outer boundary. Slavin and Cox (1992) showed that the pressure of the hot gas in a very evolved conductive bubble varies substantially with time and generally differs by factors of a few from the pressure of the surrounding $10^4\,\mathrm{K}$ gas; however, the hot gas in a highly evolved bubble is always roughly isobaric. Snowdon et al. (1990) argued that the B-band and C-band X-ray data are compatible with the Earth being surrounded by a superbubble having a uniform temperature of $10^6\,\mathrm{K}$, an electron number density of $0.0047\,\mathrm{cm}^{-3}$, and a total stored thermal energy of 10^{51} ergs which they suggested can be maintained almost indefinitely as the recurrence time for a supernova in a superbubble of the size posited is less than or comparable to the radiative lifetime at about 10^7 years. Snowden et al. (1990) attributed the observed soft X-ray anticorrelation with HI column densities to the variation of the distances, along different lines of sight, to the bubbles edge.

"Edge effects" are likely to be important for the spectrum of the superbubble if it has been re-energized recently by a single supernova as suggested by Cox and Anderson (1982) and deemed possible by Snowdon et al. (1990). For instance, the ablation of clouds distributed at and beyond the edge of a previously formed cavity will result in an adiabatic SNR emitting copiously at a wider range of energies (Dyson and Hartquist, 1987) than either a Sedov-like SNR or an old radiative superbubble which has become nearly isothermal as envisaged by Cox and Reynolds (1987) and Snowden et al. (1990).

Searches for X-ray shadowing by relatively nearby clouds have shown that substantial contributions of the soft X-ray background arise outside the local bubble or superbubble (Burrows and Mendenhall, 1991; Snowden *et al.*, 1991; Snowden, 1993; Snowden, McCammon and Verter, 1993). Significant shadows in the B-band and C-band emissions do not exist towards all clouds, but the shadow cast by the high latitude (b=40°) Draco cloud at a distance of about 300 pc. removes about half of these lower energy soft X-ray emissions. Shadowing of the M-band emission has been found in numerous directions, and possibly the M-band emissivity of the local superbubble is rather negligible.

Despite the possibility that most of the M-band emission does not arise very locally its existence remains one of the most important problems in ISM astrophysics. If thermal conductivity does occur at the "classical" (or "Spitzer") rate and leads rapidly to isothermal structures in the centers of SNRs and superbubbles the discovery of SNR and superbubble models that are compatible with the existence of M-band emission may be rather difficult. Exploration is necessary.

Observations of backscattered Lyman α and He 584 Å solar emissions provide information about an interstellar cloud through which the Earth is passing (Bertaux *et al.*, 1985). The number density is about 0.1 to 0.2 cm^{-3} and the temperature is about 8000 K. Cox and Reynolds (1987) have extensively reviewed ultraviolet absorption observations that provide additional information about this very local cloud. The C$^+$ fine structure level populations inferred in such observations towards β C Ma (Gry, York, and Vidal-Madjar, 1985) imply an electron number density of about 0.1 cm^{-3} though the absorbing material may (but probably doesn't) lie outside the local bubble. The thermal pressure of the very local cloud is significantly below that of any of the models of the local bubble or superbubble described above. Snowden *et al.* (1990) stated that the magnetic contribution to the pressure of the local cloud can easily support the very local cloud against crushing by the surrounding hot gas with higher thermal pressure and cited the conclusion of Boulares and Cox (1990) that the ISM pressure at the disk midplane must exceed $2 \times 10^4 k_B$ (in c.g.s. units) in order to support the ISM weight. However, a model of such a magnetically supported cloud surrounded by an isobaric plasma with a much higher thermal pressure has not been developed; in such a cloud, magnetic field tension probably plays as large a role as the magnetic pressure does.

In the absence of shadowing observations towards a large number of clouds at various distances but along neighbouring lines of sight, one does not know with certainty where emissions arise. This ignorance provides a major difficulty in modelling the local bubble and in determining the properties of the Galactic halo gas. It is clear that the local bubble or superbubble is unlikely to be the source of much of the CIV emission (Martin and Bowyer, 1990) and that shadowing measurements show that much of the M-band X-ray emission does not arise in the local bubble. However, those conclusions do not necessarily imply that those emissions arise in the halo. The transition region extending almost 1 kpc on either side of the Galactic mid-plane is a virtual unknown and may well be the source of many of background emissions that do not arise locally.

8   The One Free Parameter

The three phase models outlined by Cox and Smith (1974), McKee and Ostriker (1977), McKee (1990, 1993), and Slavin and Cox (1992) are not self-consistent models. Each deals with only the response of an infinite medium of specified average density to supernovae occurring at a specified rate per unit volume and to prescribed radiation and cosmic ray backgrounds. In reality, the radiation and cosmic ray backgrounds are correlated with the supernova rate per unit volume, though the exact natures of those correlations must be investigated further. In addition, the formation rate per unit volume of supernova progenitors depends on the ISM properties. One can imagine a self-consistent model of the ISM, supernova production, and the radiation and cosmic ray backgrounds. The minimum number of free parameters in such a model is unknown, but we might optimistical hope that only one quantity is necessary to parameterize the ISM-supernova progenitor-background system. A natural guess would be that the free parameter is the average density of the ISM.

However, the average density of the ISM at any point is affected by the extent to which SNRs puff up the Galactic disk. Perhaps, ultimately the most physical choice for the free parameter, if in fact one free parameter is sufficient, will prove to be the gravitational field. A goal of global ISM model-builders should be the explanation of why the disk thickness is what it is; that thickness is determined, in part, by the gravitational field but is also affected by the fact that if the disk material were to collapse to higher density the supernova rate would increase resulting in enhanced puffing.

Even if an ISM model depending only on the gravity could be constructed and were physically correct and complete, the steady state model solutions might be characterized by little activity (i.e. few supernovae). The "steady state means little activity" conjecture is supported by the prominence in spiral galaxies of spiral arms, where the gravitational field is changing, relative to interarm regions. Steady state is disrupted in the centers of some other galaxies by episodes of accretion that trigger star bursts.

Considerable evidence is given in this review to suggest that the "modest" problem of determining the response of an infinite medium of a certain average density to specified sources is extremely challenging. The goal of building a realistic self-consistent model of the type about which I am speculating in this section is even more difficult. Basic unknowns hindering progress on the "modest" problem as well as the "difficult" problem concern the means by which hot SNR gas returns to the cloud phase (c.f. the third paragraphs of section 4) and the way clouds grow to become star-forming clouds. Growth may be by collisionally induced agglomeration (e.g. Elmegreen, 1990) or by some process like conductively driven condensation. The process that stimulates a cloud to form massive stars rather than low mass stars is also unidentified.

That very basic questions must be answered before a reliable global ISM model can be constructed is evident. As mentioned in the introduction, the development of an understanding of the global structure of the ISM must precede the attainment of knowledge about a variety of other astronomical environments. Franz Kahn's work, as seen by the citations to some of it in this review, has been important in helping all of us begin to address these problems, but much remains for him and the rest of us to do.

References

Bell, A.R.: 1987, *Mon. Not. R. astr. Soc.* **225**, 615.

Benjamin, R.A. and Shapiro, P.R.: 1992, in: *The Tenth International Colloquium on UV and X-ray Spectroscopy of Astrophysical and Laboratory Plasmas*, Cambridge University Press, Cambridge.

Bertaux, J.L., Lallement, R., Kurt, V.G., and Mironova, E.N.: 1985, *Astron. Astrophys.* **150**, 1.

Borkowski, K.J., Balbus, S.A., and Fristrom, C.C.: 1990, *Astrophys. J.* **355**, 501.

Bottcher, C., McCray, R., Jura, M., and Dalgarno, A.: 1970, *Astrophys. Lett.* **6**, 237.

Boulares, A. and Cox, D.P.: 1988, *Astrophys. J.* **333**, 198.

Boulares, A. and Cox, D.P.: 1990, *Astrophys. J.* **365**, 544.

Bregman, J.N.: 1980, *Astrophys. J.* **236**, 577.

Breitschwert, D., McKenzie, J.F., and Völk, H.J.: 1987, in R. Beck and R. Grave (eds.), *Interstellar Magnetic Fields: Observations and Theories*, Springer Verlag, Heidelberg, p.131.

Burrows, D.N. and Mendenhall, J.A.: 1991, *Nature* **351**, 629.

Charnley, S.B., Dyson, J.E., Hartquist, T.W., and Williams, D.A.: 1990, *Mon. Not. R. astr. Soc.* **243**, 405.

Chevalier, R.A.: 1974, *Astrophys. J.* **188**, 501.

Chevalier, R.A. and Fransson, C.: 1984, *Astrophys. J. Lett.* **274**, L43.

Cioffi, D.F., McKee, C.F., and Bertschinger, E.: 1988, *Astrophys. J.* **334**, 252.

Cowie, L.L., McKee, C.F., and Ostriker, J.P.: 1981, *Astrophys. J.* **247**, 408.

Cox, D.P. and Anderson, P.R.: 1982, *Astrophys. J.* **253**, 268.

Cox, D.P. and Reynolds, R.J.: 1987, *Ann. Rev. Astron. Astrophys.* **25**, 303.

Cox, D.P. and Smith, B.W.: 1974, *Astrophys. J. Lett.* **189**, L105.

Dogiel, V.A.: 1991, in H. Bloemen (ed.), *The Interstellar Disk-Halo Connection in Galaxies* IAU Symp. **144**, Kluwer, Dordrecht, p. 175.

Dyson, J.E. and Hartquist, T.W.: 1987, *Mon. Not. R. astr. Soc.* **228**, 353.

Dyson, J.E., Hartquist, T.W., and Biro, S.: 1993, *Mon. Not. R. astr. Soc.* **261**, 430.

Edelstein, J. and Bowyer, S.: 1993, *Adv. Space Res.*, in press.

Elmegreen, B.G.: 1990, in L. Blitz (ed.), *The Evolution of the Interstellar Medium*, Astronomical Society of the Pacific, San Francisco, p. 247.

Falle, S.A.E.G.: 1981, *Mon. Not. Roy. astr. Soc.* **195**, 1011.

Field, G.B., Goldsmith, D.W., and Habing, H.J.: 1969, *Astrophys. J. Lett.* **155**, L149.

Gry, C., York, D.G., and Vidal-Madjar, A.: 1985, *Astrophys. J.* **269**, 593.

Hartquist, T.W.: 1983, *Mon. Not. Roy. astr. Soc.* **203**, 117.

Hartquist, T.W. and Dyson, J.E.: 1988, *Astrophys. Space Sci.* **144**, 615.

Hartquist, T.W. and Dyson, J.E.: 1993, *Quart. J. Roy. Astr. Soc.* **34**, 57.

Hartquist, T.W., Dyson, J.E., Pettini, M., and Smith, L.J.: 1986, *Mon. Not. R. astr. Soc.* **221**, 715.

Hartquist, T.W. and Morfill, G.E.: 1983, *Astrophys. J.* **266**, 271.

Hartquist, T.W. and Morfill, G.E.: 1986, *Astrophys. J.* **311**, 518.

Hartquist, T.W. and Morfill, G.E.: 1994, *Astrophys. Space Sci.*, this volume.

Hartquist, T.W., Pettini, M., and Tallant, A.: 1984, *Astrophys. J.* **276**, 519.

Innes, D.E. and Hartquist, T.W.: 1984, *Mon. Not. Roy. astr. Soc.* **209**, 7.

Kahn, F.D.: 1976, *Astron. Astrophys.* **50**, 145.
Kahn, F.D.: 1981, in F.D. Kahn (ed.), *Investigating the Universe*, Reidel, Dordrecht, p. 1.
Kahn, F.D.: 1991, in H. Bloemen (ed.), *The Interstellar Disk-Halo Connection in Galaxies* IAU Symp. **144**. Kluwer, Dordrecht, p. 1.
Klein, R.I., McKee, C.F., Colella, P.: 1990, in L. Blitz (ed.), *The Evolution of the Interstellar Medium*, Astronomical Society of the Pacific, San Francisco, p. 117.
Martin, C. and Bowyer, S. 1990, *Astrophys. J.* **350**, 242.
McKee, C.F.: 1990, in L. Blitz (ed.), *The Evolution of the Interstellar Medium*, Astronomical Society of the Pacific, San Francisco, p. 3.
McKee, C.F.: 1993, in S. Holt and F. Verter (eds.), *Back to the Galaxy*, American Institute of Physics, New York, in press.
McKee, C.F.: and Cowie, L.L.: 1977, *Astrophys. J.* **215**, 213.
McKee, C.F. and Ostriker, J.P.: 1977, *Astrophys. J.* **218**, 148.
Meaburn, J., Walsh, J.R., Clegg, R.E.S., Walter, N.A., Taylor, D., and Berry, D.S.: 1992, *Mon. Not. R. astr. Soc.* **255**, 177.
Nittmann, J., Falle, S.A.E.G., and Gaskell, P.H.: 1982, *Mon. Not. Roy. astr. Soc.* **201**, 833.
Parker, E.N.: 1992, *Astrophys. J.* **401**, 137.
Pettini, M. and West, K.A.: 1982, *Astrophys. J.* **260**, 561.
Reynolds, R.J.: 1989, *Astrophys. J. Lett.* **339**, L29.
Raymond, J.C.: 1992, *Astrophys. J.* **384**, 502.
Savage, B.D.: 1990, in L. Blitz (ed.), *The Evolution of the Interstellar Medium*, Astronomical Society of the Pacific, San Francisco, p. 33.
Scholer, M.: 1991, *Geophys. Astrophys. Fluid Dynamics* **62**, 51.
Scholer, M., Trattner, K.J., and Kucharek, H.: 1992, *Astrophys. J.* **395**, 675.
Sembach, K.R. and Savage, B.D.: 1992, *Astrophys. J. Suppl. Ser.* **83**, 147.
Shapiro, P.R. and Field, G.B.: 1976, *Astrophys. J.* **205**,
Skibo, J.G. and Ramaty, R.: 1993, *Astron. Astrophys. Suppl.* **97**, 145.
Slavin, J.D. and Cox, D.P.: 1992, *Astrophys. J.* **392**, 131.
Snowden, S.L.: 1993, *Advances Space Res.*, in press.
Snowden, S.L.: Cox, D.P., McCammon, D., and Sanders, W.T.: 1990, *Astrophys. J.* **354**, 211.
Snowden, S.L., Mebold, V., Herbstmeier, U., Hirth, W., and Schmitt, J.H.M.M.: 1991, *Science* **252**, 1529.
Snowden, S.L., McCammon, D., and Verter, F.: 1993, *Astrophys. J. Lett.* **409**, L21.
Williams, D.A. and Hartquist, T.W.: 1992, *Mon. Not. R. astr. Soc.* **251**, 351.
Woodward, P.R.: 1976, *Astrophys. J.* **207**, 484.

A Power Spectrum Description of Galactic Neutral Hydrogen

D. A. Green
Mullard Radio Astronomy Observatory, Cavendish Laboratory, Madingley Road, Cambridge CB3 0HE, United Kingdom

Abstract. Angular power spectra for neutral hydrogen towards $l = 140°$, $b = 0°$ on scales from $\approx 1°$ to $\frac{1}{10}°$ are presented. The spectra are generally well fitted by a power-law dependence on radius in the uv-plane, with an index between ≈ -2.2 and ≈ -3.0. The more distant hydrogen, which corresponds to larger physical scales, tends to have a more negative index than does the nearer material. There is no preferred angular scale apparent at any velocity. For the distant hydrogen there is no obvious difference between the slope of power spectra for emission aligned perpendicular compared with that parallel to the Galactic plane.

1 Introduction

Neutral hydrogen (HI) is an important constituent of the interstellar medium (ISM), and has been widely studied at radio wavelengths through its 21-cm line emission (e.g. Kulkarni & Heiles 1987; Burton 1988; Dickey & Lockman 1990). Within the Galactic plane, structure in HI is seen over a wide range of angular scales, although in general it cannot be readily described in terms of 'clouds'. Rather than describe HI in terms of discrete features, an alternative is to describe the emission in terms of a power spectrum over a range of angular scales. This does not provide a complete description of the emission, but it does give a concise and objective measure of the structures present, and may constrain models of the competing processes of fragmentation and coalescence at work in the ISM.

Although interferometric radio observations measure the angular power spectrum of HI emission directly, there has been as yet only limited work in this area (Crovisier & Dickey 1983). Here I present a power spectrum analysis of observations made towards $l = 140°$, $b = 0°$. I review the results for averages over all position angles, which have been presented in detail in Green (1993), and also present new results for the power spectra of distant HI averaged over different position angle ranges.

2 Observations and Analysis

The observations, which were made with the DRAO Synthesis Telescope, are described in detail in Green (1989). Continuum observations of this field show that it contains no bright SNRs or HII regions which might have HI associated with them on specific angular scales. All interferometer spacings between 3 and 141 in integer multiples of a basic increment equivalent to 20.3λ were observed, giving generally good coverage of the uv-plane. The observations covered 128 channels,

Astrophysics and Space Science **216**: 201–205, 1994.

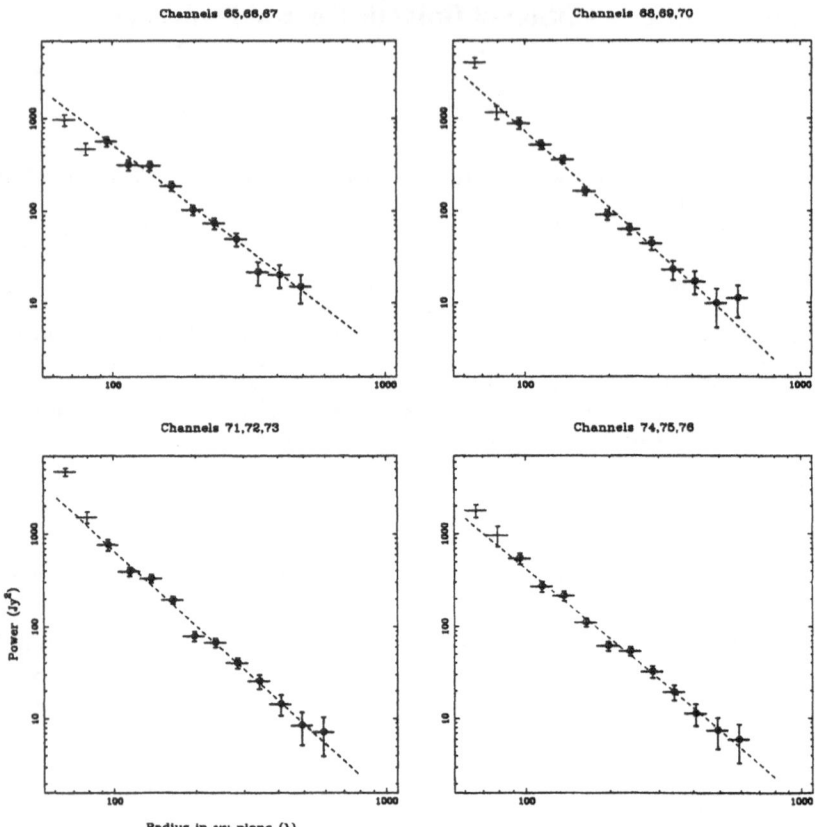

Fig. 1.   Power spectra of neutral hydrogen after removal of noise spectra. The error bars include
the statistical uncertainty in spectra plus an uncertainty in the scaling of the noise power spectrum
for removal. The dashed lines are the least-squares power-laws fitted to the data (excluding the
smallest two annuli, which are not directly comparable with the other data).

each 2.64 km s^{-1} wide spaced every 1.65 km s^{-1}. These were centred at a velocity of −41.2 km s^{-1}, and the velocity coverage ($\approx$ 210 km s^{-1}) included all
significant HI emission in the field (cf. Weaver & Williams 1973). For $l = 140°$
the velocity–distance relationship is, ideally, a single-valued function. For a simple
'flat' rotation curve with Galactocentric radius of 8.5 kpc and a circular velocity of
220 km s^{-1} the Perseus arm is at a distance of $\approx$ 4 kpc, and the extreme velocities
at which HI is seen ($\approx$ −100 km s^{-1}) correspond to a distance of $\approx$ 22 kpc.

The analysis was made in the uv-plane, rather than in the image plane, since
interferometric observations provide a direct measure of the visibility function of
the HI emission. The observations were averaged in logarithmically spaced annuli
in the uv-plane, over all or restricted ranges of position angle, and the appropriate
noise spectra were removed (see Green 1993 for full details).

TABLE I
Power-law least-squares fits to neutral hydrogen power spectra.

chan-nels	central velocity (km s^{-1})	T_B (K)	power-law index	power at 100λ (Jy^2)	scale for 100λ (pc)	comments
(1)	(2)	(3)	(4)	(5)	(6)	(7)
41–43	−3.3	57	−2.69 ± 0.27	184	3	local arm
44–46	−8.2	38	−2.57 ± 0.18	243	7	
47–49	−13.2	42	−2.18 ± 0.11	314	11	
50–52	−18.1	46	(−2.33	415)	16	intermediate arm
53–55	−23.1	37	(−2.48	433)	21	
56–58	−28.0	34	(−2.68	521)	26	
59–61	−33.0	65	(−2.36	537)	33	
62–64	−37.9	90	−2.29 ± 0.13	444	38	Perseus arm
65–67	−42.8	88	−2.28 ± 0.11	520	43	
68–70	−47.8	71	−2.75 ± 0.11	725	51	
71–73	−52.7	45	−2.69 ± 0.10	654	59	
74–76	−57.7	29	−2.52 ± 0.10	419	67	
77–79	−62.6	13	−2.06 ± 0.20	97	76	
83–85	−72.5	16	−3.02 ± 0.23	284	100	outer arm
86–88	−77.5	14	−2.67 ± 0.24	184	114	
89–91	−82.4	14	−2.79 ± 0.19	166	130	
92–94	−87.4	17	−2.78 ± 0.14	379	158	outer arm
95–97	−92.3	11	−2.82 ± 0.22	170	172	

3 Results

The results for power spectra averaged over all position angles are presented in detail in Green (1993), and they are reviewed here briefly. Data from sets of three adjacent frequency channels were averaged together to increase the signal-to-noise ratio. Fig.1 shows some of the the noise-free power spectra, together with simple power-law fits. The data for the smallest two annuli are not directly comparable to those for the other annuli, because of limited sampling of the inner uv-plane.

Table I gives the parameters of power-law fits made to the spectra, provided that at least six annuli, not including the smallest two, have signal detected at above the 2σ level. Column 1 gives the channel numbers and Column 2 the central velocity. Column 3 gives an estimate of the average brightness temperature, T_B, of the total emission in the field of the observations. Columns 4 and 5 give, respectively, the best-fitting power-law slope, m, (and usually its uncertainty), and the power at a radius of 100λ, $P_{100\lambda}$, from the least-squares fitting, for $P(r) = P_{100\lambda}(r/100\lambda)^m$, for power $P(r)$ at radius r in the uv-plane. Values in these columns in parentheses indicate poor power-law fits (see below). Column 6 gives

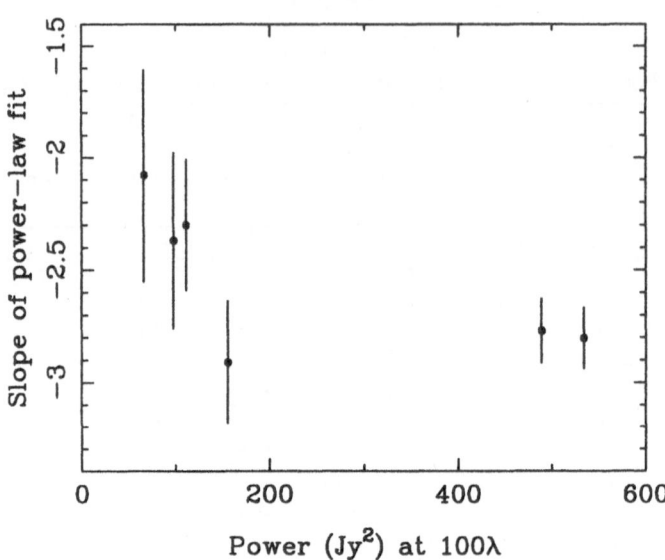

Fig. 2. Slope of best fit power-law to power spectra averaged over limited position angles ranges, against power at 100λ for distant hydrogen.

the physical scale corresponding to 100λ, based on the simple flat rotation curve. Column 7 indicates the positions of the peaks in H I emission seen in this region (cf. the data in Column 4). The power-law fits are for physical scales from slightly less than the values given in Column 6 to several times larger than these values (typically about five times). The least-squares fits to the data were generally good. For the data between ≈ −18 and ≈ −33 km s^{-1} (from the intermediate arm to near the peak of Perseus arm), however, there was poor agreement is at radii less than ≈ 200λ (≈ 17 arcmin, which corresponds to ≈ 15 pc at the expected distance of ≈ 2 to 3 kpc). As discussed by Green (1993), there are H I features on the published images (Green 1989) on these scales at these velocities, which may be associated with AFGL 437. The best-fitting power-law slopes for the distant H I are slightly steeper than those for nearer H I — this suggests that the power-law of emission on large physical scales is steeper than that for emission on smaller physical scales.

Although it is not possible to invert the observed, 2-dimensional power spectra averaged over all position angles to a unique 3-dimensional spatial structure for the H I, the power spectra provide a concise statistical description of the structure of H I emission in the outer Galaxy. There are no clearly preferred angular sizes of H I emission at any velocity. The power-law behaviour is, presumably, due to the fact that turbulent process are at work in the ISM leading to structures on a range of scales (see Scalo 1987).

In addition to the power spectra calculated by averaging over all position angles in the uv-plane, additional power spectra were also calculated for segments in the

uv-plane of restricted ranges of position angle (30° each). For most of the observed HI there were no obvious variations in the power spectra for segments at the same velocity. This is consistent with the images of HI presented in Green (1989), which do not show any preferred alignment of structures over most velocities. However, for the distant HI, the field of view of the observations is physically large, so large-scale structure aligned with the plane of the Galaxy becomes apparent. For these velocities of ≈ -71 to -92 km s^{-1} data for 14 adjacent channels were averaged together in six different segments in the uv-plane. For each segment a simple power-law fit was made. These power spectra show a large range of intensity, differing by a factor of about eight at 100λ for emission aligned perpendicular compared with that parallel to the Galactic plane. For the fainter emission — that aligned perpendicular to the Galactic plane — only a few radial bins contained significant signal, so that the slope in the derived power-law fits for these segments is relatively poorly determined. Fig.2 shows the best-fit power-law slopes plotted against the power at 100λ for these six segments. The errors in the slope do not include any systematic uncertainties due to the limited coverage of the uv-plane (see Green 1993). Fig.2 shows a tendency for the brighter emission — that aligned with the Galactic plane — to have steeper power-law fits than the other segments. However, given the uncertainties in the slopes, it is not clear that this trend is significant. The present results provide no strong evidence for differences in the power-law slopes for emission parallel compared with perpendicular to the Galactic plane.

Acknowledgements

The DRAO Synthesis Telescope is operated by the NRC of Canada as a national facility. I thank colleagues at DRAO, MRAO and elsewhere for many useful discussions, and the SERC for an Advanced Fellowship.

References

Burton, W.B.: 1988, in *Galactic and Extragalactic Radio Astronomy*, Verschuur, G.L. & Kellermann, K.I., Springer-Verlag, Berlin, p295.
Crovisier, J. & Dickey M.: 1983, *Astron. Astrophys.*, **122**, 282.
Dickey, J.M. & Lockman F.J.: 1990, *Ann. Rev. Astron. Astrophys.*, **28**, 215.
Green, D.A.: 1989, *Astron. Astrophys. Suppl.*, **78**, 277.
Green, D.A.: 1993, *Mon. Not. R. Astro. Soc.*, in press.
Kulkarni, S.R. & Heiles, C.: 1987, in *Interstellar Processes*, Hollenbach, D.J. & Thronson, H.A., Jr., Reidel, Dordrecht, p87.
Scalo, J.M.: 1987, in *Interstellar Processes*, Hollenbach, D.J. & Thronson, H.A., Jr., Reidel, Dordrecht, p349.
Weaver, H. & Williams, D.R.W.: 1973, *Astron. Astrophys. Suppl.*, **8**, 1.

A Statistical Description of Astrophysical Turbulence

A. Lazarian
DAMTP, University of Cambridge, UK.

Abstract. The properties of the ISM indicate that it is turbulent. However, the ISM turbulence is radically different from that in incompressible fluids. That is why it is so important to study it through observations. The relevant study still poses a challenging problem. In the present paper recent results based on a statistical approach to the problem are surveyed. Although it was pointed out long ago (see Kaplan *et al.*, 1970) that random 3D motions of the ISM gas result in fluctuations of the observed electromagnetic emission, it is only recently that the problem of recovering statistical properties of the ISM turbulence from the line integrated data was given an adequate mathematical treatment. Here by the example of studying turbulence in HI, it is shown that the inverse problem can be solved uniquely using a realistic model of the ISM. The application of theoretical conclusions to existing data explains some facts which used to be considered inconsistent with turbulent behaviour and reveals unexpected features of the ISM turbulence.

1   The Problem

It is generally believed that turbulence is one of the main features of the ISM (Scalo, 1987). But turbulence in astrophysics is much more complex than that in incompressible fluids, and thus, the Kolmogorov picture is likely to be inadequate (Lazarian, 1992a). Indeed, in the ISM, random motions take place in a compressible, self-gravitating magnetic medium, to mention just a few peculiarities of the large-scale astrophysical turbulence. Thus, the only rational way to deal with the problem is to use observational material.

For the statistical analysis discussed in the paper, it is important that turbulence is never completely chaotic. On the contrary, turbulent motions are correlated and can be characterised, for example, by structure functions:

$$d(\mathbf{r}, \mathbf{R}) = \langle (n(\mathbf{x}_1) - n(\mathbf{x}_2))^2 \rangle \qquad (1)$$

where $d(\mathbf{r}, \mathbf{R})$ is a structure function of density $n(\mathbf{x})$, which depends on two vectors $\mathbf{r} = \mathbf{x}_1 - \mathbf{x}_2$ and $\mathbf{R} = \frac{1}{2}(\mathbf{x}_1 + \mathbf{x}_2)$.[1] A similar description is applicable to other random fields, both scalar fields, like temperature and pressure, and vector fields, like velocity and magnetic. However, unlike in the case of the laboratory turbulence, we are unable to measure pointwise characteristics $n(\mathbf{x})$, and only quantities proportional to line integrals, e.g. intensity $I(\mathbf{e}) = æ \int n(\mathbf{x}) dl$, are available. 3D turbulence causes fluctuations of $n(\mathbf{x})$ and therefore those of $I(\mathbf{e})$. This means that one can apply a statistical description to $I(\mathbf{e})$ and use, for example, structure

---

[1]   Note that here and elsewhere in the paper we use angular brackets $\langle \ldots \rangle$ to denote ensemble averaging.

Astrophysics and Space Science **216**: 207–212, 1994.
© 1994 *Kluwer Academic Publishers.*

functions of intensity

$$D(\theta, \varphi) = \langle (I(\mathbf{e}_1) - I(\mathbf{e}_2))^2 \rangle \tag{2}$$

where θ is the angle between unit vectors $\mathbf{e}_1$ and $\mathbf{e}_2$, and φ is the positional angle over the sky.

In what follows, recent results concerning the experimental study of turbulence are briefly discussed. A direct use of statistics available through observations (e.g. $D(\theta, \varphi)$) is addressed in section 2, while the inversion in order to recover statistical properties of turbulence (e.g. $d(\mathbf{r}, \mathbf{R})$) is discussed in section 3.

2   Information Directly Available

It would not be an exaggeration to state that a major effort of researchers in the field has been concentrated on the direct use of structure functions of observable parameters (see Eq. (2)) in order to obtain information about the ISM turbulence (see Dickman, 1985; Scalo, 1987). For example, one can estimate a characteristic size of the 3D turbulence (i.e., the size over which the turbulent motions are correlated) from 2D structure functions of intensity. However, the actual geometry of the lines of sight becomes crucial when such estimates are performed (Lazarian, 1992a). Our analysis has shown that not accounting for the geometry of crossing lines of sight can, on the one hand, result in the overestimation of the actual characteristic size of turbulence. On the other hand, this geometry leads to peculiar behaviour of structure functions (e.g. their growth at all θ from 0 to π). The latter behaviour, when observed (e.g. Kaplan et al., 1970), usually causes embarrassment. But this is not the only puzzle that structure functions can present to the observer. Apart from the geometry of lines of sight, one has to take into account the additive properties of the measurable parameters. Indeed, one can use not only intensities, which are always positive, but, for example, structure functions of the degree of interstellar polarization (Kaplan et al., 1970). Such quantities averaged along the crossing lines of sight can result in "inexplicable" decrease of the appropriate structure functions (Lazarian 1992a).

2.1   STRUCTURE AND CORRELATION FUNCTIONS

Apart from structure functions similar to the ones given by Eq. (2), one can use correlation functions:

$$C(\theta, \varphi) = \langle (I(\mathbf{e}_1) - \langle I \rangle)(I(\mathbf{e}_2) - \langle I \rangle) \rangle \tag{3}$$

The latter characteristics were advocated in Dickman (1985) despite their sensitivity to linear gradients. We feel that one should be very cautious about using correlation functions as, for example, averaging over a strip of the sky results in the correlation function $C_\Omega(\theta)$ which is related to the ensemble averaged quantity $C(\theta)$ in the following way

$$C_\Omega(\theta) = C(\theta) - \frac{1}{|\Omega|} \int_\Omega C(\theta) \mathrm{d}\theta \tag{4}$$

and if the size of the strip Ω is not much greater than the characteristic angular scale of $C(\theta)$, the error may be considerable. In other words, if one wants to use the correlation function and compare observational results with the theoretical predictions, one has to solve an inverse problem to find $C(\theta)$ from Eq. (4) first. This problem will be discussed in more detail in other papers. In the present paper, we will use only structure functions and reformulate all the results in terms of these functions.

2.2   ANISOTROPIES OF STRUCTURE FUNCTIONS

It is possible to show that anisotropies of turbulence should result in the φ dependence of D. However, there are cases when isotropic 3D turbulence produces anisotropic 2D structure functions of intensity. This happens, for instance, for structure functions of synchrotron intensity of the Galactic halo (Chibisov *et al.*, 1980; Lazarian *et al.*, 1990)

$$D(\theta, \varphi) = D_0(\theta) + D_2(\theta) \cos 2\varphi \tag{5}$$

where the anisotropic part $D_2(\theta)$ emerges due to the existence of the regular magnetic field. This is why the statistical analysis brings the information not only about the random magnetic field but also about the value and the direction of the regular magnetic field (Dagkesamanskii *et al.*, 1987, Lazarian *et al.*, 1990).

2.3   SIZE OF THE EMITTING VOLUME

In addition to their direct application, structure functions can be used to study geometrical properties of emitting astrophysical objects. For example, it was shown in Chibisov *et al.*, (1991) that by using HII regions as distance indicators, it is possible to estimate the size of the Galactic synchrotron halo. The idea of such measurements is very simple and is based on the scaling relations for structure functions:

$$D_i(\theta, \varphi) = æ_i \frac{R}{L} \bar{D}_i \left(\frac{R}{L} \theta \right) \tag{6}$$

where indices $i = 0$, 2 are used to denote isotropic and anisotropic parts of the structure function (see Eq. (5)), respectively, $æ_i$ is a coefficient, R is the size of the halo, and L is the distance to the HII region. Note, that the scaling relation (6) enables one to perform averaging using a number of different HII regions. Similar measurements are applicable not only to the synchrotron halo but also to the HI disc etc.

2.4   TOMOGRAPHY OF THE ISM TURBULENCE

The easiest possibility for ISM turbulence tomography may be performed using HII regions as the opaque screens in the decameter region of wavelengths. From elementary geometrical considerations it is clear that effective contributions to the structure function (see Eq. (2)), where one of the vectors **e** is directed towards the

HII region, come due to emission from points lying on the second line of sight out to distances $\sim L + \lambda$, where λ is the characteristic scale of the turbulence. For $\lambda \ll L$, one can solve the inverse problem for the MHD turbulence using different HII regions. Then, in general, different characteristics of the turbulent spectrum will be obtained, and these differences will correspond to actual variations of the parameters of turbulence in different places in our Galaxy. Here, HII regions and MHD turbulence are mentioned just to exemplify the procedure, which is applicable in a number of other cases, when we are able to resolve an object which screens the emission behind it.

2.5 OTHER STATISTICAL CHARACTERISTICS

To characterise the distribution of diffuse matter not only the structure function D can be used. Crovisier *et al.*, (1983) and Green (1993) applied the averaged signal of the radiointerferometer with the diagram $\frac{1}{\alpha}$

$$\langle \bar{S}(t) \rangle = \int\int \langle I(x,y)I(x',y') \rangle \cos(2\pi t(x - x'))e^{-\frac{x^2+x'^2+y^2+y'^2}{\alpha^2}} dx\,dx'\,dy\,dy' \qquad (7)$$

for studying HI. We expect that this characteristic of turbulence will soon become very popular for two reasons. First, interferometers provide relatively high resolution, and second, it will be shown below that a simple relation exists between the measured 3D spectrum and the interferometric signal.

3 The Inverse Problem

It was realised rather long ago (Kaplan *et al.*, 1970) that, given a statistical description of the transparent astrophysical medium, it is possible to predict statistical properties of the observable diffuse emission. However, a more important problem is to be able to deduce the properties of the ISM turbulence through observations. To do this one should solve an inverse problem in the framework of an appropriate physical model (Gough, 1984). The turbulence, isotropic on its characteristic scale, seems to be a good model for the purpose as a major part of energy in the ISM comes from the smallest scales (due to supernova explosions and stellar winds) rather than from the largest scales, where motions are influenced by the mean flow. Still, when dealing with observational data, one is advised to put the assumption of isotropy under test. The simplest one includes a study of D as a function of φ (Lazarian, 1992a). The results of such a study can be used to evaluate the ambiguity of the inversion (Lazarian, in preparation).

Consider, as an example, a structure function of HI intensity (Kaplan *et al.*, 1970):

$$D(\theta) = \mathit{æ}^2 \int\int_0^L \{d\left(\sqrt{(x_1 - x_2)^2 + R^2\theta^2}\right) - d(|x_1 - x_2|)\}dx_1 dx_2 \qquad (8)$$

where R is the distance towards the particular slice of HI, being under study. The transformations similar to those presented in Lazarian (1991a, 1992b) result in

$$d(r) - d(L) = -\frac{1}{\pi æ^2 LR} \int_r^L \frac{d\theta}{\sqrt{\theta^2 - \frac{r^2}{R^2}}} \frac{d}{d\theta} D(\theta) \tag{9}$$

where L is the thickness of the HI slice, which can be found using the rotation curve. Eq. (9) gives a unique opportunity to study 3D turbulence using the statistics of intensity fluctuations available through observations (e.g. using $D(\theta)$).

The structure function $d(r)$ is related to the turbulence spectrum $E(k)$ in the following way (Monin et al., 1975):

$$d(r) = 2 \int_0^\infty \left(1 - \frac{\sin kr}{kr}\right) E(k) dk \tag{10}$$

A strikingly simple relation exists between the signal of the interferometer (see Eq. (7)) and the 3D spectrum $E(k)$. Namely, (Lazarian, 1992b; Lazarian, unpublished paper):

$$\left\langle \bar{S}_i \left(\frac{u}{2\pi}\right) \right\rangle \approx æ_1 \frac{E(u)}{u^2} \tag{11}$$

where $æ_1 = 16\pi^2 \alpha^2 æ^2$. The latter relation makes the interferometer a unique tool for studying 3D turbulence.

In spite of the fact that the relations are more complex in the case of studying a random magnetic (Lazarian et al., 1990) or velocity (Lazarian, in preparation) fields, one can still obtain the corresponding 3D statistical measures.

3.1   FIRST RESULTS OF THE INVERSION

It happened that the first object for the inversion was the MHD turbulence in the Galactic halo. The corresponding analysis enabled us to obtain statistics of the random magnetic field (Lazarian et al., 1990), though ambiguities were large, partially due to poor quality of the initial data. The application of Eq. (11) to the data of Green (1993) reveals a shallow 3D spectrum of turbulence with $k^{-\alpha}$ and $\alpha < 1$. More details will be supplied elsewhere (Lazarian, in preparation).

4   Discussion

The observational study of astrophysical turbulence is a subject in its infancy. Despite the fact that an infinite number of tensors is required for a complete description of the ISM turbulence, simple structure functions of intensity and density can be taken as a starting point in this study. In the present paper, we have been dealing only with intensities of diffuse radiation, though we believe that in future more complex characteristics, involving line profiles and polarisation properties of the radiation as well as information from different emission lines, will be used in order to get an adequate description of the ISM turbulence (see Lazarian, 1992). The increase of resolving power of ground and space-based instruments will stimulate the development of this branch of science. But even now much valuable information can be obtained.

Acknowledgements

The author is grateful to B. Elmegreen and N. Weiss for their encouragement. Discussions with D. Green, D. Gough, E. Falgarone and J.-L. Puget are acknowledged. The research is supported by an Isaac Newton Scholarship.

References

Chibisov, G. V. and Lazarian, A.: 1991, *Sov. Astron. Lett.* **17**(3), 208.
Chibisov, G. V. and Ptuskin, V. S.: 1980, *Proc. 17th Int. Cosmic Ray Conf.*, Paris, vol. 2, p. 233.
Crovisier, J. and Dickey, M.: 1983, *Astron. Astrophys.* **122**, 282.
Dagkesamanskij, R. D. and Shutenkov, V. R.: 1987, Sov. Astron. Lett. **13**.
Dickman, R. L.: 1985, *Turbulence in Molecular Clouds*, in: Black D.C., Mathews M.S. (eds) *Protostars and Planets II*, Tucson: University of Arizona, p. 150.
Green D.A.: 1993, *Mon. Not. Roy. Astr. Soc.* **262**, 327.
Gough, D. O.: 1984, *Phil. Trans. Roy. Soc. London*, Ser. A., **313**, 27.
Kaplan, S. A. and Pickelner, S. B.: 1970, *The Interstellar Medium*, Harvard University Press.
Lazarian, A. and Shutenkov, V. R.: 1990, *Sov. Astron. Lett*, **16**(4), 297.
Lazarian, A.: 1991, in E. Falgarone, F. Boulanger, G. Duvert (eds) *Fragmentation of Molecular Clouds and Star Formation*, Kluwer, Dordrecht, p. 65.
Lazarian, A.: 1992a, *Astron. and Astrophys. Transactions*, **3**, p. 33.
Lazarian, A.: 1992b, in *Proc. of the ESA Colloquium Targets for Space-Based Interferometry*, October 13-16, 1992, Côte d'Azur, France, p. 177.
Lazarian, A.: 1993, *A Statistical Method for Astrophysical Turbulence Investigation*, in F. Krause, K. H. Rädler (eds) Cosmic Dynamo, Kluwer, Dordrecht.
Scalo, J. M.: 1987, *Theoretical Approaches to Interstellar Turbulence*, in D. F. Hollenbach and H. A. Thronson (eds.), *Interstellar Processes*, Reidel, Dordrecht, p. 349.

Rosat Wide Field Camera Data And The Temperature Of The Interstellar Medium

J. J. Quenby, T. J Sumner, S. D. Sidher and S. Immler
Astrophysics Group, Blackett Laboratory,
Imperial College, London SW7 2BZ, UK

February 16, 1994

Abstract. The shadowing effects of the molecular clouds in the nearby interstellar medium on the soft x-ray background has been investigated, using ROSAT WFC data in conjunction with previous rocket B and C band surveys. Shadowing over a 5° extent occurs only for a few percent of the sky, but the mixed model of the ISM is supported.

Detailed modelling of the 'Draco' shadowing region shows little evidence for a multi-temperature, hot ISM component.

1 Introduction

Three models of the soft x-ray background in relation to the properties of the interstellar medium have been extensively discussed.

These are:

(a) Extra-galactic absorption. It is possible to consider a halo of hot gas around the galaxy at about $3 \ 10^6$ K which is seen through the intervening absorbing column. However the predicted SXRB-N_H correlation is too steep and a local, hot component is required to produce the observed, flat equatorial region of this correlation.

(b) The Snowden *et al* (1990) cavity model assumes all the SXRB arises in a local cavity with walls at varying distances from the sun, so that the actual x-ray intensity is proportional to this distance. The cavity boundary map produced as a result of this model does not correspond in detail to the Frisch and York (1983) contours of N_H, based upon ultraviolet stellar data, however.

(c) The 'mixed' model of Jakobsen and Kahn (1986) places a random distribution of the clouds within the hot emitting region. These authors find that the SXRB depends only on the ratio, R, of the absorbing cloud scale height to that of the x-ray emitting regions and also η, the 'clumping' factor in the x-ray absorption cross-section. Again there is inconsistency between different wavelength observations with the SXRB results suggesting $\eta \sim 0.5$ and $R \sim 0.1$ while radio measurements of N_H angular fluctuations imply $\eta \sim 1.0$ and $R \sim 1.0$.

If the x-ray shadow of a known HI cloud can be found, such that the SXRB can be described by:-

$$I_x = I_o + I_1 e_{exp}\left(-\sigma n_H\right) \tag{1}$$

where I_o is due to the foreground and I_1 is due to emission behind the cloud of column N_H, absorption cross-section σ, then the above models can be distinguished. The cloud may separate two hot components. Kahn (1991) distinguished a super-

Astrophysics and Space Science **216**: 213–217, 1994.
© 1994 *Kluwer Academic Publishers.*

nova generated ISM with an energy content of 10^{14} erg g^{-1} from an early stellar wind generated ISM with an energy content of 10^{15} erg g^{-1}. Magnetic field 'walls' could also separate the two hot media.

2   Shadowing Experiments

The primary data used in this work is the ROSAT WFC S1 filter survey (Lieu et al, 1992) in the 90-188 eV energy range encompassing 20% of the sky and the Wisconsin rocket survey (McCammon et al, 1983) B band (130-188 eV) and C band (160-284 eV) full sky surveys. Note the Wisconsin work in C band is verified by SAS 3 observations to within 10%. The Landini-Fossi (1990) hot plasma emission code is employed.

Estimates of the properties of the local ISM include those of Lieu et al (1992) who find $\log T = 5.8$ and a low galactic latitude emission measure of 0.0086 cm^{-6} pc, implying $p/k = 8.9 \, 10^3$ cm^{-3} K, for an average cavity size (Snowden et al, 1990) of 100pc, where p is pressure. Warwick et al (1992), from EUV source counts, find a local gas density of 0.05 atoms cm^{-3} within a local bubble $\sim$ 80 pc radius. Cox and Snowden (1986) suggest the local $< n_e^2 > = 2 \, 10^{-5}$ cm 10^{-6} while combining the Lieu et al (1992) results with those of Warwick et al (1992) yields $< n_e^2 > = 1.1 \, 10^{-4}$ cm^{-6}.

In an x-ray based search for regions of shadowing or anomalous ISM temperature, Sumner et al (1993) identify regions where the C/B band Wisconsin counts ratio is $\geq$ 3.7 : 1 rather than the mean 2.4 : 1 ratio. While a significant high C : B ratio is found mainly near the south galactic pole, other isolated clouds are seen especially corresponding to the DRACO shadow (Snowden et al, 1991). Shadowing searches were also made in the direction of 57 'MBM' CO molecular clouds at $\mid b \mid >$ 25° at 50–100 pc distance, 4 IRAS high cirrus dust clouds, Chameleon-Musca and a Be–B band cloud at $(l,b) = (132°, -69°)$. With a 5° resolution available, only MBM17, MBM41 (DRACO), Chameleon-Musca and an object at $(l,b) = (309°, -15°)$ were detected as shadows. Hence on this angular scale, the 'mixed' model receives limited support.

3   Slab Model for Draco

Shadowing in the Draco region has been investigated with a 3 or 5 slab model of alternate hot and cold ISM components, arranged so that the nearest and most distant slabs both contain hot ISM. Within each slab, we use the transport equation:

$$I(\lambda) = I_0(\lambda)exp\left(-n_H\sigma(\lambda)l\right) + \frac{n_e^2\rho(\lambda,T)}{n_H\sigma(\lambda)}\left[1 - exp(-n_H\sigma(\lambda)l)\right] \qquad (2)$$

where $\rho(\lambda,T)$ = emission coefficient,
$\sigma(\lambda)$ = absorption coefficient,

TABLE I
Draco Observations

Detector	On Cloud	Off Cloud
C cps	238	262
B cps	57	90
S1 cps	1.7	1.8

TABLE II
Draco Simulations

Slab Number	$\log T$	EM in $10^{16} cm^{-5}$	N_H in 10^{18} cm^{-2}
1	6.0	3.8	10
2	<4.0	0.0	20
3	6.0	1.8	30

$I_0(\lambda)$ = Intensity on entry into slab.

The ROSAT Draco shadow is deepest over a 1° field of view, although we detect it with 5° averaging. It is either due to the cloud at 300 pc–1500 pc distance, or to a 'finger filament' at 60 ± 20 pc.

Table 1 lists the WFC, B and C count rates per second, for the on-cloud and off-cloud SXRB in the Draco direction.

The temperatures emission measures and absorbing columns for the best fit, 3 slab model are given in Table 2, where increasing slab number corresponds to increasing nearness to the sun.

With the slab parameters of Table 2, the predicted count rates of B, C and S1 are given in Table 3

Using our fit for slab 3 column density and the Warwick *et al* (1992) local density of 0.05 atoms cm^{-3} we obtain a DRACO distance of 194 pc but this can reduce to 81 pc if we take into account the Warwick *et al* (1992) error bars which allow a local density of 0.12 cm^{-3}. However, the Lieu *et al* (1992) derived value of $< n_e^2 > \sim 1.1\ 10^{-4}$ cm^{-6} taken with the emission measure for slab 3 gives a DRACO distance of 53 pc. Finally the slab 1 emission measure suggests a cavity

TABLE III
Draco Simulations

Detector cps	On Cloud	Off Cloud
C	241	266
B	56	67
S1	2.6	2.5

TABLE IV
Draco 5 Slab Simulation

Slab	$\log T$	em (10^{16} cm^{-5})	N_H (10^{18} cm^{-2})
1	6.0	2.5	0
2	<4.0	0.0	20
3	6.5	2.5	10
4	<4.0	0.0	20
5	6.0	1.8	30

Detector cps	On -Cloud
C	244
B	86
S1	1.91

TABLE V
Draco 5 Slab Simulations

Slab	$\log T$	em (10^{16} cm^{-5})	N_H (10^{18} cm^{-2})
1	6.0	1.0	0
2	<4.0	0.0	25
3	6.5	8.0	10
4	<4.0	0.0	25
5	6.0	0.6	0.4

Detector cps	On-Cloud
C	214
B	62
S1	1.9

boundary at 165 pc. Hence the x-ray data can be reconciled with the position of the 'finger filament' as source of the shadowing and a 'mixed' ISM model.

Less success was obtained with 5 slab simulations of the DRACO, on-cloud counting rates where an attempt was made to find a multi-temperature hot ISM. Our best fits to date are given in Tables 4 and 5.

With a 5 slab, multi-temperature model, either the B or C rates tend to be too high for a given S1 rate.

4 Conclusions

Most known 'MBM' and IRAS molecular clouds do not produce noticeable x-ray shadowing, so either they are near the edge of the local, hot bubble or are significantly less than 5° in angular extend or they contribute an absorption significantly less than that of a column $N_H \sim 10^{18}$ cm^{-2}.

Shadowing occurs over a few percent of the sky within our local hot bubble, believed to be ~ 100 pc in radius. Hence the 'mixed' model receives limited support.

The DRACO shadow is best fitted by a single, absorption model involving a single, hot ISM component. X-ray evidence alone cannot locate the position of the absorber better than to within 50–200 pc distance.

References

Cox, D. P. and Snowden, S. L.: 1986,*Adv. Sp. Res.* **6**, No.2, 97.

Frisch, P. C. and York, D. G.: 1983,*Astrophys. J.* **271**, L59.

Jacobson, P. and Kahn, S. M.: 1986,*Astrophys. J.* **309**, 682.

Landini, M. and Fossi, Monsignori B. C.: 1990, *Astron. Astrophys. Suppl.* **82**, 229.

Lieu, R., Quenby, J. J., Sidher, S. D., Sumner, T. J., Willingale, R., West, R. G., Harris, A. W., Snowden, S. L. and Bickert, K.: 1992, *Astrophys. J.* **397**, 158.

McCammon, D., Burrows, D. N., Saunders, W. T. and Kraushaar, W. C.: 1983, *Astrophys. J.* **269**, 107.

Snowden, S. L., Cox, D. P., McCammon, D. and Saunders, W. T.: 1990, *Astrophys. J.* **354**, 211.

Snowden, S. L., Mebold, A., Hirth, W., Herbstmeier, J. and Schmitt, J. H. M. M.: 1991, *Science* **252**, 1529.

Sumner, T. J., Sidher, S. D., Immler, S. and Quenby, J. J.: 1993, *Proc. Calgary 23rd Int. Cos. Ray Conf.*

Warwick, R. S., Barber, C. R., Hodgkin, S. T. and Pye, J. P.: 1993, *Mon. Not. Roy. Astr. Soc.* **262**, 289.

Hierarchial Galactic Dynamo and Seed Magnetic Field Problem

A. Lazarian
DAMTP, University of Cambridge, UK

Abstract. A new approach to the galactic seed magnetic field problem is briefly discussed. It is shown that, in early stages of galactic evolution, the hierarchial agglomeration and fragmentation processes can account for the generation of a dynamically important magnetic field. The amplification of this field follows an inverse cascade since a non-zero average value of the field amplified on a smaller scale serves as a seed field on the next (earlier) hierarchial scale. In such a scenario, a problem of how to get things started never occurs as any infinitesimally small battery generated seed field (Lazarian 1992a) can be efficiently amplified passing by through a sufficient number of amplification cascades.

1 Seed Field

According to Rees (1987), the galactic seed magnetic field problem has two aspects. For one thing, a finite magnetic field $B_0 > 10^{-21}$ G should exist for any galactic dynamo to feed on. For another thing, in early galaxies, physical processes can be different due to the absence of dynamically important magnetic field. Both of the aspects are addressed in the paper. Unlike our previous attempt to explain the seed magnetic field generation by a large-scale battery process (Lazarian, 1992a), here, we sketch the alternative approach: we consider the hierarchial dynamo amplification of magnetic field. Since a magnetic field generated by battery processes can more readily emerge on small scales, we discuss a possibility of the inverse cascade for the magnetic energy. Indeed, not only the battery generation of magnetic field, but also a dynamo can be shown to be more efficient at small scales (see Kulsrud *et al.*, 1992). Therefore, if there is a hierarchy of motions in the discs of early galaxies, one should expect to observe magnetic energy coming to equipartition with kinetic energy at the smallest scales first. Let the scale with the equipartition reached be l (the field intensity is B_l, the size of loops is l). At a scale $L \gg l$, magnetic loops are distributed randomly. However, their volume averaged value is proportional to $B_l \left(\frac{l}{L}\right)^{3/2}$, where the power index $\frac{3}{2}$ appears instead of the expected $\frac{1}{2}$ due to the fact that magnetic loops are approximated by a set of the δ-function derivatives (Ruzmaikin *et al.*, 1988). This non-zero averaged value of magnetic field will serve as a seed magnetic field for the scale L. Such amplification proceeds up to the largest scale of hierarchical motions. To exemplify the qualitative arguments presented above we consider a Kolmogorov turbulence[1] with $E(l) \sim l^{\frac{5}{3}}$. For the scale l, the velocity $v(l) \sim l^{\frac{1}{3}}$ and the characteristic time

---

[1] This type of turbulence is unlikely to be present in interstellar conditions (Lazarian, 1992b), and is used here merely to exemplify the process discussed.

Astrophysics and Space Science **216**: 219–221, 1994.
© 1994 *Kluwer Academic Publishers.*

of the magnetic field growth is of the order of $\frac{l}{v(l)} \sim l^{\frac{2}{3}}$. Then the non-zero field at the scale L, which emerges due to the magnetic field generation at the scale l, is $\sim B_0 \left(\frac{l}{L}\right)^{\frac{3}{2}} \cdot \frac{\exp(\alpha l^{-\frac{2}{3}} t)}{1 + \beta l^{-\frac{2}{3}} \exp(\alpha l^{-\frac{2}{3}} t)}$, where α is a coefficient inversely proportional to the fluid velocity $\beta \sim \frac{\alpha}{\sqrt{2\pi}} \varrho^{-\frac{1}{2}} B_0$, ϱ is the fluid density, and B_0 is the initial seed field on the scale l, which may have a statistical nature. Therefore, the scales responsible for the formation of the non-zero seed field at the scale L depend not only on the value and scale of the battery generated magnetic field (Lazarian, 1992a), but also on the time t of initiation of the dynamo process. More details will be supplied elsewhere.

2   Simple Model

Our previous calculations were done for a very simple model where the magnetic field was generated due to a dynamo in molecular clouds (Lazarian, 1993a,b). Assuming the average size of a molecular cloud to be $\sim$ 10 pc, the following estimate for the seed large-scale magnetic field can be obtained: $\sim 10^{-12}$ G. This model is essentially a two scale model, where there exists a considerable separation between the small scale of the initial magnetic field and the scale at which the mean field dynamo operates. Taking into account the existence of large-scale instabilities like the gravitational or the Parker instabilities, one may increase the estimate for the characteristic field at least by one order of magnitude (i.e., to $\sim 10^{-11}$ G). The latter estimate for the seed magnetic field can be easily fitted in the framework of modern mean field dynamo theories (see Ruzmaikin et al., 1988). The argument that the magnetic field reaches equipartition with kinetic energy first at small scales were put forward by Kulsrud et al. (1992) in attempt to show that the turbulence at the smallest scales will be quelled, and consequently galactic dynamo impossible. We do not share such an extreme point of view since we believe that turbulence on the smallest scales should survive, though in a modified form. The reconnection of the magnetic field is likely to be an additional source of small-scale turbulence, activating turbulent diffusion.

3   Summary

To summarise, it may be possible to solve the galactic seed field problem without much reference to a particular battery process responsible for the generation of the initial field by including the hierarchial motions into consideration. Another important conclusion is that a dynamically important magnetic field is likely to appear at the earliest stages of the galactic evolution and influence the formation of the first generation of stars.

Acknowledgements

The author is grateful to M. Rees for formulating the problem and for helpful discussions. The research is supported by an Isaac Newton Scholarship.

References

Kulsrud, R. M. and Anderson, S. W.: 1992, *Astrophys. J.* **396**, 606.
Lazarian, A. 1992a, *Astron. Astrophys.* **264**, 32.
Lazarian, A.: 1992b, *Astron. and Astrophys. Transactions*, **3**, p. 33.
Lazarian, A.: 1993a, *Generation of the Seed Magnetic Field*, in F. Krause, K.H.Rädler and G. Rüdiger (eds.), *Cosmic Dynamo*, Kluwer, Dordrecht.
Lazarian, A. 1993b, *Magnetic Field Generation within Molecular Clouds* in F. Krause, K.H.Rädler and G. Rüdiger (eds.), *Cosmic Dynamo*, Kluwer, Dordrecht.
Rees, M. J.: 1987, *Quart. Journ. Roy. Astr. Soc.* **28**, 197.
Ruzmaikin, A. A., Sokoloff, D. D. and Shukurov, A. M.: 1988, *Magnetic Fields of Galaxies*, Kluwer Acad. Publ., Dordrecht.

Cosmic Ray Diffusion at Energies of 1 MeV to 10^5 GeV

T. W. Hartquist and G. E. Morfill
Max-Planck-Institut für extraterrestrische Physik
D-85740 Garching, Germany

Abstract. A supernova remnant accelerates cosmic rays to energies somewhat above 10^5 GeV by the time that the free expansion phase of its evolution has come to an end. As the remnant's outer shock slows, these highest energy cosmic rays diffuse away from the shock along a magnetic flux tube with a radius comparable to that of the remnant at the end of its free expansion phase and which eventually (over a distance of the order of a kiloparsec) bends into the Galactic halo. A similarity solution exists for the temporal and spatial variations, in such a tube, of both the number density for these $\sim 10^5$ GeV cosmic rays and the energy density of the waves on which they resonantly scatter. Wave-wave interactions probably do not dominate the evolution of the energy density of these lowest frequency waves, but we assume that they do establish a Kraichnan wave spectrum at higher wavenumber. Although we cannot rigorously justify this assumption, it does receive some support from the analysis of pulsar signals. There is a large body of observations to which such a model can be applied, yielding constraints that must be met. With the model that we develop here we obtain the following results:

1. The local intensity of $\sim 10^5$ GeV cosmic rays implies that the flux tube which currently surrounds the Solar System last contained a remnant in the free expansion phase several times 10^7 years ago. We comment on the rough agreement between this age and that inferred from Be^{10} data.

2. The theoretical value of the cosmic ray diffusion coefficient at ~ 1 GeV in the tube corresponding to that time is in harmony with the value of the diffusion coefficient inferred from cosmic ray composition and synchrotron measurements.

In the light of our inhomogeneous cosmic ray acceleration/propagation model we re-examine our earlier work on the evidence for second order acceleration in a very old remnant. Such evidence is provided by the molecular compositions along several lines of sight to the Perseus OB2 association. We find as a third significant result that the model value of the diffusion coefficient at energies in the range of 1 MeV agrees within about an order of magnitude with that which we infer from the molecular data.

1 Introduction

As stressed by Hartquist (1994) in this volume, the value of the cosmic ray diffusion coefficient is important in governing the effects that cosmic rays have on supernova remnant evolution and the global structure of the interstellar medium. This article should be considered to be a companion to that review.

The interpretation of Galactic radio synchrotron emission data and of the elemental and isotropic cosmic ray composition suggests that the Galactic cosmic ray diffusion coefficient is roughly 10^{29} cm^2 s^{-1} for GeV particles within 10 kpc of the disk (e.g. Morfill, Meyer and Lüst, 1985; Dogiel, 1991). A variety of self-consistent steady state cosmic ray propagation models, in which waves are generated by cosmic ray streaming which is in turn limited by scattering on the waves, have been developed (e.g. Kulsrud and Pearce, 1969; Skilling, 1975c; Lerche and Schlickeiser, 1982). In principle, these approaches yield values of the cosmic ray diffusion coefficient, provided that the cosmic ray anisotropy and the wave amplitudes are

Astrophysics and Space Science **216**: 223–234, 1994.
© 1994 *Kluwer Academic Publishers.*

small enough that the quasi-linear theory applies and that the wave dissipation mechanism is known.

The temporal fluctuations in the physical conditions at any point in the supernova moderated interstellar medium are not considered in such steady state models. An extreme, though reasonable, cosmic ray propagation model in which local variations in the properties of the interstellar medium are included was advanced by Streitmatter *et al.* (1985) who proposed that all except the highest energy cosmic rays reaching the Earth were produced in a superbubble, powered by multiple supernovae, which engulfs the Sun and which emits much of the observed soft X-ray background (Innes and Hartquist, 1984; Cox and Reynolds, 1987; Hartquist, 1994). Streitmatter *et al.* (1985) and Morfill and Hartquist (1985) argued that data restricting the past variation in cosmic ray intensity at the Earth are not inconsistent with superbubble modification of the local cosmic ray background. (See also Sonett, Morfill and Jokipii, 1987.)

In this paper, we argue that the diffusion properties of many cosmic rays that reach the Earth are determined by the physics of the transport of the most energetic ($\sim 10^5$ GeV) cosmic rays injected by a supernova remnant at the end of its free expansion phase into a magnetized tube in which the Solar System is now located. These $\sim 10^5$ GeV cosmic rays diffused away from the remnant at a typical speed of about 2000 km s^{-1} as the remnant slowed, and their streaming generated resonant waves throughout the tube, which has a cross section comparable to that of the remnant at the end of its free expansion phase.

In section 2 we review Bell's (1978) solution of the coupled cosmic ray transport and resonant wave generation and convection problem. We use that solution to estimate the highest energy of the cosmic rays that can be accelerated in the remnant; that estimate depends on the comparison of a lengthscale in Bell's solution to the remnant radius, a comparison suggested by Bell who used cosmic ray and remnant properties inferred in interpretations of observations. We have used simply model properties which permit an explicit expression of the dependence of the highest energy on supernova and ambient medium characteristics. In section 2 we also show that the coupled time dependent equations used by Bell possess a similarity solution. Its nature suggests that the leading front of the $\sim 10^5$ GeV cosmic rays propagates through the tube at a constant velocity. From this we are able to estimate the number density of the highest energy cosmic rays and the energy density of the waves with which they are resonant—as functions of the time elapsed since the occurrence of the supernova that "activated" the tube. From the locally measured energy density of cosmic rays with energies of 2×10^5 GeV we estimate the time elapsed since the most recent local "activation" to be several times 10^7 y.

In section 3 we simply *assume* that at late times the wave spectrum in the tube develops into a Kraichnan (1965) spectrum at wavenumbers greater than the wavenumber of the waves resonant with the 10^5 GeV cosmic rays. The existence of a spectrum for compressive interstellar waves similar to a Kraichan-type spectrum has been advocated by Armstrong, Cordes and Rickett (1981) on the basis of their analysis of scintillations, angular broadening, and other behavior of pulsar signals. Perhaps, the overlapping of supernova remnants in a McKee and Ostriker (1977) type of interstellar medium leads to lower energy cosmic rays having a relatively uniform spatially distribution and, consequently, in them giving rise to little resonant wave generation by streaming. However, that sort of argument is highly speculative. The analysis of Armstrong *et al.* and the agreement between other observationally based results and the model results that we obtain provide more reliable support for our assumption that higher frequency waves are maintained by a Kraichnan cascade. For instance, we show that a Kraichnan wave spectrum, normalized to the energy density of waves resonant with 2×10^5 GeV cosmic rays, in a tube that was most recently activated by a supernova several times 10^7 yr ago gives a diffusion coefficient at 1 GeV of about 10^{29} cm^2 s^{-1} in harmony with that inferred from observational data.

The arguments given in sections 2 and 3 include modifications to some that we employed in a paper (Hartquist and Morfill, 1983) in which we showed that molecular data for clouds near an old radiative supernova remnant in the Per OB2 association imply that second order Fermi acceleration and spatial diffusion result in a peak at 2–3 MeV in the cosmic ray spectrum in that remnant. In section 4 we discuss the Per OB2 remnant within the context of the model developed in sections 2 and 3.

Specifically, we show that a Kraichnan spectrum up to the wavenumbers of waves resonant with MeV cosmic rays gives a spatial diffusion coefficient close to that required to account for the Per OB2 molecular data.

2   The Acceleration and Propagation of 10^5 GeV Cosmic Rays

Bell (1978) considered the acceleration and propagation of cosmic rays and the generation and convection of Alfvén waves in and near a plane-parallel shock propagating parallel to a large scale magnetic field, $B_o\hat{x}$, by obtaining a steady state solution to the following set of equations:

$$\frac{\partial f}{\partial t} + (u \pm v_A)\frac{\partial f}{\partial x} = \frac{\partial}{\partial x}\left(D(x,t)\frac{\partial f}{\partial x}\right) \tag{1a}$$

$$\frac{\partial J}{\partial t} + (u \pm v_A)\frac{\partial J}{\partial x} = \sigma \tag{1b}$$

$$D(x,t) = \frac{4}{3\pi}\frac{\beta r_g c}{J} \tag{1c}$$

$$\sigma = \mp\frac{4\pi}{3}\frac{v_A \beta c}{U_M}p^4\frac{\partial f}{\partial x}. \tag{1d}$$

(See also e.g. Skilling, 1975a,b,c.) t is time, x is the position coordinate, and $u\hat{x}$ is the bulk velocity of the thermal plasma. $\pm v_A\hat{x}$ are the velocities of the Alfvén waves on which the cosmic rays scatter (We will assume that only forwardly or only backwardly propagating waves are present in each situation that we consider until we note otherwise.) c and $\beta c, p$, and r_g are the speed of light, the particle speed, the magnitude of the momentum, and the gyroradius of the cosmic rays being considered. U_M is the energy density of the large scale uniform component of the magnetic field. $k\hat{x}$ will signify the wavenumber of the Alfvén waves resonant with a cosmic ray of momentum p. $U_M J(k, x)$ is the wave energy density per unit logarithm of wavenumber. $f(p, x)$ is the cosmic ray (proton) distribution function, and $p^3 f(p, x)$ has the dimension of a number density in coordinate space.

Bell's steady state solution is for a shock positioned at $x = 0$ with the thermal plasma velocity at $x < 0$ being $+v_s\hat{x}$ and the Alfvén waves at $x < 0$ having a velocity of $-v_A\hat{x}$ with respect to the thermal plasma. The solutions for $f(p, x < 0)$ and $J(k, x < 0)$ are

$$f(p, x < 0) = f(p, x = -\infty) + \frac{a}{x_0 - x} \tag{2a}$$

$$J(k, x < 0) = \frac{b}{x_0 - x} \tag{2b}$$

with

$$a = \frac{c\,U_m}{\pi^2 e B_0 v_A p^3} \tag{2c}$$

$$b = \frac{4cp^2}{3\pi e B_0 \gamma m(v_S - v_A)} \tag{2d}$$

$$x_0 = \frac{cU_m}{\pi^2 e B_0 v_A} \frac{1}{p^3[f(p, 0) - f(p, -\infty)]} \tag{2e}$$

where we have used c.g.s. units rather than the M.K.S. units which Bell employed, and we have taken $\gamma = (1 - \beta^2)^{-1/2}$ and m to signify the proton rest mass.

The highest energy to which cosmic rays are accelerated in the vicinity of a supernova remnant's leading shock at any time can be estimated by equating the supernova remnant's radius, R_s, to x_0. (This condition is, of course, quite crude and is not intended to replace a nonplanar shock geometry analysis. On the other hand, a scale length for the accelreation region larger than the SNR scale is clearly not acceptable in a diffusive transport picture.) We take

$$f(p, 0) = \frac{\epsilon_c}{4\pi c(\ln\left(\frac{2p_H}{mc}\right) - 1)} p^{-4} \tag{3}$$

where p_H is the momentum of the most energetic cosmic ray, ϵ_c is the spatially averaged cosmic ray kinetic energy density within the remnant, and a p^{-4} law has been assumed, consistent with diffusive shock acceleration at a planar shock without losses. Then setting x_o and R_s equal, we find

$$\left(\frac{R_s}{30 \text{ pc}}\right)^2 = 1.0 \times 10^5 \left(\frac{n_e}{5 \times 10^{-3} \text{ cm}^{-3}}\right)^{-\frac{1}{2}} \left(\frac{p_H}{mc}\right)^{-1}$$

$$\left(\frac{E_c}{10^{50} \text{ erg}}\right) \left(\frac{13.5}{\ell n \left(\frac{2p_H}{mc}\right) - 1}\right) \tag{4}$$

where E_c is the total kinetic energy of cosmic ray protons contained in the remnant. E_c probably grows (at least) as rapidly as R_s^3 until the supernova ejecta have swept up a mass comparable to their own by which time the kinetic energy of the cosmic rays may be assumed to be a substantial fraction of the total energy of the remnant. The end of the free expansion phase occurs when $R_s = R_E$ where

$$\left(\frac{R_E}{30 \text{ pc}}\right) \approx 0.8 \left(\frac{M_E}{10 \, M_\odot}\right)^{\frac{1}{3}} \left(\frac{n_e}{5 \times 10^{-3} \text{ cm}^{-3}}\right)^{-\frac{1}{3}} \tag{5}$$

and M_E is the mass of the supernova ejecta.

Substituting (5) into (4) we find that the maximum momentum to which a supernova remnant can accelerate a cosmic ray is given roughly by

$$p_H \approx 2 \times 10^5 mc \left(\frac{M_E}{10 \, M_\odot}\right)^{-\frac{2}{3}} \left(\frac{n_e}{5 \times 10^{-3} \text{ cm}^{-3}}\right)^{\frac{1}{6}}$$

$$\left(\frac{E_c}{10^{50} \text{ erg}}\right) \left(\frac{13.5}{\ell n \left(\frac{2p_H}{mc}\right) - 1}\right) \tag{6}$$

Of course, most of a remnant's shock's surface is orientated obliquely to the ambient magnetic field. If Θ is the angle between the shock propagation velocity and the magnetic field, $p_H \propto (\cos\Theta)^{-1}$. We neglect this factor of order unity. By considering the acceleration timescale other authors (e.g. Drury, 1990 and references therein) have obtained similar estimates for p_H, but usually they have assumed Bohm diffusion (rather than calculating $J(k)$ with a self-consistent model) and have specified the remnant age and shock speed. When the assumption of Bohm diffusion is made, significant further acceleration after the end of the free expansion phase is limited by the time dependence of the decrease of the shock speed, whereas in Bell's and our picture the increase in the diffusion coefficient (due to the drop in the cosmic ray energy density) acts to end acceleration more abruptly as evolution beyond the free expansion phase occurs.

As the remnant expands beyond R_E, it slows and the cosmic rays with the highest momenta ($\sim p_H$) diffuse out of the remnant along those magnetic field lines that passed through the remnant when $R_s = R_E$. If the ambient field were uniform, the diffusion of most of those high energy cosmic rays would be contained within a cylinder of radius R_E.

We consider the propagation of the cosmic rays with $p = p_H$ in the stationary ambient gas ($u = 0$) contained in that cylinder. We take $x = 0$ to be the position of the supernova and treat the diffusion of the cosmic rays in the $+\hat{x}$ direction. The resonant waves are assumed to be propagating at the velocity $+v_A\hat{x}$. Equations (1) then possess a similarity solution with

$$f = \frac{1}{t}q(\chi) \tag{7a}$$

$$J = \frac{1}{t}s(\chi) \tag{7b}$$

$$\chi = \frac{x}{t} \tag{7c}$$

where $q(\chi)$ and $s(\chi)$ are governed by the coupled ordinary differential equations

$$-q - \chi\frac{dq}{d\chi} + v_A\frac{dq}{d\chi} = -\frac{4}{3\pi}\beta r_s c\left(\frac{1}{s^2}\frac{ds}{d\chi}\frac{dq}{d\chi} + \frac{1}{s}\frac{d^2q}{d\chi^2}\right) \tag{7d}$$

$$-s - \chi\frac{ds}{d\chi} + v_A\frac{ds}{d\chi} = -\frac{4}{3\pi}\frac{v_A\beta c}{U_M}p^4\frac{dq}{d\chi} \tag{7e}$$

Clearly, the expressions (7a) and (7c) imply that the leading front of the high energy cosmic ray distribution propagates at a constant velocity along the tube with radius R_E and that the spatially averaged (between the supernova site and the leading front) density of cosmic rays with $p \approx p_H$ falls as the inverse of the time since the supernova occurred.

It is reasonable to suppose that the leading front of the 2×10^5 GeV cosmic rays moves at a speed equal to that of the supernova remnant when $R_s = R_E$. That speed is about $2000\,\text{km s}^{-1}$. When the leading front has propagated a distance of the order of a kiloparsec, it will probably have encountered a region in which a substantial magnetic field component out of the Galactic plane occurs. (For instance, such lengthscales are thought to be associated with the formation of giant molecular cloud complexes (e.g. Mouschovias, 1974) through the Parker (1966) instability which naturally gives rise to gentle bending of field lines towards the halo.) The existence of such regions in which the field is so oriented is consistent with the presence of diffuse gas in the Galactic halo, the source of which is most likely the disk. Furthermore, the distribution of the high latitude synchrotron radiation, from which estimates of the diffusion coefficient in the halo are made, requires that the halo magnetic field strength be of the same order of magnitude as that in the disk. However, it is not improbable that a single "high energy cosmic ray tube" remains in the disk for several kiloparsecs and contains several sources of 10^5 GeV cosmic rays. The basic argument will be made for the case that such a tube has a magnetic field of roughly constant magnitude, though it bends, and contains only one high energy cosmic ray source, but the more general case can be described in a similar fashion.

In the restricted case on which we are focussing we can estimate the time variation of the density of 2×10^5 GeV cosmic rays near the site of the supernova by simply assuming that the bending of the field lines occurs over such a long distance that the flux tube can be treated as a cylinder and that the region filled with the 2×10^5 GeV cosmic rays maintains a constant cross section of πR_E^2. If v_p is the propagation speed of the leading front of the 2×10^5 GeV cosmic rays the time that must have passed since the most recent supernova 2×10^5 GeV activation of the flux tube currently containing the Sun must be

$$t_A \approx 9 \times 10^7 \text{ y} \left(\frac{E_c}{10^{50} \text{ erg}} \right) \left(\frac{M_E}{10 \, M_\odot} \right)^{-\frac{2}{3}}$$

$$\left(\frac{n_e}{5 \times 10^{-3} \text{ cm}^{-3}} \right)^{\frac{2}{3}} \left(\frac{v_p}{2 \times 10^8 \text{ cm s}^{-1}} \right)^{-1} \left(\frac{13.5}{\ell n \left(\frac{2 p_H}{mc} \right) - 1} \right) \tag{8}$$

in order for the spatially averaged 2×10^5 GeV cosmic ray proton energy density to have dropped to the measured (e.g. Fichtel and Linsley, 1986) value in the solar vicinity.

A value for t_A of 9×10^7 y is several times larger than the estimated age of the Be^{10} cosmic rays. (Garcia-Munoz, Mason and Simpson, 1977). The Be^{10} cosmic rays are of lower energy than 2×10^5 GeV, and their age should not be thought of as a direct measure of the local "2×10^5 GeV activation age" or "tube age", since lower energy cosmic rays can be accelerated by the passage of a remnant that has evolved well past the free expansion phase. However, the Be^{10} age does provide a rough measure of how frequently a typical point in the interstellar medium is passed by a remnant shock, and if a tube occupies a volume in the disk comparable to that of a typical remnant when its expansion stalls the Be^{10} age is also a rough measure of how frequently a new source of high energy cosmic rays occurs in a flux tube.

From (2b), (2d), (2e), (3), and (5) and assuming J to fall as the inverse of the volume between the site of the supernova and the 2×10^5 GeV cosmic ray propagation front, we find that a typical value of $J(k_{\min})$, where $k_{\min}$ is the wavenumber of waves resonant with the highest energy cosmic rays to be accelerated in the supernova remnants, is

$$J(k_{\min}) \approx 1 \times 10^{-3} \left(\frac{E_c}{10^{50} \text{erg}} \right) \left(\frac{M_E}{10 \, M_\odot} \right)^{-\frac{2}{3}}$$

$$\left(\frac{n_e}{5 \times 10^{-3} \text{ cm}^{-3}} \right)^{\frac{1}{6}} \left(\frac{B_0}{1 \times 10^{-6} \, G} \right)^{-1} \left(\frac{v_s(R_s = R_E)}{2 \times 10^8 \text{ cm s}^{-1}} \right)^{-1}$$

$$\left(\frac{v_p}{2 \times 10^8 \text{ cm s}^{-1}}\right)^{-1} \left(\frac{13.5}{\ell n\left(\frac{2p_H}{mc}\right) - 1}\right) \left(\frac{t_A}{10^7 \text{ y}}\right). \tag{9}$$

The analysis in this section has been restricted by a number of assumptions. The primary one is that no damping mechanisms affect $J(k_{\min})$. We now show that some regularly invoked wave damping mechanisms probably do not affect $J(k_{\min})$. From the work of Arnaud and Rothenflug (1985) we see that at temperatures above 5×10^5 K the fractional abundance of neutrals is about 10^{-6} and below; in gas with $2n_eT \approx 5000 \text{ cm}^{-3}$ K, ion-neutral collisions will damp waves only on timescales longer than about 10^9 yr. Random phase wave-wave interactions (e.g. Sagdeev and Galeev, 1969; Chin and Wenzel, 1972; Skilling, 1975b; Schwartz, 1977) transfer energy to different frequency waves on a timescale of about $(v_A k J(k))^{-1}$ if $v_A \gg c_s$, where c_s is the adiabatic sound speed. The transfer occurs on a timescale of about $(v_A k J(k))^{-1} c_s^2/v_A^2$ if $c_s \gg v_A$. Wave-wave interactions would then contribute a loss term on the right hand side of (1b) and in this model effectively requires that

$$J(k_{\min}) \lesssim 7 \times 10^{-4} \left(\frac{p_H}{2 \times 10^5 mc}\right) \left(\frac{B_0}{1 \times 10^{-6} \text{ G}}\right)^{-2}$$

$$\left(\frac{t_A}{10^7 \text{ y}}\right) \left(\frac{n_e}{5 \times 10^{-3} \text{ cm}^{-3}}\right)^{1/2} \left(\frac{c_s}{v_A}\right)^2 \tag{10}$$

since in the hot interstellar gas $c_s > v_A$ is expected. Random phase wave-wave interactions probably do not cause $J(k_{\min})$ to evolve significantly differently than the expression given in (9) indicates, but as we will argue in the next section they may play an important role in establishing a Kraichnan (1965) spectrum at larger k. Boulares and Morfill (1992) have calculated the rates at which waves are damped by cosmic ray viscosity. Their result for the damping rate of a transverse wave with a wavelength short compared to the mean free path of the cosmic rays is

$$\Gamma_v = 2 \times 10^{-15} \text{ s}^{-1} \left(\frac{f_o m^3 c^3}{10^{-7} n_e}\right) \left(\frac{D_0}{10^{29} \text{ cm}^2 \text{s}^{-1}}\right)^{-1} \left(\frac{p_{\min}}{mc}\right)^{-1} \tag{11}$$

if $f(p) = f_o p^{-4.5}$ (the observed dependence) for $p \geq p_{\min} \gg mc$ and $f(p) = 0$ for $p < p_{\min}$ and $D = D_0(p/mc)^{1/2}$. Hence, the cosmic ray viscous damping of waves resonant with cosmic rays with $p \leq p_H$ is of the order of several times 10^7 y, only marginally long enough for us to neglect it.

Our analysis in this section has also been restricted by the assumption that diffusion of the highest energy cosmic rays in momentum space is negligible unlike the diffusion in coordinate space. A comparison of the spatial and second order Fermi diffusion coefficients (Skilling, 1975a) gives the timescale for second order Fermi processes to be of the order of

$$\tau_2 \approx \frac{D}{v_A^2}. \tag{12}$$

From (1c), (6), and (9) we find that for $p = p_H$

$$\tau_2(p_H) \approx 3 \times 10^{10}\,\text{y}\left(\frac{B_0}{1 \times 10^{-6}G}\right)^{-2}\left(\frac{n_e}{5 \times 10^{-3}\,\text{cm}^{-3}}\right)$$

$$\left(\frac{v_s(R_s = R_E)}{2 \times 10^8\,\text{cm s}^{-1}}\right)\left(\frac{v_p}{2 \times 10^8\,\text{cm s}^{-1}}\right)\left(\frac{t_A}{10^7\,\text{y}}\right). \qquad (13)$$

Clearly, the timescale for second order Fermi processes to affect the momentum of a particle with $p \approx p_H$ is longer than t_A.

3   The Diffusion Coefficient at Energies Below 10^5 GeV

In a plasma in which $J(k_{min})$ is maintained at a constant value as time passes and in which the generation and the damping of waves with $k > k_{min}$ are due solely to random phase wave-wave interactions, the Kraichnan (1965) wave spectrum corresponding to

$$J(k > k_{min}) = J(k_{min})(k/k_{min})^{-1/2} \qquad (14)$$

eventually will be established. As stated in the Introduction, we will *assume* that this wave spectrum obtains in the tube. In fact, because $J(k_{min})$ decreases with time due both to the propagation of the highest energy cosmic rays and due to the random phase wave-wave interactions (A comparison of the value for $J(k_{min})$ given by (9) and its upper bound given by (10) shows that the propagation and random phase wave-wave interactions affect $J(k_{min})$ on comparable timescales.) equation (14) at best slightly underestimates $J(k)$ at $k \gg k_{min}$ and the spectrum at k slightly greater than k_{min} is flatter than indicated by (14).

We leave largely untouched the question of why the streaming of lower energy cosmic rays can be neglected as a source of waves with $k > k_{min}$ but do wish to reiterate two suggestive points: 1) As noted in the Introduction, properties of pulsar signals have been invoked as evidence for a spectrum of compressive waves similar to that which we are assuming for the Alfvén waves. 2) As also mentioned in the Introduction, in an interstellar medium of overlapping supernova remnants the cosmic rays at lower energies may, by their diffusion from remnant to remnant, establish a relatively uniform spatial distribution giving rise to little streaming of low energy cosmic rays in overlapping remnants. Certainly, if the diffusion coefficient of 1 GeV cosmic rays is $10^{29}\,\text{cm}^2\,\text{s}^{-1}$, their remnant to remnant transit time is very short compared to a remnant's total lifetime.

We cannot rigorously justify our assumption of a Kraichnan spectrum. However, we will show in this section and the next one that the results of section (2) together with equation (14) lead to model properties that agree well with those inferred from observations.

We use the particle-wave resonance condition (Skilling, 1975c)

$$0.41kp = \gamma m\Omega \qquad (15)$$

to calculate $k_{\min}$ from p_H; here Ω is the gyrofrequency $eB_o/\gamma mc$. Use of equations (6), (15), (9), (14), and (1c) yields

$$D \approx 2 \times 10^{28} \text{ cm}^2 \text{ s}^{-1} \gamma^{\frac{1}{2}} \beta^{\frac{3}{2}} \left(\frac{E_c}{10^{50} \text{ erg}} \right)^{-\frac{1}{2}}$$

$$\left(\frac{M_E}{10 \, M_\odot} \right)^{\frac{1}{3}} \left(\frac{n_e}{5 \times 10^{-3} \text{ cm}^{-3}} \right)^{-\frac{1}{12}} \left(\frac{v_s}{2 \times 10^8 \text{ cm s}^{-1}} \right)$$

$$\left(\frac{v_p}{2 \times 10^8 \text{ cm s}^{-1}} \right) \left(\frac{\ell n \left(\frac{2p_H}{mc} \right) - 1}{13.5} \right)^{\frac{1}{2}} \left(\frac{t_A}{10^7 \, y} \right) \qquad (16)$$

Substitution of the expression given by (8) for t_A into (16) yields

$$D \approx 2 \times 10^{29} \text{ cm}^2 \text{ s}^{-1} \gamma^{\frac{1}{2}} \beta^{\frac{3}{2}} \left(\frac{E_c}{10^{50} \text{ erg}} \right)^{\frac{1}{2}} \left(\frac{M_E}{10 \, M_\odot} \right)^{-\frac{1}{2}}$$

$$\left(\frac{n_e}{5 \times 10^{-3} \text{ cm}^{-3}} \right)^{\frac{7}{12}} \left(\frac{v_s}{2 \times 10^8 \text{ cm s}^{-1}} \right)$$

$$\left(\frac{13.5}{\ell n \left(\frac{2p_H}{mc} \right) - 1} \right)^{\frac{1}{2}} \left(\frac{p_H^4 f_{\text{obs}}(p_H)}{1 \times 10^{-27} \text{ g cm s}^{-1}} \right)^{-1} \qquad (17)$$

Expression (17) gives a diffusion coefficient of about $2 \times 10^{29} \text{ cm}^2$ for a cosmic ray with a kinetic energy of 1 GeV if the normalization values of various parameters are assumed. This is in remarkably good agreement with values inferred from measurements given that the extrapolation has been over five orders of magnitude in p.

4   Stochastic Acceleration in the Vicinity of the Perseus OB2 Association

Hartquist and Morfill (1983) argued that various molecular data for several lines of sight through a slowly ($\approx 5 \text{ km s}^{-1}$) propagating shell of an old radiative supernova remnant near the Perseus OB2 cluster imply that cosmic rays with kinetic energies of about 2 MeV are entering the shell from the ionized parts of the remnant and inducing the ionization that drives the chemistry in the mostly neutral shell. Some observed variations in the chemistry would not be expected if the cosmic rays driving that part of the chemistry had energies much above 2 MeV while lower energy cosmic rays would suffer too many losses before penetrating to sufficient depths to induce the formation of the molecules used in the diagnosis.

In the light of the inhomogeneous cosmic ray transport model developed above, we now reexamine the Perseus OB2 observations interpreted earlier by Hartquist and Morfill (1983).

The dominant loss mechanism in the Per OB2 remnant is spatial diffusion which occurs on a timescale of

$$\tau_L \approx \frac{R_{\mathrm{Per}}^2}{\pi^2 D} \tag{18}$$

where R_{Per} is the radius of that remnant. By equating τ_2, given by equation (12), and τ_L and specifying D we can estimate the energy to which second order Fermi scattering will accelerate cosmic rays in the Per OB2 remnant. Consistent with the arguments given in section 3, we will assume that the Per OB2 remnant is sufficiently old that low energy cosmic ray streaming no longer affects the wave spectrum in the vicinity of the molecular material. Furthermore, we assume that the cosmic ray energy density near Per OB2 is not significantly different from the "near-Earth" value. Thus, equation (17) should give a reasonable approximation to the diffusion coefficient throughout most of the Per OB2 remnant. Using (17) and setting τ_2 and τ_L equal, we find that

$$\beta^2 \approx 2 \times 10^{-4} \left(\frac{v_{A,\mathrm{Per}}}{3 \times 10^6 \text{ cm s}^{-1}} \right)^{\frac{4}{3}} \left(\frac{R_{\mathrm{Per}}}{20 \text{ pc}} \right)^{\frac{4}{3}} \left(\frac{v_s}{2 \times 10^8 \text{ cm s}^{-1}} \right)^{-\frac{4}{3}}$$

$$\left(\frac{E_c}{10^{50} \text{ erg}} \right)^{-\frac{2}{3}} \left(\frac{M_E}{10 \, M_\odot} \right)^{\frac{4}{9}} \left(\frac{n_e}{5 \times 10^{-3} \text{ cm}^{-3}} \right)^{-\frac{7}{9}}$$

$$\left(\frac{\ln\left(\frac{2p_H}{mc}\right) - 1}{13.5} \right)^{\frac{2}{3}} \left(\frac{p_H f_{\mathrm{obs}}(p_H)}{1 \times 10^{-27} \text{ g cm}^{-3}} \right)^{\frac{4}{3}} \tag{19}$$

typifies the cosmic rays accelerated by second order Fermi scattering in the Per OB2 remnant. ($v_{A,\mathrm{Per}}$ is the Alfvén speed in the hot gas in that remnant.) This is within about an order of magnitude of the β^2 of 4×10^{-3} inferred from the molecular data to characterize the low energy cosmic rays in that remnant. Given that the use of the normalization parameters in equation (17) lead to a diffusion coefficient for GeV cosmic rays that may be a factor of a few larger than the Galactic value, that the value for t_A in the Per OB2 region may be somewhat smaller than the local value, and that the Per OB2 ionizing cosmic rays have kinetic energies that are 10^8 times smaller than those of the 2×10^5 GeV cosmic rays, the agreement between the theoretical β^2 and the inferred β^2 to about an order of magnitude can be considered encouraging. This implies, as originally suggested by Hartquist and Morfill (1983), that the Per OB2 data provide evidence for second order Fermi acceleration. The difference to our previous work lies in the fact that the inhomogeneous cosmic ray transport model developed here gives a quantitative estimate of the wave power level required for this acceleration, derived independently from general considerations of energy densities, "tube age" etc. In turn, the Per OB2 data may then be invoked to support the inhomogeneous transport model.

5   References

Armstrong, J.W., Cordes, J.M., and Rickett, B.J.: 1981, *Nature* **291**, 561.
Arnaud, M. and Rothenflug, R.: 1985, *Astron. Astrophys. Suppl.* **60**, 425.
Bell, A.R.: 1978, *Mon. Not. R. astr. Soc.* **182**, 147.
Boulares, A. and Morfill, G.: 1992, *Astrophys. J.* **400**, 622.
Chin, Y.C. and Wentzel, D.G.: 1972, *Astrophys. Spac. Sci.* **16**, 465.
Cox, D.P. and Reynolds, R.J.: 1987, *Ann. Rev. Astron. Astrophys.* **25**, 303.
Dogiel, V.A.: 1991, in: J.B.G.M. Bloemen (ed.), *The Interstellar Disk-Halo Connection in Galaxies - IAU Symposium No. 144*, Kluwer, Dordrecht, p. 175.
Drury, L.O'C.: 1990, *Proceedings of the 21st International Cosmic Ray Conference* **12**, 85.
Fichtel, C.E. and Linsley, J.: 1986, *Astrophys. J.* **300**, 474.
Garcia-Munoz, M., Mason, G.M. and Simpson, J.A.: 1977, *Astrophys. J.* **217**, 859.
Hartquist, T.W.: 1994, *Astrophys. Spac. Sci.*, this volume.
Hartquist, T.W. and Morfill, G.E.: 1983, *Astrophys. J.* **266**, 271.
Innes, D.E. and Hartquist, T.W.: 1984, Monthly Notices Roy. Astron. Soc. **209**, 7.
Kraichnan, R.H.: 1965, *Phys. Fluids* **8**, 1385.
Kulsrud, R.M. and Pearce, W.P.: 1969, *Astrophys. J.* **156**, 445.
Lerche, I. and Schlickeiser, R.: 1982, *Astron. Astrophys.* **107**, 148.
McKee, C.F. and Ostriker, J.P.: 1977, *Astrophys. J.* **218**, 148.
Morfill, G.E. and Hartquist, T.W.: 1985, *Astrophys. J.* **297**, 194.
Morfill, G.E., Meyer, P. and Lüst, R.: 1985, *Astrophys. J.* **216**, 670.
Mouschovias, T.Ch.: 1974, *Astrophys. J.* **192**, 37.
Parker, E.N.: 1966, *Astrophys. J.* **145**, 811.
Sagdeev, R.Z. and Galeev, A.A.: 1969. *Nonlinear Plasma Theory*, Benjamin, New York.
Schwartz, S.J.: 1977, *Mon. Not. R. Astr. Soc.* **178**, 399.
Skilling, J.: 1975a, *Mon. Not. R. Astr. Soc.* **172**, 557.
Skilling, J.: 1975b, *Mon. Not. R. Astr. Soc.* **173**, 245.
Skilling, J.: 1975c, *Mon. Not. R. Astr. Soc.* **173**, 255.
Sonett, C., Morfill, G.E., and Jokipii, J.R.: 1987, *Nature* **330**, 458.
Streitmatter, R.E., Balasubrahmanyan, V.K., Protheroe, R.J., and Ormes, J.F.: 1985, *Astron. Astrophys.* **143**, 249.

Alfvénic Waves and Alignment of Large Grains

A. Lazarian
DAMTP, University of Cambridge, UK

Abstract. The alignment of grains under the influence of the Alfvenic waves is discussed. It is shown that even small deviations from grain uniformity result in the alignment of large ($l > 6 \cdot 10^{-5}$ cm) grains. The latter result is important for the interpretation of the IR polarization data.

1 Grains in a Gaseous Flow

The alignment of grains in gaseous flows may take place when the grain centre of mass does not coincide with its centre of pressure. If the radius vector from the centre of pressure to the centre of mass is $\Delta \mathbf{R}$, the torque applied to the grain is given by

$$M(\theta) = S \varrho v_T u \, \Delta R \sin \theta \tag{1}$$

for the velocity $\mathbf{u}$ of relative motion of the grain in respect to the gas less than the thermal velocity of gas atoms v_T. Note, that S is the grain cross-section, ϱ is the density of the gas, and θ is the angle between vectors $\Delta \mathbf{R}$ and $\mathbf{u}$. Due to this momentum and the influence of frictional forces the grain performs oscillations according to the following equation

$$\frac{d^2\theta}{dt^2} + q\frac{d\theta}{dt} - \omega_0^2 \sin\theta = 0 \tag{2}$$

where $\omega_0 = \sqrt{\frac{S \varrho u \Delta R}{I_z}}$, I_z is the z-moment of inertia, $q = \frac{2 \varrho v_T l^2 S}{3 I_z}$, l is the grain size. For the grain motion not to be determined by friction the following inequality should be fulfilled: $|\omega_0^2 \sin\theta| \gg \left|q\frac{d\theta}{dt}\right|$, which means that $u\frac{\Delta R}{l} \gg \frac{\varrho}{m_g} v_T \sim 10^{-8} v_T$, where m_g is the grain mass. In other words, for a grain moving with velocity $u < v_T$, even small irregularities of shape or chemical composition result in the grain alignment if the temperature of the gas is close to zero.

2 Temperature Effects

As a result of a random bombardment by atoms the grain obtains a temperature T. We can also introduce a potential $\phi_{\theta_i} = \int_0^{\theta_i} M(\theta) d\theta$. It is easy to see that, for an axially symmetric grain, $\phi_{\theta_i} = W \, \Delta R \cos\theta$, where $W = S \varrho u v_T$. Considering an idealised situation when $l = \beta \, \Delta R$, it is possible to introduce the mean squared moment of alignment

$$\langle l_z^2 \rangle = \frac{\int_0^{\frac{\pi}{2}} l^2 \cos^2\theta \exp(\kappa \cos\theta) \sin\theta d\theta}{\int_0^{\frac{\pi}{2}} \exp(\kappa \cos\theta) \sin\theta d\theta} = \frac{l^2 \frac{\partial^2}{\partial \kappa^2}\left(\frac{1}{\kappa}\exp\kappa - \frac{1}{\kappa}\right)}{\frac{1}{\kappa}\exp\kappa - \frac{1}{\kappa}} \tag{3}$$

Astrophysics and Space Science **216**: 235–237, 1994.
© 1994 *Kluwer Academic Publishers.*

As $W \ll \beta kT$, then $\kappa = \frac{W}{\beta kT} \ll 1$, and an expansion of $\exp \kappa$ can be used. Keeping terms up to the order of κ, one has

$$\frac{\langle l_z^2 \rangle}{l^2} \approx \frac{\frac{1}{3} + \frac{\kappa}{12} + \frac{\kappa}{6}}{1 + \frac{\kappa}{2}} \approx \frac{1}{3} + \frac{\kappa}{12} = \frac{1}{3} + \frac{W}{12\beta kT} \tag{4}$$

which means that

$$\frac{\langle l_z^2 \rangle}{l^2} \sim 0.1 \frac{l^3 n}{\beta} \left(\frac{u}{v_T} \right)^i \frac{m_g}{m} + \frac{1}{3} \sim \frac{1}{2\beta} \left(\frac{u}{v_T} \right)^i \left(\frac{l}{5 \cdot 10^{-5}} \right)^6 \left(\frac{n}{10^3} \right) + \frac{1}{3} \tag{5}$$

where n is the gas concentration, while $i = 1$, for $u < v$, and 2, for $u \gg v$. Thus, the parameter of the alignment (see Purcell, 1979)

$$Q_A = \frac{1}{2} \left\langle 3\frac{\langle l_z^2 \rangle}{l} - 1 \right\rangle \tag{6}$$

can be estimated as

$$Q_A \approx \frac{1}{\beta} \left(\frac{u}{v_T} \right)^i \left(\frac{l}{5 \cdot 10^{-5}} \right)^6 \left(\frac{n}{10^3} \right) \tag{7}$$

It can be seen from Eq. (7) that the mechanism of alignment discussed implies the dependence of Q_A on the characteristic size (mass) of the grain. If $\beta \sim 10-10^2$, the alignment can be expected for the grains with $l \sim 6 \cdot 10^{-5}$ cm and $u \sim v_T$.

3   Alfvénic Perturbations

Uncharged grains can obtain the velocity u with respect to the gas due to Alfvénic perturbations with $v_A = v_0 \sin \omega_A t$. The corresponding equation for the grain motion can be written as follows:

$$\frac{\mathrm{d}p}{\mathrm{d}t} + \omega_A k p = \omega_A e^{i\omega_A t} \tag{8}$$

where $p = \frac{u}{v_0}$, ω_A is the frequency of oscillations, $k = \frac{\varrho S v_T}{4 m_g \omega_A}$. The solution of Eq. (8) is $u = u_0 \sin(\omega_A t \varphi_{sh})$, where $u_0 = \frac{v_0}{\sqrt{1+k^2}}$, and $\mathrm{tg}\varphi_{sh} = k$. For typical conditions in the ISM $k \sim 1$ and $u_0 \sim v_0 \sim 10^5$ cm s^{-1}, is $l > 6 \cdot 10^{-5}$ cm. This means that the large grains will be aligned by the Alfvénic perturbations with their long axes perpendicular to the magnetic field lines.

The introduction of charge on the grain surface does not alter the result for the grains with $l \sim 10^{-5}$ cm. Indeed, in a steady state, the grains become negatively charged to equalise the impact rates of electrons and positive ions. For a gas of temperature T, the number of electrons is $\sim 1.4 \cdot 10^{-3} lT$ (Spitzer, 1978). Then, for typical conditions in the ISM, the charge does not exceed a few electrons, and the ratio of the number of elementary charges to the number of atoms composing the grain is $\sim 10^{-9}$. Comparing this value with a typical value of the ratio ($\sim 10^{-4}$), one can see that the "inertia per charge" is considerably greater for grains than for gas. Therefore the direct influence of the oscillating magnetic field on the grains can be ignored, unless the grains are so tiny that their size is $\sim 10^{-6}$ or 10^{-7} cm.

4 Summary

It has been shown that the Alfvénic waves can be responsible for the alignment of large ($l > 6 \cdot 10^{-5}$ cm) grains. The alignment of smaller grains will be discussed in another paper (Lazarian, in preparation).

Acknowledgements

Valuable comments of M. Rees are acknowledged. The research is supported by an Isaac Newton Scholarship.

References

Purcell, E.M.: 1979, *Astrophys. J.* **231**, 404.
Spitzer, L.Jr., McGlunn, T.A.: 1979, *Astrophys. J.* **231**, 417.

Summary

Acknowledgement

References

An Interstellar Thermostat: Gas Temperature Regulated by Grain Charge

J. A. Turner and A. P. Whitworth
Department of Physics and Astronomy, University of Wales, Cardiff CF2 3YB, U.K.

Abstract. The temperature of interstellar gas at low optical depths, is determined by the balance of heating by photoejected electrons from grains and cooling by fine structure transitions of ionized Carbon. The heating rate is dependant on the UV radiation and the grain charge. The grain charge is determined by the intensity of the UV radiation, and the increase in grain charge with increasing UV intensity, acts to reduce the heating rate, and regulate the gas temperature with respect to increases in the UV intensity. This thermostat is novel compared to others in the interstellar medium, because it does not depend upon the cooling rate of the gas.

1 Temperature of Interstellar Gas

The equilibrium temperature of interstellar gas is defined as the balance of the heating and cooling rates. The heating of the gas at low optical depths is caused by energetic electrons emitted from dust grains via the photoelectric effect (de-Jong, 1977). The charge of the dust grains acts to reduce this heating rate, by both decreasing the fraction of incident photons that can cause this effect, and by decreasing the average energy of the emitted electron. The cooling at low optical depths is through the fine structure lines of ionized Carbon, which are collisionally excited by electrons and hydrogen atoms (Dalgarno and McCray, 1972).

Figure 1 plots the temperature as a function of ψ, for the two cases of including and ignoring the grain charge. The graph clearly demonstrates the thermostatic property of the grain charge, as the UV intensity changes over 4 orders of magnitude, the temperature only varies by 1 order of magnitude.

2 Summary

The temperature of the interstellar gas in regions of low optical depth, i.e. where the UV intensity is equal to the background value or above, is determined by the balance of heating by electrons emitted from grains via the photoelectric effect of the UV radiation, and the cooling by fine structure lines of ionized Carbon. Increasing the grain charge, increases the minimum energy of the photon that can cause the photoelectric effect, thereby quenching the emission of electrons and the heating rate. The grain charge is increased by increasing the UV intensity and therefore as the UV radiation intensity increases the temperature of the gas does not increase in proportion (due to quenching by the grain charge). This thermostat is interesting as the temperature of the interstellar gas is normally regulated by the cooling rate, which is strongly dependant upon the temperature of the interstellar gas, through the exponential terms in the cooling rate equations.

Astrophysics and Space Science **216**: 239–240, 1994.
© 1994 *Kluwer Academic Publishers.*

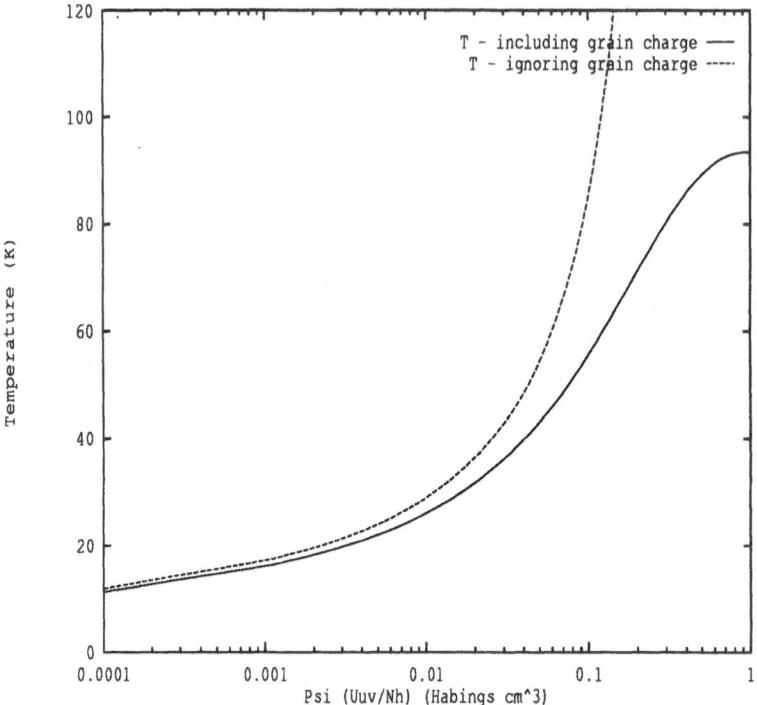

Fig. 1.   The equilibrium temperature as a function of ψ (Habings cm^3)

The inclusion of other heating agents does alter the this maximum, so in regions of high UV intensity, near to an OB star. The gas near to the star would be cold ($\psi > 1$), then further from the star the gas would heat up until $\psi = 1$, at which the temperature would reach its maximum value, and even further from the star ($\psi << 1$), the gas would become cold again. Therefore for a uniform gas density surround a new OB star, there would be a zone of cold gas then hot gas then cold gas again. This temperature structure would lead to expansion in the hot region causing it to heat up. Whilst the cold regions would become denser and so cool, however this would also have the effect of blocking more of the UV radiation from the OB star. Therefore it is not clear what the time evolution of this structure would be, however it could lead to expansion and contraction waves around the HII region, and so cause an instability in the PDR surrounding the HII region, and thus trigger star formation.

References

Dalgarno, A. and McCray, R.: 1972, *Ann. Rev. Astr. Ap.* **10**, 375.
deJong, T.: 1977, *Astron. Astrophys.* **55**, 137.

Recent Optical Observations of Circumstellar and Interstellar Phenomena

J. Meaburn
Department of Astronomy, The University of Manchester
Manchester, M13 9PL, England

Abstract. Complex motions are found over the shells which comprise the Honeycomb nebula. Evidence is presented for an episodic jet from Eta Carinae. The compact globules with radial spokes in the Helix nebula have both been shown to have dusty, molecular cores. Finally, high-speed flows of ionized gas are shown to be associated with the compact knots in the vicinity of the Trapezium cluster in the Orion nebula.

1 Introduction

Several interstellar and circumstellar phenomena have recently aroused considerable interest: there is the curious elongated cluster of around twenty ≈ 2 pc diam shells, the Honeycomb nebula, in the vicinity of SN1987A discovered by Lifan Wang (1992): a jet-like feature has been found by Hester *et al* (1991) projecting from the eruptive, super-luminous star Eta Carinae: the origin of the compact knots with molecular cores and radial spokes in the nearest (NGC) planetary nebula, the Helix nebula (NGC 7293) remains the subject of speculation: and the compact, ionized knots found originally by Laques and Vidal (1973) in the core of the Orion nebula (NGC 1976) are increasingly relevant to the understanding of low-mass star formation. Revealing, spatially-resolved, spectral and CCD imaging observations of these phenomena have now been made.

2 The Honeycomb Nebula

Wang (1992) has discovered that one of the nebulous filaments in the vicinity of SN1987A (see Meaburn, 1990) is composed of around twenty, ≈ 7" diam ($\equiv$ 2pc), interlocking shells. Consequently, this extraordinary region, ≈ 45 pc from SN1987A, has been aptly named the 'Honeycomb' nebula. Its morphology, as revealed by Wang's CCD images, appears to be unique but this is most likely a consequence of the relatively low angular resolutions employed in previous optical imagery of the LMC.

Any definitive theories concerning the origin and nature of the Honeycomb nebula have awaited knowledge of its kinematics. Spatially resolved, long-slit, arrays of Hα and [NII]6584Å profiles have now been obtained (reported in Meaburn *et al*, 1993a) over the nebula with the Manchester Echelle spectrometer (MES-Meaburn *et al*, 1984) combined with the Anglo-Australian telescope (AAT).

The slit positions 1-3 are shown in Fig. 1 against a sketch of the nebulosity and a contour map of the position-velocity (pv) array of Hα profiles with $\log_{10}$ contour

Astrophysics and Space Science **216**: 241–252, 1994.

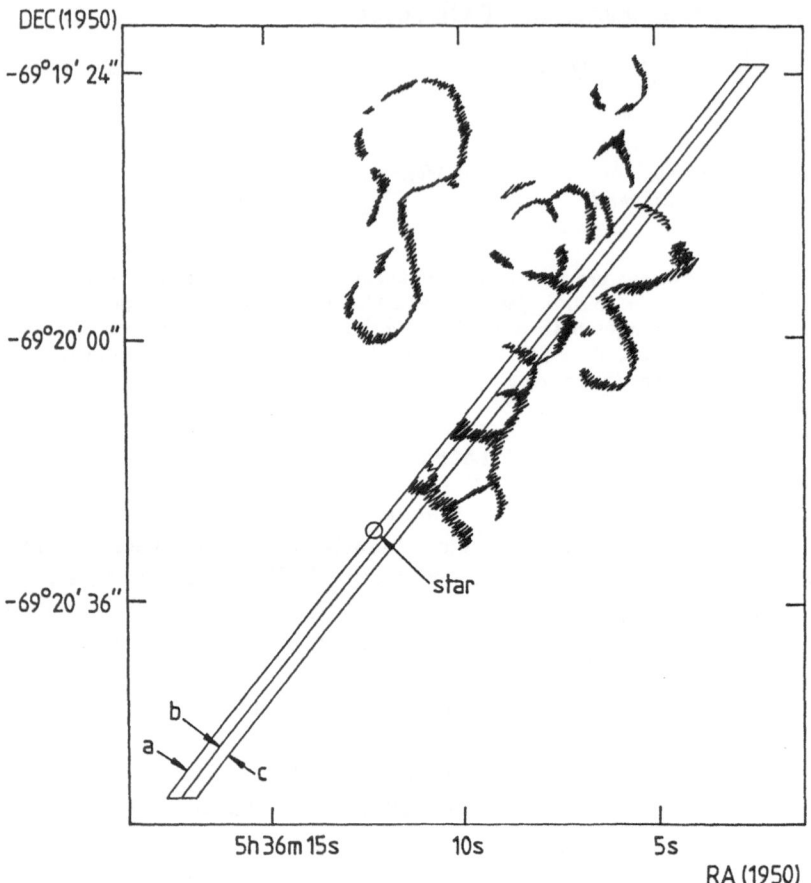

Fig. 1.   Slit positions along which Hα profiles were obtained are shown against a sketch of the
Honeycomb nebula.

intervals is shown in Fig. 2 for slit position c. Sharp 'spikes' project to ≈ −250
km s⁻¹ with respect to the systemic radial velocity in all of the pv arrays over the
filamentary edges of the Honeycomb shells. There is some evidence of aspheric,
approaching, radial expansion of each shell.

Individual supernova explosions, Wolf-Rayet ejecta or stellar winds of any origin
are easily ruled out as mechanisms for forming each of the shells that comprise the
Honeycomb nebula: it is unlikely that twenty stars exploded at the same time and
there are no obvious WR or OB stars in each shell.

Any explanation of the Honeycomb nebula must be considered in the context
of the well-established structure and kinematics of the halo of 30 Dor. Typical
kinematics (Meaburn, 1981, 1984 and 1988a and see Chu (this conference) and

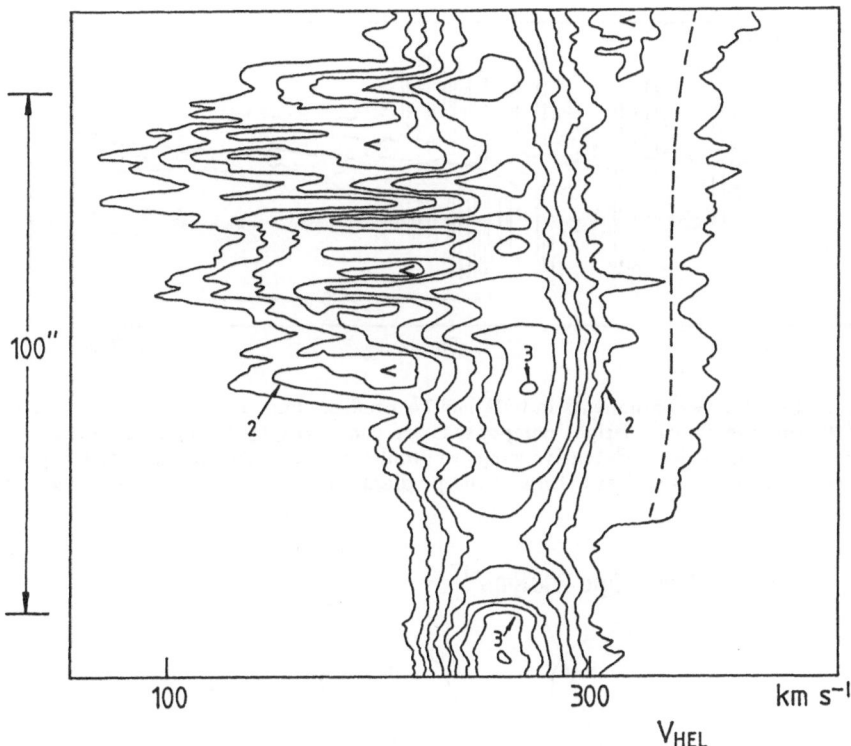

Fig. 2. A contour map of the Hα along slit position c in Fig. 1 is shown. The data were converted to log values. The Hα surface brightnesses for the 2 and 3 contours are 9.6 10^{-6} and 1.73 10^{-4} erg s^{-1} cm^{-2} sr^{-1} $\AA^{-1}$ respectively. The edges of the Honeycomb shells are characterised by spikes to negative radial velocities with respect to the systemic radial velocity (≈ 270 km s^{-1}).

Chu and Kennicutt (1993)) across the halo of 30 Dor are shown in Fig. 3 and a schematic model of the halo is presented in Fig. 4. The halo appears to be composed of giant shells expanding radially at 30-50 km/s whose diameters increase up to 100pc with distance from the dense nebular core surrounding the R136 cluster. These giant shells are most likely in momentum conserving phases and predominantly generated by the successive supernova explosions within the enclosed OB associations over periods of 10^6-10^7 yrs. Localised regions of high-speed (Meaburn, 1988a) gas indicate the presence of ≈ 30, $\leq 10^4$ yr old, supernova remnants in the 30 Dor halo.

It is now proposed that the Honeycomb nebula is the consequence of a single supernova explosion which has occurred just inside the dense perimeter of the giant shell surrounding SN1987A. Here the blast wave expands unhindered until its collision with the perimeter. The 'overstability' model of Mac Low and Nor-

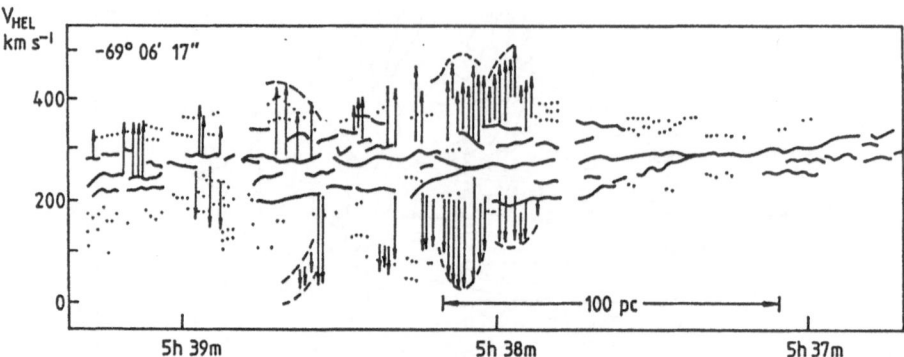

Fig. 3. Separate velocity components in [OIII]5007Å profiles along a line of measurements over the halo of 30 Dor are shown. Bright components are connected by heavy lines, whereas faint components are shown by dots. Extensive wings to profiles are arrowed and indicate the presence of young supernova remnants. Giant shells ≈ 100 pc diam. can be seen to be expanding at ≈ 30 km s^{-1}.

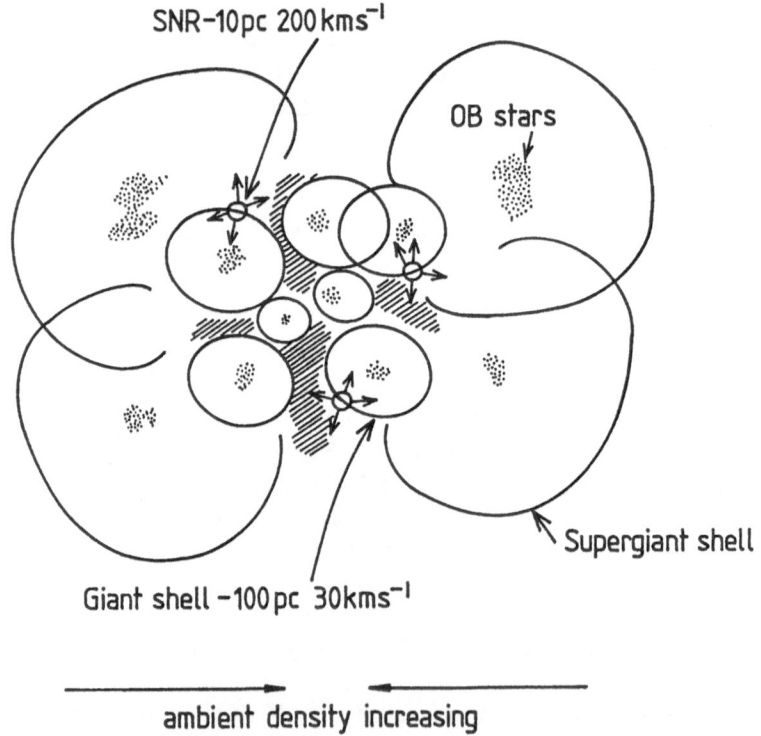

Fig. 4. A schematic model of the halo of 30 Dor is shown. Giant shells expanding at around 30 km s^{-1} have young supernova remnants within their perimeters.

man (1993) where such a blast wave encounters a dense, comparatively stationary, medium, promises to predict motions similar in detail to those depicted in Fig. 2 over the Honeycomb shells. It is perhaps relevant that very similar motions are found within the galactic supernova remnant IC443 (Meaburn et al, 1990) which appears to have expanded into pre-existing cavities. Moreover, Chu and MacLow (1990) and Wang and Helfand (1991a and b) have all found soft X-ray sources, presumably generated by the collisions of young supernova remnants, just within the perimeters of giant LMC shells.

3 The Jet from Eta Car

At a distance of 2800pc (Walborn and Hesser, 1975) Eta Car is arguably the most luminous ($10^{6.6}L_{\odot}$ — Davidson et al, 1986) star in the Galaxy and thought to be a massive star on the verge of a supernova explosion (Davidson, Walborn and Gull, 1982). Its irregular outbursts, observed over the last 300 yrs (Feinstein and Marraco, 1974, Walborn and Liller, 1977) manifest themselves in the shells of expanding nebulosity which are sketched in Fig. 5. Of particular interest is the jet-like feature found by Hester et al (1991) pointing away from the central star and culminating in the collisionally ionized knot NN whose proper motion from Eta Car had been measured as 9.45″/century (Walborn, Blanco and Thackeray, 1978).

With MES on the AAT, spatially resolved profiles of the [NII]6584Å emission line have now been obtained (see Meaburn et al, 1993d) along the two slit positions (marked 1 and 2) in Fig. 5. Slit pos. 1 is along the 'jet' and through knot NN whereas slit pos. 2 is just off the jet but still through the edge of knot NN.

The pv arrays of profiles of the [NII]6584Å line along these positions are shown in Figs. 6a and b respectively.

The brightest emissions from knot NN and the 'jet' in Fig. 6a have remarkably similar heliocentric radial velocities, $V_{HEL}= -637 kms^{-1}$ which for a systemic radial velocity of $V_{HEL}= -13$ km s^{-1} indicates velocities of 1374 km s^{-1} away from Eta Car at an angle of 27° to the sky when the proper motion of knot NN is considered.

Consequently, only one possibility plausibly explains the pv map in Fig. 6b. The features marked NN and 'jet' must both be produced by shocks driven at 140 km s^{-1} (Meaburn et al, 1988) into separate unseen bullets, travelling at 1510 km s^{-1} for 130 and 100 yrs respectively from Eta Car. If knot NN was simply the working surface of a shock being driven into the 'jet' then a significant radial velocity difference would have been present between these two features in Fig. 6a.

Episodic jets are being emitted by a wide range of galactic phenomena e.g. from YSO's which produce HH objects, planetary nebulae such as Fleming 1 (Lopez et al, 1993a and b) and here the luminous blue variable star Eta Car. Perhaps the common feature of these diverse phenomena is that a low mass star is present in the envelope of the eruptive companion, in which case the mechanism proposed

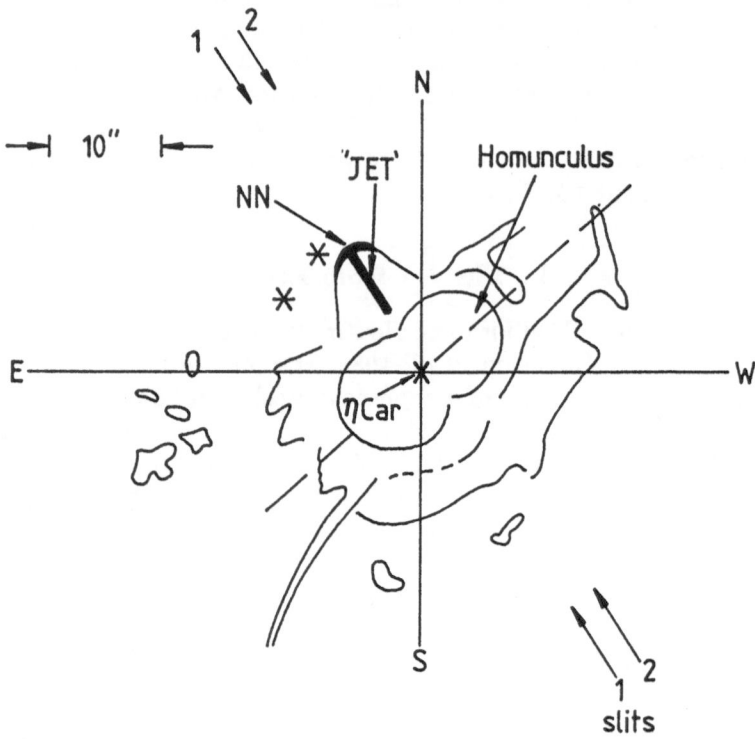

Fig. 5. A sketch of the expanding gas surrounding Eta Car is shown. The slit positions 1 and 2 are on and off the 'jet' respectively.

by Soker (1992) may apply in modified forms to all cases where episodic jets are observed.

Incidentally, the bow-shaped envelope sketched in Fig. 5 and marked as ENV in Figs. 6a and b is more likely produced by matter being 'squirted ' from the working surface of knot NN than by a shock being driven into the ambient gas (Falle and Raga, 1993). The cooling times would be far too large in the latter case.

4   The Helix globules

The most interesting features within the closest (distance $\approx$ 130 pc) planetary nebula, the Helix (NGC 7293) are surely the ionized knots, $\approx 1''$ across, with comet-like tails (Voronstev-Velyaminov, 1968) pointing in radial spokes away from the central star. These knots are found on the inside edge of the prominent low-ionization/molecular helical structure, yet on the outer edge of the highly ionized

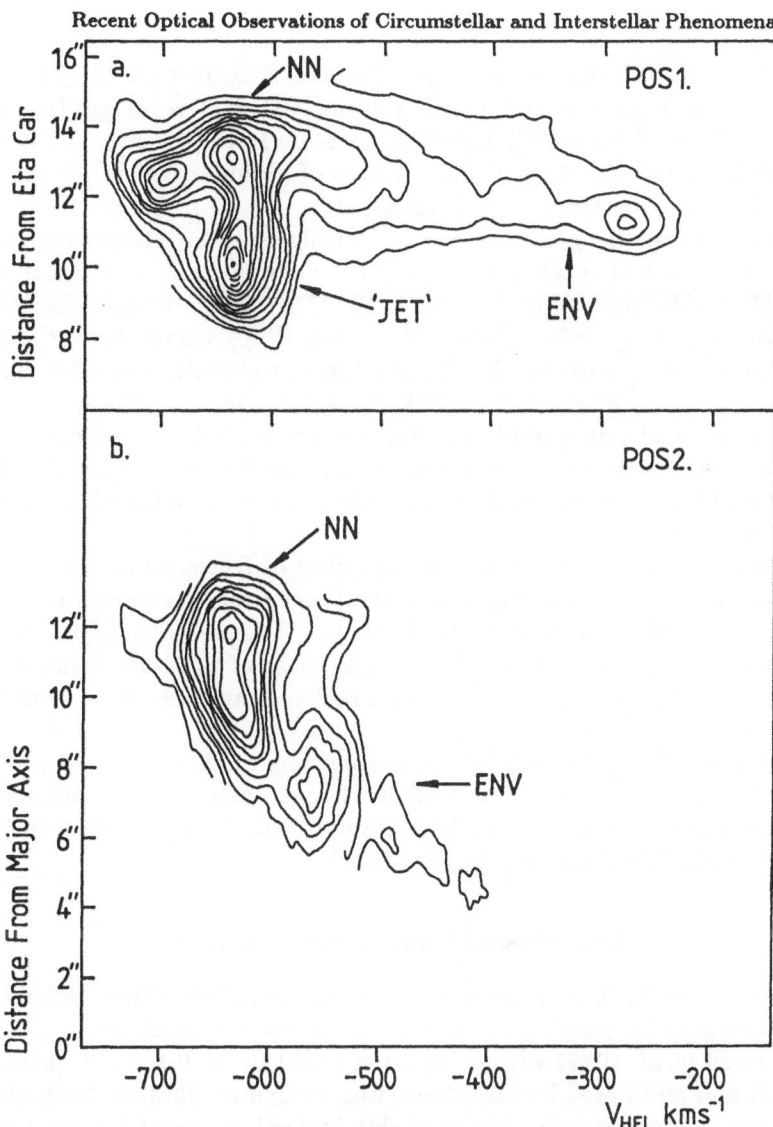

Fig. 6.   Position-velocity arrays of [NII]6584Å profiles along slit positions 1 and 2 have been contoured with linear intervals. The motions of knot NN, the envelope (ENV) and the 'jet' sketched in Fig. 5 are identified.

nebular core.

It is the presence of the central, spherical, shells which emit the [OIII]5007Å line strongly (Meaburn and White, 1982) that has permitted the detection of the dense, dusty, molecular cores of the Helix knots (see Meaburn et al, 1992, Walsh and Meaburn, 1993). The knots with their absorbing and scattering dust, on the nearside surface of the nebular core, are silhouetted against its bright emission. Consequently these appear as dark patches in the [OIII]5007Å images (see Fig. 7a) whereas, in the light of $H\alpha$ and [NII]6584Å emission lines (see Fig. 7b) the bright rims of ionized gas can be seen on the surfaces of globules which are exposed to the Lyman photons from the central ionizing star.

The parameters of the globules can be calculated from the fractions of the background light that they absorb. A typical globule mass and mean number density (of HI and H_2) derived in this way is $10^{-5} M_\odot$ and $\geq 10^5$ cm^{-3} respectively. Angular resolutions (e.g. with COSTAR on the HST) better than the present 0.8″ will be required to explore the densities in the globular cores by this same absorption technique. These are expected to reach far higher values.

CO observations of one globule by Huggins et al (1992) closely confirm the present 'optical' measurements of the mass and mean density of its molecular gas. Moreover, the CO measurements show that the globule has a velocity close to the systemic radial velocity.

These observations are consistent with the view of Dyson et al (1989) that the globules were formed in the atmosphere of the progenitor star during its red giant phase and ejected in the subsequent slow (10 km s^{-1}) dense wind only to be overrun by the 30 km s^{-1} shell ejected in the later AGB phase of the star. A more fanciful suggestion is that the knots and tails form a real cometary cloud in orbit around the central star.

Incidentally, neutral globules with similar parameters have now been found in the Dumb-bell planetary nebula (NGC6853) though their irregular shapes, along with the absence of cometary tails make them unlike their distinctive counterparts in the Helix nebula (Meaburn and Lopez, 1993).

5   High-speed ionized flows in M42

The close vicinity of the Trapezium cluster in the core of the Orion nebula (M42) is swept by the energetic particle wind of the O6 star θ^1C Orionis of $\dot{M} \leq 10^{-7} M_\odot$ yr^{-1} (Churchwell et al, 1987) with a terminal speed of 1650 km s^{-1} (Franco and Savage, 1982) and irradiated by an intense flux of Lyman photons from all of the Trapezium stars. Consequently, the six highly ionized, compact (diams from 1.3–2.7 10^{13} cm with electron densities $\approx 3 \times 10^6$ cm^{-3}) LV knots found by Laques and Vidal (1973) within 9″ of the Trapezium cluster (see Fig. 8), each of which contains an unresolved K-band source and presumably a low mass star (Meaburn, 1988), are expected to exhibit evidence of an interaction as their surfaces are directly exposed to this extreme environment. Moreover, if the central source in each knot is a young stellar object (YSO), collimated outflows of ionized gas may occur.

Both localised and extensive outflows of high-speed, ionized gas have now been found in the vicinity of the LV knots. Longslit observations (Meaburn, 1988b; Meaburn et al, 1993c) with MES on the AAT, INT and WHT of the profiles of the [OIII]5007Å line have revealed collimated receding outflows in the vicinity of

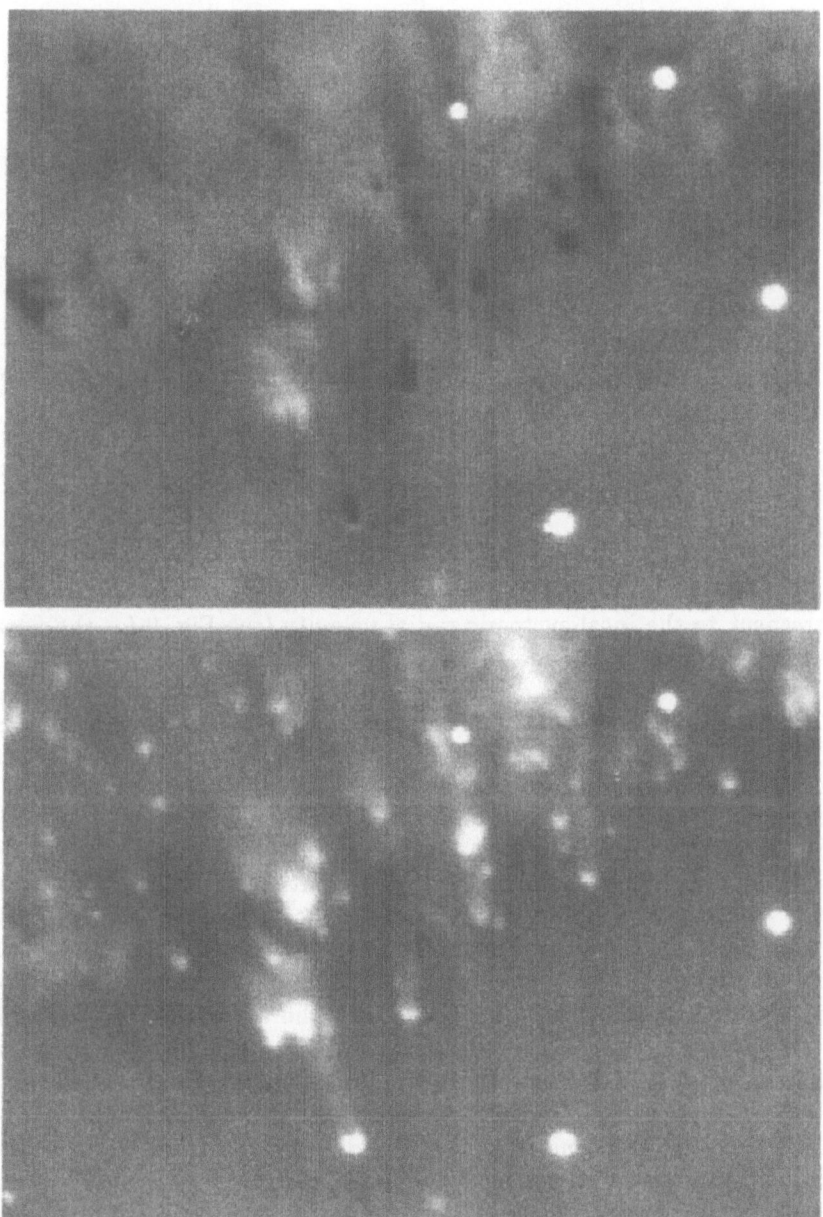

Fig. 7.   Images of a region of the Helix nebula in the light of [OIII]5007Å and Hα+[NII] lines are shown in a and b respectively (Meaburn et al 1992, Walsh and Meaburn, 1993).

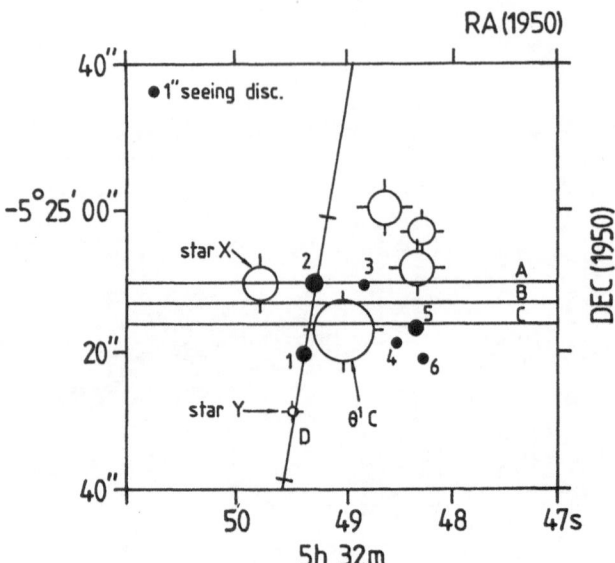

Fig. 8.   Slit positions are shown against a sketch of the Trapezium stars and the LV knot 1–6.

LV 1, 2, 3 and 5. These flows are from regions ≈ 1″ across and out to ≈ 130 km s^{-1} with respect to the local systemic radial velocity (see Fig. 9 for the flow from LV 2). Also some kind of ionized shell of 1′ radius appears to be centred on the Trapezium cluster (Meaburn *et al*, 1993; Massey and Meaburn, 1993 and see the Poster Paper at this conference). The approaching side of this 'shell' is clearly detected with radial velocities of up to 100 km s^{-1} with respect to the systemic radial velocity.

An extended (7″ across) 'loop' of ionized gas, receding at 94 and 134 km s^{-1}, is similarly found in the vicinity of LV 1 (Meaburn, 1988b; Meaburn *et al*, 1993c).

A plausible, unified, mechanism for the generation of all of these kinematic phenomena was proposed by Meaburn *et al* (1993c). Here a shell, expanding at 95 km s^{-1}, driven by the pressure of the shocked wind from θ^1C Orionis (see Dyson and de Vries, 1972), overruns the LV knots to cause localised flows of ionized gas from their surfaces. This model has the attraction of producing the two flow velocities that are observed. However, the later observations (Massey and Meaburn, 1993) which show that the 1′ radius expanding 'shell' is not a simple, single, radially expanding structure have lead to the possibility that all of the localised flows are the consequence of the direct interaction of the unshocked stellar wind with evaporating gas from the ionized surfaces of the globules by the momentum-conserving mechanism of Dyson (1975). This possibility is depicted schematically in Fig. 10 as an explanation of the present kinematic phenomena. True jets from YSO's, though, may occur in some cases.

The six LV knots have their counterparts in the 16 further compact thermal

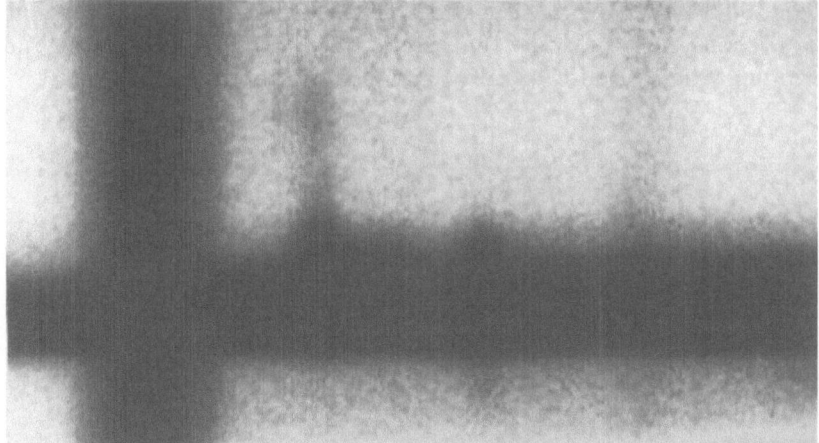

Fig. 9.   A greyscale representation of part of the position–velocity array of [OIII]5007Å profiles along slit position A in Fig. 8 is shown. The horizontal dimension is along the slit and positive radial velocities are to the top. The dark vertical band is the continuous spectrum of star X in Fig. 8. Just to its right the collimated flow from LV 2 can be seen out to receding radial velocities of $\approx$ 130 km s^{-1} with respect to the systemic radial velocity. The horizontal band is formed by the [OIII]5007Å profiles from the bulk of the nebular gas near this systemic value.

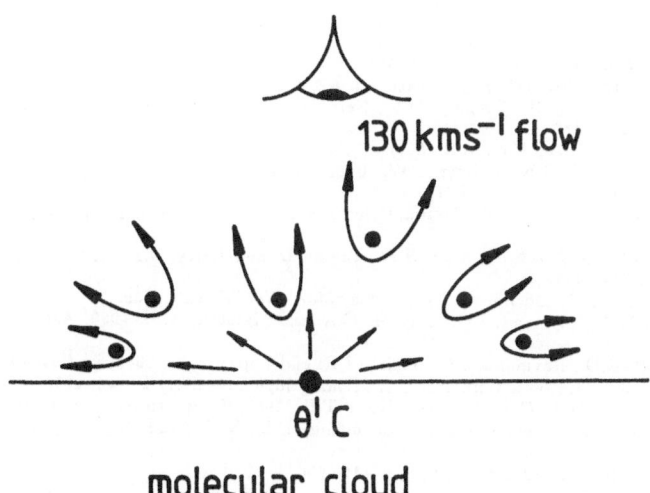

Fig. 10.   A schematic model to explain both the localised flows and the extensive 1′ radius feature is shown. Here the particle wind fro θ^1C Orionis mixes with the evaporating globular gas to sweep the flows away.

radio sources found in the 4' core of M42 by Moran *et al* (1982) and Churchwell
et al (1987) and those imaged with the HST in such a spectacular fashion by
O'Dell *et al* (1993) and shown at this conference. A nursery of YSO's, still in their
interstellar cocoons, is revealed by the scouring action of the energetic winds from
the Trapezium stars (O'Dell *et al*, 1993 and this conference).

References

Churchwell, E., Felli, M., Wood, D.O.S. and Massi, M.: 1987, *Astrophys. J.* 321, 516.
Chu, Y.H., MacLow, M.M., Garciasegura, G., Wakker, B. and Kennicutt, R.C.: 1993, *Astrophys. J.* 414, 213.
Chu, Y.H. and MacLow, M.M.: 1990, *Astrophys. J.* 365, 510.
Davidson, K., Dufour, R.J., Walborn, N.R. and Gull, T.R.: 1986, *Astrophys. J.* 305, 867.
Davidson, K., Walborn, N.R. and Gull, T.R.: 1982, *Astrophys. J.* 254, L47.
Dyson, J.E.: 1975, *Astrophys. Space Sci.* 35, 299.
Dyson, J.E. and de Vries, J.: 1982, *Astron. Astrophys.* 20, 233.
Dyson, J.E., Hartquist, T.W., Pettini, M. and Smith, L.J.: 1989, *Mon. Not. Roy. astr. Soc.* 241, 635.
Falle, S.A.E.G. and Raga, A.C.: 1993, *Mon. Not. Roy. astr. Soc.* 261, 573.
Feinstein, A. and Marraco, H.G.: 1974, *Astron. Astrophys.* 30, 271.
Franco, J. and Savage, B.D.: 1982. *Astrophys. J.* 225, 541.
Hester, J.J., Light, R.M., Wetsphal, J.A., Currie, D.G., Groth, E.J., Holtzman, J.A., Lauer, T.R. and O'Neil, E.J.: 1991, *Astron. J.* 102, 654.
Huggins, P.J., Bachiller, R., Cox, P. and Forveille, T.: 1992, *Astrophys. J.* 401, L43.
Laques, P. and Vidal, J.L.: 1979, *Astron. Astrophys.* 73, 97.
Lopez, J.A.L., Meaburn, J. and Palmer, J.W.: 1993, *Astrophys. J. Letts.* 415 L135.
Lopez, J.A.L., Roth, M. and Tapia, M.: 1993, *Astron. Astrophys.* 267, 194.
MacLow, M.M. and Norman, M.L.: 1993, preprint 013 University of Illinois at Urbana-Champaign.
Massey, R.M. and Meaburn, J.: 1993, *Mon. Not. Roy. astr. Soc.*
Meaburn, J.: 1981, *Mon. Not. Roy. astr. Soc.* 196, 19P.
Meaburn, J.: 1984, *Mon. Not. Roy. astr. Soc.* 251, 521.
Meaburn, J.: 1988a, *Mon. Not. Roy. astr. Soc.* 235, 375.
Meaburn, J.: 1988b, *Mon. Not. Roy. astr. Soc.* 233, 791.
Meaburn, J.: 1990, *Mon. Not. Roy. astr. Soc.* 244, 551.
Meaburn, J. and White, N.J.: 1982, *Astrophys. Space Sci.* 82, 423.
Meaburn, J., Blundell, B., Carling, R., Gregory, D.F., Keir, D. and Wynne, C.G.: 1984, *Mon. Not. Roy. astr. Soc.* 210, 436.
Meaburn, J. and Lopez, J.A.: 1993b, *Mon. Not. R. astr. Soc.*, in press.
Meaburn, J., Gehring, G., Walsh, J.R., Palmer, J.W., Lopez, J.A., Bryce M. and Raga, A.C.: 1993d, *Astron. Astrophys. (Letts)* 276, L21.
Meaburn, J., Massey, R.M., Raga, A.C. and Clayton, C.A.: 1993c, *Mon. Not. R. astr. Soc.* 260, 625.
Meaburn, J., Walsh, J.R., Clegg, R.E.S., Walton, N.A., Taylor, D. and Berry, D.S.: 1992, *Mon. Not. Roy. astr. Soc.* 255, 177.
Meaburn, J., Wang, L., Palmer, J. and Lopez, J.A.: 1993a, *Mon. Not. R. astr. Soc.,* ,.
Meaburn, J., Whitehead, M.J., Raymond, J.C., Clayton, C.A. and Marston, A.P.: 1990, *Astron. Astrophys.* 227, 191.
Meaburn, J., Wolstencroft, R.D., Raymond, J.C., Walsh, J.R. and Lopez, J.A.: 1988, in *Dust in the Universe*, eds. Bailey M.E. and Williams, D.A., (Cambridge University Press), 381.
Moran, J.M., Garay, G., Reid, M.J., Genzel, R. and Ho, P.T.P.: 1982, Symp. to Honour Henry Draper, eds. Glasgold, A.E., Huggins, P.J. and Schucking, E.L., N.Y. Acad. Science., 395, 64.
O'Dell, C.R., Wen, Z. and Hu, X.: 1993, *Astrophys. J.* 410, 696.
Soker, N.: 1992, *Astrophys. J.* 389, 628.
Vorontsov-Velyaminov, B.A.: 1968, in: *Planetary Nebulae*, IAU Symp. No 34, p 256, eds. Osterbrock, D.E. and O'Dell C.R., D. Reidel Publ. Co., Dordrecht.
Walborn, N.R., Blanco, B.M. and Thackeray, A.D.: 1978, *Astrophys. J.,*2 19, 498.
Walborn, N.R. and Hesser, J.E.: 1975, *Astrophys. J.* 199, 535.
Walborn, N.R. and Liller, M.H.: 1977, *Astrophys. J.* 211, 181.
Walsh, J.R. and Meaburn, J.: 1993, *ESO Messenger* No. 73, 35.
Wang, L.: 1992, *ESO Messenger*, No 69, 34.
Wang, Q. and Helfand, D.J.: 1991a, *Astrophys. J.* 370, 541.
Wang, Q. and Helfand, D.J.: 1991b, *Astrophys. J.* 373, 497.

Internal Motions of HII Regions and Giant HII Regions

You-Hua Chu
Astronomy Department, University of Illinois, 1002 W. Green Street, Urbana, IL 61801, USA

and

Robert C. Kennicutt, Jr.
Steward Observatory, University of Arizona, Tucson, AZ 85721, USA

Abstract.
We report new echelle observations of the kinematics of 30 HII regions in the LMC, including the 30 Doradus giant HII region. All of the HII regions possess supersonic velocity dispersions, which can be attributed to a combination of turbulent motions and discrete velocity splitting produced by stellar winds and/or embedded supernova remnants (SNRs). The core of 30 Dor is unique, with a complex velocity structure that parallels its chaotic optical morphology. We use our calibrated echelle data to measure the physical properties and energetic requirements of these velocity structures. The most spectacular structures in 30 Dor are several fast expanding shells, which appear to be produced at least partially by SNRs.

Key words: kinematics, HII regions, giant HII regions, stellar winds, supernova remnants

1   Introduction

The basic knowledge of internal motions of HII regions and giant HII regions was established by Smith and Weedman (1970, 1971, 1972, 1973) using Fabry-Perot scanner observations. They measured the velocity dispersions of $\sim$100 HII regions in the Magellanic Clouds, and found the typical velocity dispersion to be $\beta \sim$15 km s^{-1} (or FWHM $\sim$ 25 km s^{-1}). The giant HII region 30 Doradus on the other hand has velocity dispersions nearly twice as large. The giant HII regions in M33 and M101 also have similarly large velocity dispersion, $\beta \sim$20–34 km s^{-1} (or FWHM $\sim$ 33–57 km s^{-1}).

Observations of the integrated velocity profiles of large samples of giant HII regions confirm that violent internal motion is characteristic of giant HII regions in general (*e.g.*, Melnick 1980; Hippelein 1986; Roy *et al.* 1986). These observations also show that nebular velocity dispersions are loosely correlated with the diameters and luminosities of the HII regions, suggesting that giant HII regions may be useful as extragalactic distance indicators (Melnick 1977, 1980; Arsenault and Roy 1988). Several physical mechanisms have been proposed to account for the supersonic motions and the L–σ and D–σ correlations, including self-gravity, large-scale rotation or champagne flows, turbulence, stellar winds, and SNRs (see references above). Long-slit spectroscopy for the central parts of a few objects provide support for local champagne flows or stellar winds (Gallagher and Hunter 1983; Meaburn 1984; Rosa and Solf 1984; Skillman and Balick 1984) and SNRs (Chu and Kennicutt 1986), but the relation of these local features to the global kinematics of the HII regions is unclear.

The Large Magellanic Cloud (LMC), at a distance of only 50 kpc, provides a unique opportunity to study the kinematic structure of HII region over a range of sizes, with sufficient spatial resolution to isolate these individual dynamical processes. In a series of important papers, Meaburn has used echelle spectroscopy to reveal the diverse range of kinematic structures in the LMC HII regions, especially the 30 Doradus giant HII region (Meaburn 1981, 1984, 1988). We have used the echelle CCD spectrograph on the CTIO 4 m telescope to map the nebular velocity field in ~30 HII regions around OB associations in the LMC, including extensive observations of the 30 Dor region (Chu *et al.* 1992; Chu and Kennicutt 1993a, b). Kinematics alone often cannot distinguish between energetic phenomena, such as fast stellar winds and SNR shocks, so we have obtained additional information at X-ray wavelengths. In this paper we describe the kinematic structure of 30 Dor, compare it to smaller HII regions, analyze the dynamic processes, and use its integrated velocity profile to decipher those for distant, unresolved giant HII regions.

2   Internal Motions of LMC HII Regions

We study HII regions around OB associations because they are expected to show combined effects of stellar winds and SNRs that are similar to those in giant HII regions only at a smaller scale. Over one hundred OB associations are cataloged in the LMC (Lucke and Hodge 1970). Ideally, we would like to include a large number of HII regions for a wide range of stellar ages, so that we may examine how the internal motions of HII regions evolve with the central OB associations. However, the stellar contents of only a few OB associations have been studied in detail (*e.g.*, Conti *et al.* 1986; Massey *et al.* 1989); the majority of LMC OB associations have just photographic photometry available (Lucke 1971). With limited stellar information available, we have to use the nebular morphology as a rough indicator of their evolutionary status.

Since fast stellar winds and supernovae from an OB association will sweep up the ambient gas into a shell, a superbubble (McCray and Kafatos 1987), we expect the shell HII regions be more evolved than the amorphous HII regions. We selected 29 HII regions, with ~2/3 being amorphous and ~1/3 being shells, and observed a total of ~50 positions with a 4′-long slit. Several spectra were extracted along each slit position. Figure 1 shows the distribution of the observed velocity FWHM (including instrumental and thermal broadenings) of all spectra extracted. The data have not been fully analyzed, but the preliminary results indicate that the shell HII regions indeed have higher kinetic energy density than the amorphous objects. The apparent velocity FWHM of the amorphous HII regions are 32–38 km s^{-1}, corresponding to 15–25 km s^{-1} "turbulent" FWHM after the correction for the instrumental FWHM of 18.5 km s^{-1} and the thermal width of 20 km s^{-1}. The shell regions show expansion velocity of 15–45 km s^{-1}. The broadest unresolved line profile is seen in the diffuse nebulosity around the OB association

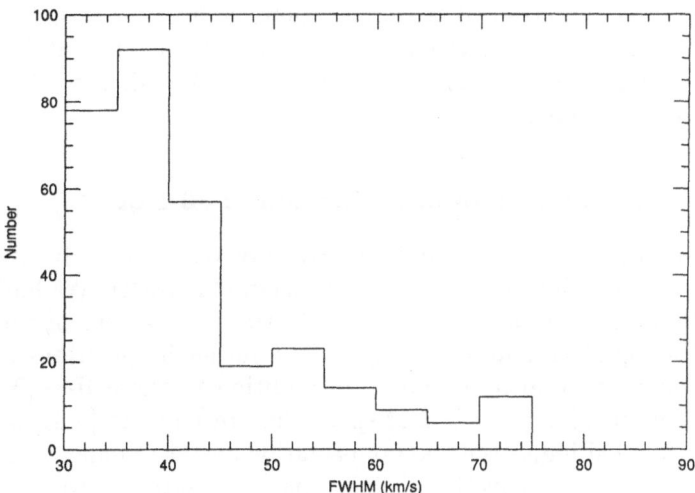

Fig. 1.   Distribution of the observed velocity FWHM (including instrumental and thermal broadenings) of all spectra extracted in 29 LMC HII regions, not including 30 Dor.

LH79, apparent FWHM of 47±5 km s^{-1}. If the stellar contents had been studied, it would be possible to estimate the energy input rates from the OB associations and compare them to the kinetic energy density growth rates. Such investigation is currently undertaken by S. Oey as a Ph.D. thesis project.

Two HII regions show very high velocity (Δ V $>$ 100 km s^{-1}) features – N44 and N159. Bright X-ray emission has been reported in the main shell of N44 (Chu et al. 1993), and it is suggested that SNRs near the shell walls hit the dense shell gas and emit X-rays (Chu and Mac Low 1990). The high velocity component in N44 is detected at $\Delta V = -125$ km s^{-1} near the "B" knot. Since no HeIIλ4686 emission is detected in N44B, this high-velocity Hα feature cannot be low-velocity HeIIλ6560 emission, as in the case of N44C (Pakull 1990, private communication). This high velocity feature is probably associated with the same SNR shocks which produced the bright diffuse X-ray emission. In the HII region N159 we detected a 23 pc shell expanding at nearly 250 km s^{-1}. This is probably a SNR. The [S II]/Hα ratio is enhanced, but this shell is superposed on the luminous HII region N159 and is only 100$''$ from LMC X-1, so the radio and X-ray signatures of the SNR cannot be confirmed. For both cases, N44 and N159, the high velocity features appear to be related to SNR shocks.

This apparent association of SNRs with large OB/HII complexes is by no means unique in the LMC. The Crab-like SNR 30 Dor B is located in an active star forming region, and echelle and Fabry-Perot observations reveal kinematic features which can be ascribed to both stellar winds and SNR shocks (Chu et al. 1992). A complete study of the environments of the 32 known SNRs in the LMC revealed that 7 lie within the boundaries of OB associations, and 17 lie within HII regions (Chu and Kennicutt 1988b). Given these statistics we should not be surprised if the 30 Dor nebula, which contains a quarter of all the OB stars in the LMC, should contain several young SNRs, as we show later.

3 Violent Internal Motion of 30 Dor

The turbulent internal motion in 30 Dor is readily testified by its chaotic filamentary structure and bright diffuse X-ray emission (Figure 2). We have mapped the velocity field of the core $9' \times 9'$ region of 30 Dor with a grid of slit positions separated by $3'$ along E-W and $45''$ along N-S. Another 53 positions in the halo or regions of interest were also observed. In addition to these Hα+[N II] observations, we also observed 5 positions of special interest in the [S II]$\lambda\lambda$6717,6731 doublet, which provided high resolution information on the nebular excitation and kinematics in the shock-sensitive [S II] lines, and kinematically resolved electron density information.

The large-scale kinematic structure of 30 Dor is characterized by a complex, energetic core and a relatively quiescent halo. Most of the fast expanding shells and distinct high velocity features are confined within the core. At regions $9'-13'$ from the center, the Hα line image becomes quite smooth and featureless, although a few small fast expanding shells still appear. To demonstrate the contrast between the halo and the core of 30 Dor, the spatial variation of velocity dispersion in 30 Dor is plotted in Figure 3. The data were obtained by selecting slit positions at large radial distances from R136 (but avoiding a large expanding network in the NW), and measuring the Hα line widths over regions $10-70''$ in length. There is a clear upper envelope to the distribution, and the halo shows a distinctly lower velocity dispersion than the core. Hence the halo shows not only fewer numbers of organized high velocity structures, but a more quiescent "turbulent" velocity field overall. A similar radial variation in dispersion is also seen in NGC 604, a comparable giant HII region in M33 (Melnick 1980; Hippelein and Fried 1984). Note however that the velocity dispersions in the halo of 30 Dor are still higher than the median of those in smaller HII regions, as can be seen by comparing Figures 1 and 3.

In order to study the range of kinematic structure in 30 Dor we combined our grid of 37 positions in the $9' \times 9'$ core region into echellogram mosaics, and a mosaic of the Hα line is shown in Figure 4. Each horizontal cut in Figure 4 covers $9'$ along an E-W line, with dispersion running vertically. The 13 E-W spectra are separated by 45" in declination, and the velocity range in each spectrum is ± 300 km s^{-1}. The bright continuum source near the center of Figure 4 is R136. The mosaic illustrates the complexity of internal motions in 30 Dor. Line splitting by $20-100$ km s^{-1} in the low-velocity gas is frequently observed; high-velocity features with $100-300$ km s^{-1} offset from the systemic velocity are

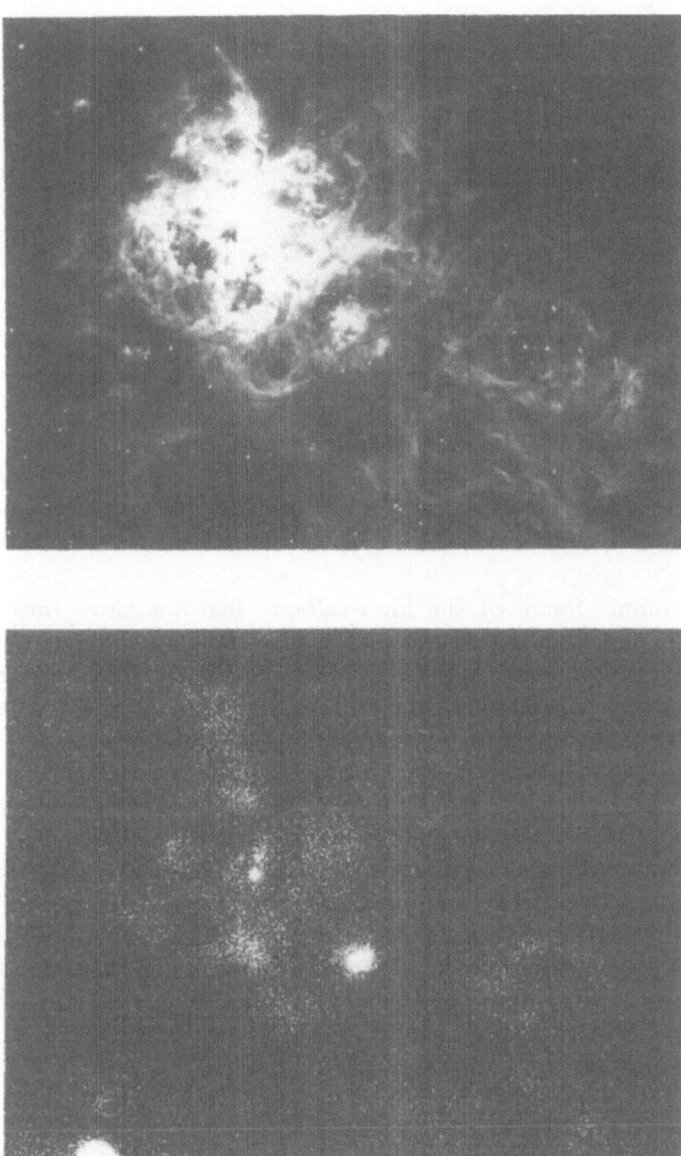

Fig. 2.   The 30 Doradus giant HII region. Curtis Schmidt Hα image (top) and ROSAT PSPC soft X-ray image (bottom)

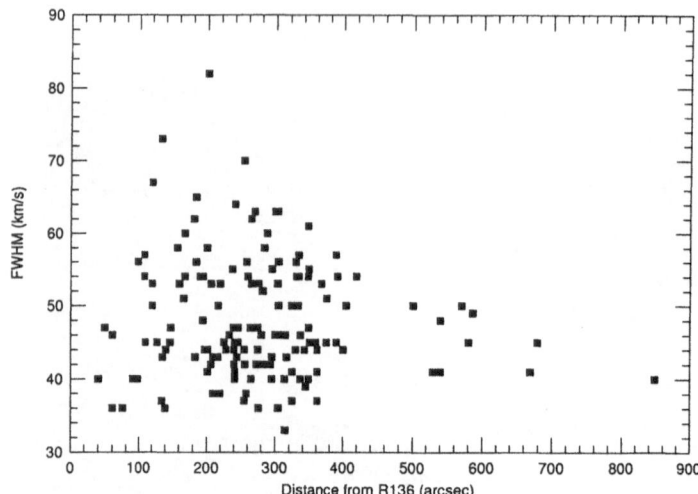

Fig. 3.   FWHM velocity widths at various positions in 30 Dor, plotted as a function of distance from R136.

also quite common. Some of the high-velocity features show bow-shaped line images, indicating an expanding-shell structure, while others remain detached from the low-velocity components, indicating discrete clouds, filaments, or sheets of gas crossing the slit. In many cases discrete high-velocity features with similar kinematic characteristics congregate into large "networks."

Despite the violent internal motions, the systemic velocity field in 30 Dor is surprisingly simple. In halo regions as far out as 9'E, 10'S, 12'W, and 13'N of R136, the peak velocities all lie within a small range $V_{hel} = 265–275$ km s^{-1}. The mean velocity measured at 13 points in the outermost halo is $V_{hel} = 273\pm2$ km s^{-1}, while the mean velocity of the core region is $V_{hel} \simeq 266\pm2$ km s^{-1}. The simple systemic velocity field of 30 Dor suggests that the HII region originated from one large cloud, as opposed to a collision of two or more clouds. The velocity difference between the core and the halo is expected as a result of thermal expansion in ionized gas. No global rotation is evident in the velocity field. No large systematic velocity drift across the whole nebula is evident, either.

4   Expanding Shell Structures in 30 Dor

The most spectacular kinematic features in 30 Dor are the many different types of expanding shells. Many such shells can be readily seen in the mosaic in Figure 4. The smallest shells are only a few pc across and their expansion velocities are up to ~100 km s^{-1}. Line splitting of up to 100 km s^{-1} is also observed over larger regions up to 100 pc in diameter (*cf.* Meaburn 1984), including a massive

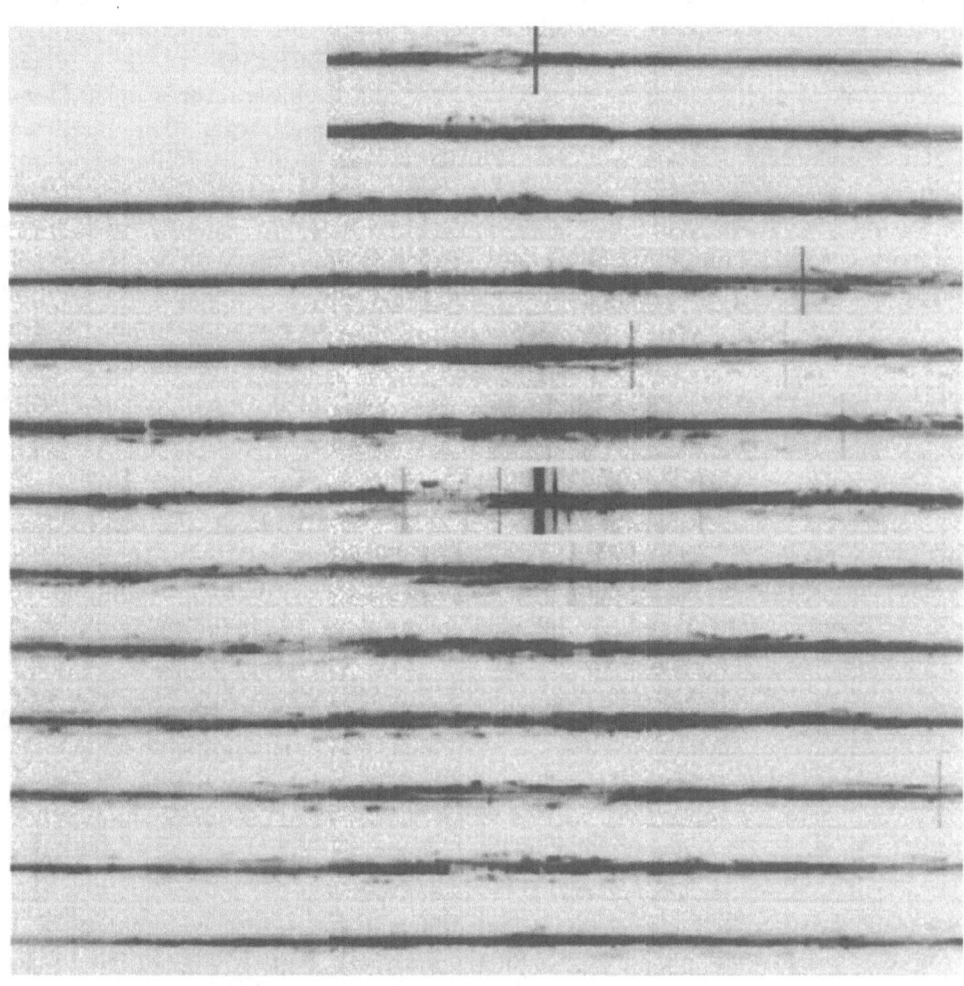

Fig. 4.   Mosaic of echellograms of the central $9' \times 9'$ core of 30 Dor. See text for scale and explanation of features.

shell surrounding the central R136 cluster, which contains some of the brightest nebulosity in 30 Dor.

The most energetic structures are large fast shells with diameters of 2–20 pc and expansion velocities of 100–300 km s^{-1}. Several such shells can be seen in our echellogram mosaics in Figure 4. Since our echelle data are intensity calibrated, and we have [S II] densities for several objects, we have been able to estimate the physical parameters of a number of prototype shells. The shell masses range from 100–5000 M$_\odot$, and the kinetic energy in the *visible* ionized shells ranges over $10^{49} - 10^{51}$ ergs. The energy required to accelerate the gas, either from winds or supernovae, is roughly an order of magnitude higher. In some cases a number of shells and more isolated high velocity features are clearly organized into larger expanding networks, *e.g.*, the region at $2' - 4'$ south of R136.

Analysis of our observations indicates that the shell structures in 30 Dor are produced by a combination of stellar winds and supernovae. The small shells with expansion velocities of less than 100 km s^{-1} are probably stellar wind-blown bubbles, rather than SNRs, because SNRs of such sizes would have much higher expansion velocities, and their X-ray emission would have shown up as bright, unresolved sources in Figure 2. At the other extreme, most of the fast expanding shells must be produced by supernovae, probably in combination with stellar winds. Several lines of evidence support this interpretation. On energetic grounds alone, the large energetic requirements of the largest shells, combined with their short expansion time scales (order $10^4 - 10^5$ yr) require far more stars than are present within the shell boundaries, if stellar winds were to be the sole source of kinetic energy (the shell located immediately around the R136 cluster could be an exception). In addition most of the fast shells are coincident with extended X-ray sources, which can best be understood as originating from an SNR (or SNRs) hitting the walls of a superbubble (Chu and Mac Low 1990). The resemblence of the velocity structure of the 30 Dor shells to isolated LMC SNRs (Chu and Kennicutt 1988a; Chu *et al.* 1992) lends further support to this interpretation. The physical mechanisms responsible for the slower large shells are less clear, though it would not be surprising if a combination of winds and supernovae, analagous to what was inferred for N44 and N159, is responsible. This would make the 30 Dor shells perhaps the most accessible observed examples of the interstellar supershells envisioned theoretically by Mac Low and McCray (1988) and others.

5   Integrated Velocity Profile of 30 Dor

The spatial coverage of our echelle map is sufficient that we can combine the data to produce a pseudo-integrated profile for the giant HII region. This profile is shown in Figure 5. Despite the chaotic velocity structure of the nebula on small scales, the integrated profile is remarkably smooth. The lesson to be learned from this result is that the nearly Gaussian velocity field observed for more distant extragalactic HII regions offers virtually no physical insight into the physical mechanisms which

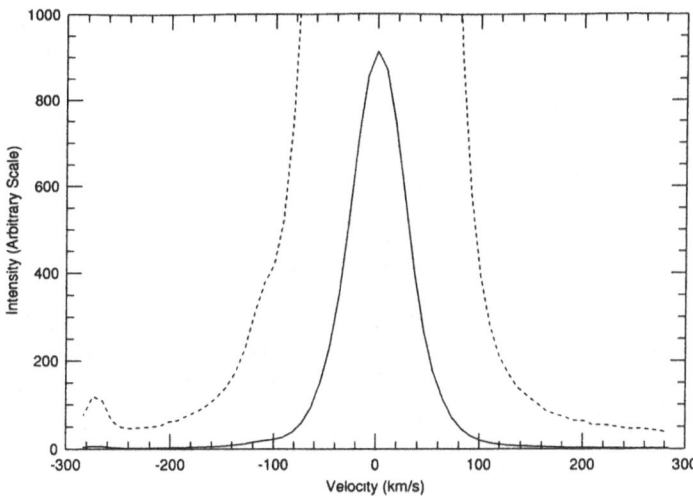

Fig. 5. Integrated profile of the Hα emission line in 30 Dor, derived by summing the line profiles in Fig. 4. Top plot (dashed line) shows an expanded scale for the line wings.

accelerate the gas; they are artifacts of the poor spatial and spectral resolution of those observations.

As discussed in the introduction, previous authors have attributed the supersonic velocity dispersions of giant HII regions to a variety of physical mechanisms, ranging from self-gravity and rotation to turbulence, champagne flows, and bulk input from stellar winds and supernovae. Our spatially resolved maps allow us to measure the relative contributions of these different mechanisms. In 30 Dor it can easily be shown that most of the observed line width can be attributed to three sources, the expansion velocities of individual shells (in particular the massive central shell around R136, which by itself is responsible for roughly half the line width in Fig. 5), the variation in central velocity between different shells, and the underlying "turbulent" line width of the smooth gas component (see Fig. 3). Photoevaportive motions such as champagne flows may contribute to the motions in the shells, but do not appear to contribute as much to the global velocity dispersion of the nebula. By contrast self-gravity ("virial" motions), rotation, and any other bulk motion are of minor consequence, contributing less than 10% of the observed integrated velocity dispersion.

It is notable that two of the three mechanisms of importance are associated with the large, discrete velocity structures. Hence it appears that the physical mechanisms which are responsible for the expanding shells, stellar winds and/or SNRs, are the dominant contributors to the global kinematics of the giant HII region. Indeed the total kinetic energy contained in the fast shells ($V_{exp} > 100$

km s^{-1}) represents up to half of all the kinetic energy in the nebula. This energy exceeds by severalfold the gravitational binding energy of the region, and indicates that 30 Dor is destined to evolve into a supergiant shell not unlike the large supergiant shells observed elsewhere in the LMC (Meaburn 1980). Hence 30 Dor may offer a unique opportunity to observe self-regulating star formation in action, and a precursor to the large shell structures which dominate the morphology and energetics of the interstellar medium in late-type galaxies.

Acknowledgements

YHC would like to thank Profs. F. Kahn and J. Dyson for their invitation to the conference and their hospitality during her visit to Manchester. YHC acknowledges the support of NASA grants NAG 5-1900 and NAG 5-2112. RCK acknowledges the support of NSF grant AST90-19150.

References

Arsenault, R., & Roy, J.-R. 1988, A&A, **201**, 199
Chu, Y.-H. and Kennicutt, R.C. 1986, ApJ, **311**, 85
Chu, Y.-H. and Kennicutt, R.C. 1988a, AJ, **95**, 1111
Chu, Y.-H. and Kennicutt, R.C. 1988b, AJ, **96**, 1874
Chu, Y.-H. and Kennicutt, R.C. 1993a, submitted to ApJ
Chu, Y.-H. and Kennicutt, R.C. 1993b, in preparation
Chu, Y.-H., Kennicutt, R.C., Schommer, R.A., and Laff, J. 1992, AJ, **103**, 1545
Chu, Y.-H. and Mac Low, M.-M. 1990, ApJ, **365**, 510
Chu, Y.-H., Mac Low, M.-M., Garcia-Segura, Wakker, and Kennicutt 1993, ApJ, in press
Conti, P.S., Garmany, C.D., and Massey, P. 1986, AJ, **92**, 48
Gallagher, J.S., and Hunter, D.A. 1983, ApJ, **274**, 141
Hippelein, H. H. 1986, A&A, **160**, 374
Hippelein, H. and Fried, J.W. 1984, A&A, **141**, 49
Lucke, P.B. 1971, ApJS, **255**, 73
Lucke, P.B. and Hodge, P.W. 1970, AJ, **75**, 171
Mac Low, M.-M. and MCray, R. 1988, ApJ, **324**, 776
McCray, R. and Kafatos, M.C. 1987, ApJ, **317**, 190
Massey, P., Garmany, G.D., Silkey, M., and Degioia-Eastwood, K. 1989, AJ, **97**, 107
Meaburn, J. 1980, MNRAS, **192**, 365
Meaburn, J. 1981, MNRAS, **196**, 19p
Meaburn, J. 1984, MNRAS, **211**, 521
Meaburn, J. 1988, MNRAS, **235**, 375
Melnick, J. 1977, ApJ, **213**, 15
Melnick, J. 1980, A&A, **86**, 304
Rosa, M., & Solf, J. 1984, A&A, **130**, 29
Roy, J.-R., Arsenault, R., and Joncas, G. 1986, ApJ, **300**, 624
Skillman, E.D., & Balick, B. 1984, ApJ, **280**, 580
Smith, M.G. and Weedman, D.W. 1970, ApJ, **161**, 33
Smith, M.G. and Weedman, D.W. 1971, ApJ, **169**, 271
Smith, M.G. and Weedman, D.W. 1972, ApJ, **172**, 307
Smith, M.G. and Weedman, D.W. 1973, ApJ, **179**, 461

High-speed flows in the vicinity of the Trapezium stars

J. Meaburn and R. M. Massey
Department of Astronomy
University of Manchester
Manchester M13 9PL

20th June 1993

Abstract. [OIII] 5007Å line profiles at high spectral and spatial resolution have been obtained at a single slit position near the Trapezium cluster in the Orion Nebula (M42, NGC1976) using the Manchester Echelle Spectrometer (MES). The very long integration time at this position confirms the earlier tentative identification of a shell on the nearside of the Trapezium cluster with a relative velocity of $-100\,\mathrm{km\,s^{-1}}$ and a radius of 1 arcminute. No receding counterpart is found. We believe this is the first detection of this feature at optical wavelengths, previous spectroscopic work (O' Dell *et al.*, 1993),(Baldwin *et al.*, 1991),(Castaneda, 1988) having concentrated on the main nebular material at relatively low velocities.

Key words: line:profiles - ISM:individual:Orion nebula - ISM:shells

The shell observed in this work was tentatively detected by Meaburn *et al.* (1993), [OIII] 5007Å line profiles having been obtained at slit positions in the vicinity of the knots LV1, 2, 3 and 5 (Laques & Vidal, 1979). High-speed material was observed faintly at velocities of $-100\,\mathrm{km\,s^{-1}}$ with respect to the systemic nebular velocity.

Twelve 600-second integrations were made at a single slit position (centred on $-5°24'40''$, 5h 32m 48.9s (1950.0) at p.a. 350°) near the Trapezium stars and through the highly-ionized knots LV1 and LV2 using the Manchester Echelle Spectrometer (MES (Meaburn *et al.*, 1984)) at the $f/15$ focus of the 2.5m Isaac Newton Telescope (INT) on the nights of 5 and 6 November 1992. A 70Å interference filter isolated the 114th echelle order which contains the [OIII] 5007Å line. The slit width was 150 μm ($\equiv 9\,\mathrm{km\,s^{-1}}$ and 0.7 arcseconds). The data was wavelength-calibrated to an accuracy of $\pm 2\,\mathrm{km\,s^{-1}}$ using the spectrum from a Thorium-Argon arc lamp and intensity-calibrated using the previous data (Meaburn *et al.*, 1993).

The resultant co-added position-velocity (pv) array is shown in *Fig 1*.

Following arguments applied in the previous work (Meaburn *et al.*, 1993), a $7'' \times 2''$ extent was assumed for the brightest clump X and applying Menzel's case B, we find the clump mass to be $\approx 3 \times 10^{28}$kg and its inferred momentum (if viewed along the axis) to be $P_{\mathrm{HII}} \approx 3 \times 10^{35}$g cm s^{-1}.

The star $\Theta_1 C$ is known to have a wind (Franco & Savage, 1982) with a terminal velocity of $1650\,\mathrm{km\,s^{-1}}$ and a mass-loss rate of $\leq 10^{-7}\,M_\odot\,\mathrm{yr^-}$ Assuming that the clump lies at a distance of 4.5×10^{17} cm from this star then the intercepted momentum $\delta P_W \geq 3.1 \times 10^{33}$g cm s^{-1} may be sufficient to drive a shell of this nature, although the published values are extremely uncertain.

A more detailed description is required to explain the complicated sub-structure, further frames not shown here suggest the shell has a sharper southern edge and a slightly elliptical shape. A more comprehensive analysis is given elsewhere (Massey

Astrophysics and Space Science **216**: 263–265, 1994.
© 1994 *Kluwer Academic Publishers.*

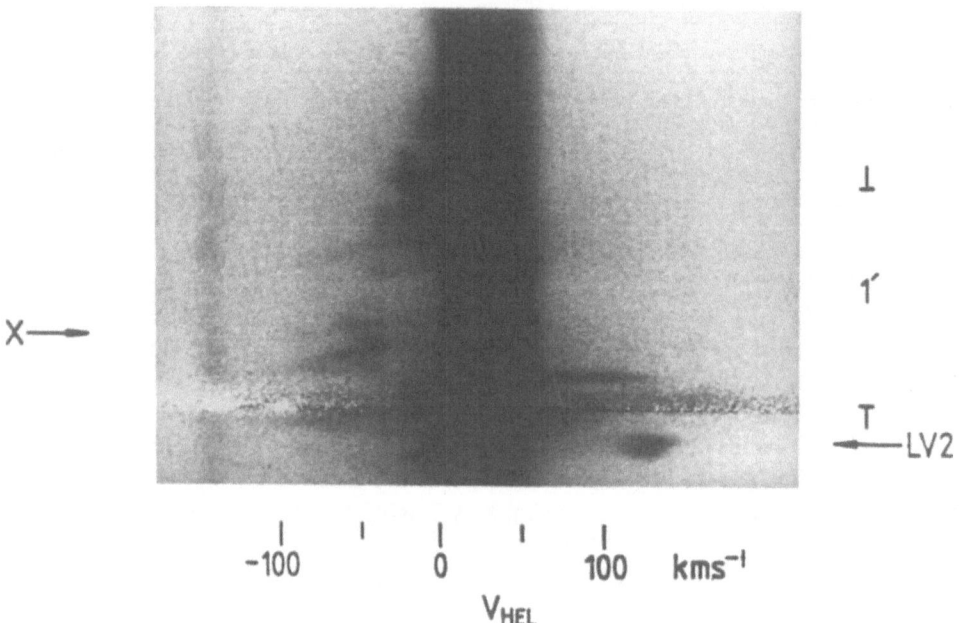

Fig. 1.   The resultant co-added [OIII] 5007Å deep negative print of the position-velocity array obtained. This frame has been continuum-subtracted with only a faint residual from the star $\Theta_1 C$ still visible. By far the most dramatic feature on this frame is the clumpy shell which covers the entire slit length and reaches velocities of $-100\,\mathrm{km\,s^{-1}}$ with respect to the bulk of the nebular material ($v_{HEL} \approx +18\,\mathrm{km\,s^{-1}}$). The shell itself appears to be made up of several smaller clumps or 'bubbles', most clearly seen at the top of the frame, typically on $\approx 10''$ scales, (7.5×10^{16} cm at 450 pc) and with extensions down to the systemic nebular velocity. Also labelled is the jet from the knot LV2 and part of the S-feature in the vicinity of the knot LV1.

& Meaburn, 1993).

We are grateful to the staff at the INT and to Bill Garner for printing the images so well. RMM is grateful for an SERC studentship.

References

Baldwin J. A., Ferland G. J., Martin P. G., Corbin M. R., Cota S. A., Peterson B. M., Slettebak A., 1991, *Astrophys. J.*, **374**, 580.

Castaneda H. O., 1988, *Astrophys. J.*, **67**, 93.

Franco J., Savage B.D., 1982, *Astrophys. J.*, **225**, 541.

Laques P., Vidal J. L., 1979, *Astr. Astrophys.*, **73**, 97.

Massey R.M., Meaburn J., 1993, *Mon. Not. R. Astr. Soc.*, **262**, L48.

Meaburn J., Blundell B., Carling R., Gregory D.F., Keir D., Wynne C.G., 1984, *Mon. Not. R. Astr. Soc.*, **210**, 436.

Meaburn J., Massey R. M., Raga A. C., Clayton C. A., 1993, *Mon. Not. R. Astr. Soc.*, **260**, 625.

Meaburn J., 1988, *Mon. Not. R. Astr. Soc.*, **233**, 291.

O' Dell C. R., Walk J.H., Wen Z., 1993, *Astrophys. J.*, **403**, 678.

The Orion Nebula: Structure, Dynamics, and Population

C. R. O'Dell
Rice University, Houston, Texas, USA

30 June 1993

Abstract.
    The physical nature and evolution of the Orion Nebula has begun to be revealed by calibrated emission line images and high resolution spectroscopy. We review the evidence that the nebula is a thin wall of emission on the near side of the Orion Molecular Cloud and that its separation from the dominant ionizing star is about 0.3 pc. The density of the nebula decreases rapidly away from the ionization front and the ionized gas is moving at 8 km s^{-1} away from the front. A three dimensional model of the surface is presented and its peaks and valleys interpreted as due to irregularities in the density of the molecular cloud. The front is moving rapidly into the molecular cloud, so that objects previously shielded from ionization are continuously being revealed. Recent Hubble Space Telescope images are reviewed and they indicate that protoplanetary disks around pre-main sequence stars are both common and rendered visible by ionization and projection against the bright nebula. A large body of velocity data is discussed and it is seen that Kolmogorov type turbulence seems to only apply to material in the ionization front and the statistical fluctuations become less correlated away from the front.

Key words: Orion Nebula, HII Region, Planetary Disks

1 Introduction

In spite of its antiquity and proximity, the Orion Nebula has only recently begun to yield information about its true physical structure. Small as compared with many other Galactic HII regions and the giants found in extragalactic systems, Orion remains the Rosetta Stone of this discipline owing to its apparent brightness and closeness. Although it may not be a typical HII region, the Orion complex manifests the major structural features and physics that are found in the larger, but more incompletely observed nebulae. Dominating the long early winter nights in the northern hemisphere, it is a prime target for optical telescopes and has been the subject of a series of investigations by my students and colleagues at Rice Univeristy. This paper summarizes and integrates the knowledge gained in these studies and those from the infrared and radio regimes.

2 Structure and Dynamics of the Principal Ionization Front

The Orion Nebula (M42, NGC 1976) is a zone of ionization by the O6 star θ^1C Ori close to the surface on the near side of a giant molecular cloud lying at a distance of about 500 pc in the Galactic anticenter direction. It satisfies the usual criteria for being called a blister nebula, since it is dominated by a region of surface ionization of the neutral molecular cloud, the ionized material being open to the inter-cloud medium and therefore freely flowing away from the cloud. In this case the ionizing star is moving with a large spatial velocity with respect to the parent

Astrophysics and Space Science **216**: 267–280, 1994.
© 1994 *Kluwer Academic Publishers.*

molecular cloud, so it seems likely that this is a chance encounter of a star formed elsewhere which happens to be moving near the surface of the molecular cloud and producing the high surface brightness HII region (Goudis 1982).

The height of the star above the ionization front can be estimated in two ways, through the surface brightness in an HII recombination line and the reflection of visual starlight. We know that the in the direction of the ionization front all of the stellar Lyman continuum photon flux (F_{LyC}) will be absorbed in the column of hydrogen. In an equilibrium situation, which applies because of the short recombination time that applies, this number of ionizing photons must equal the number of recombinations of HII, which will then be directly related to the surface brightness ($S_{H\alpha}$) of a recombination line like Hα, when viewed along the same line of sight. Originally pointed out by Baldwin et al. (1991), this relation needs to also consider the role of reflection of the Hα emission by the dust of the material lying near the ionization front. If this amount is 50%, then the relation for the distance r from a star of total Lyman continuum luminosity Q_{LyC} is

$$S_{H\alpha} = \frac{1.5 Q_{LyC} \alpha_{H\alpha} h\nu_{H\alpha}}{16\pi^2 r^2 \alpha_B}$$

for the simple case of a very thin absorbing layer and $\alpha_{H\alpha}$ and α_B are the recombination coefficients for hydrogen (Osterbrock 1989). If one adopts the extinction corrected $S_{H\alpha}$ of Hester et al. (1991a), and $Q_{LyC} = 1.5 \times 10^{49}$ s^{-1} (Osterbrock 1989), then the distance from the star to the ionization front at the substellar point is 0.24 pc. This value actually results from a more complex and accurate relation that assumes an exponential decay in the density and that the absorbing layer is not necessarily thin as compared with the distance of the front from the ionizing star, as described later in this paper. If one assumes that all four of the Trapezium stars forming θ^1C Ori are grouped and that the the albedo of the scattering particles is 0.5, then the distance determined from the surface brightness of the scattered visual starlight is 0.4 ± 0.2 pc. Throughout this paper I will use the value 0.3 pc for the substellar distance.

The ionized material is confined on the side of the molecular cloud by the dense, shocked Photo-Dissociation-Region (PDR) but free to expand away from the front, basically representing the Champagne phase of flow modeled by Tenorio-Tagle and his collaborators (Yorke 1986), which would produce an exponential fall in density away from the ionization front and acceleration of the gas, in this case towards the observer and thus blueshifted.

This flow of material away from the ionization front has been characterized in a series of papers coming out of our group at Rice, using the Kitt Peak National Observatory coude feed telescope, which offers a modest spatial resolution of about 4" but high velocity resolution of about 4 km s^{-1}. Castañeda (1988) determined velocity maps of [OIII], O'Dell and Wen (1992) [OI], Jones (1992) [OII], and Wen and O'Dell (1993) [SIII]. More recently Zheng Wen (1993) has measured central region velocities for Hα, [NII], [SII], and HeI. These velocities can be grouped according to the ionization structure that is expected. [OI] and [SII] will come from directly within the ionization front, while [NII] and [SIII] will arise in the thin region where hydrogen is ionized and helium is neutral, and the [OIII] and HeI emission will come from even further away. The grouped velocities are given in Table I, where we also present the velocities for the CO emission from the

TABLE I
Velocities Near the Main Ionization Front.

Emission Line Region	Heliocentric Velocity(km s^{-1})
Molecules (CO)	28±1
CII (PDR)	28±1
Ionization Front ([OI]+[SII])	25.5±1.1
Low Ionization Region ([NII]+[OII]+[SIII])	18.8±1.5
Moderate Ionization Region (HeI+[OIII])	17.9±1.3
HII (All, Low Temperatures)	16.0±1.2

molecular cloud and CII radio lines from the PDR (Goudis 1982). Hα emission will arise from all of the ionized region but will show a preferential emission from low electron temperature regions along the line of sight, which stands in contrast with the collisionally excited forbidden lines that prefer the hot regions. However, the main term in the emissivity is the square of the density.

Examination of Table I reveals that the material is accelerated very rapidly away from the ionization front. The velocity difference is about 6 km s^{-1} by the low ionization zone, which is only about 2×10^{-3} pc away from the front (O'Dell 1993), and the velocity is only slightly larger in the moderate ionization zone which is further out. This velocity gradient is certainly of the type expected for free expansion. It may be a challenge for the theoreticians to explain how this much change in velocity can occur over such a small distance.

The determination of the density gradient is much more problematic since one is looking at emission coming from along a line of sight with widely varying conditions. [SII] certainly characterizes the electron density N_e in the ionization front itself, but this region is only partially ionized (Hester 1991). The [OII] 372.7 nm doublet will arise in the thin low ionization zone, which may or may not have a strong density gradient. Comparison of [SII] densities (Pogge et al. 1992) and [OII] densities (Jones 1992) on a point by point basis by Wen (1993) shows that the [SII] densities are about 0.9 times smaller, which probably reflects the incomplete ionization of the ionization front and argues for a higher total density. density. A recent study of the [Cl III] doublet at 551.8+553.8 nm by Walter (1993) is of interest as it arises from an ion requiring 23.8 eV, which means that it should basically coexist with the moderate ionization material lying outside of the low ionization zone sampled by [OII]. Walter's study shows that the [Cl III] densities are about

0.6 that of the [OII] along the same line-of-sight near the brightest region. The density increase towards the ionization front must continue into the PDR that lies on the molecular cloud side of the front. A study of the emission from this region by Tielens and Hollenbach (1985) and Escalante, *et al.* (1991) indicated densities of $10^5 cm^{-3}$, which could be due to either compression by a shock front preceding the ionization front, or the natural increase in density towards the middle of the molecular cloud, or both. Together these data indicate a general agreement with the density of material rapidly falling away from the ionization front and the situation is summarized in Table II.

TABLE II
Properties of Substellar Regions Near the Orion Ionization Front.

Region	PDR	Front	Low Ionization	Medium Ionization
Markers	CO,CII	[OI],[SII]	[OII],[NII],[SIII]	[OIII],[Cl III],HII,HeI
$V_\odot(kms^{-1})$	28	25.5	18.8±1.5	17.9±1.3
Density(cm^{-3})	10^5	$\geq$6,000	7,000	4,000
Depth (pc)	?	10^{-4}	$2x10^{-3}$	0.1

The role of scattered light in Orion has not been fully exploited, even though it was recognized very early that stellar scattered light was dominate over the atomic continuum (Baldwin *et al.* 1991). The most thorough modeling study is probably that of Schiffer and Mathis (1974), but they did not treat the blister model and could draw no firm conclusions about the nature and distribution of the particles. In light of the well known stellar scattered light and the blister geometry, we should have been concerned with reflection of the emission lines by dust in the ionized region and the PDR. The reality of this reflection became obvious only through examination of the high velocity resolution emission line studies, where we noted a common component, often a significant fraction of the flux from the velocity component coming from the main nebular emission, but always more redshifted and much broader. This is now interpreted (O'Dell *et al.* 1992) as the reflected nebular emission. In effect it is a moving mirror, located behind the emission source. The relative velocity of the source and the scattering layer will double the velocity difference and since the light scattered from any one point will be incident from a variety of angles and velocity components, the scattered emission line will be broadened. The inferred velocities of the various reflecting layers for different ions are listed in Table III. This reflection component is interesting not only in its own right, but also because it means that corrections for this must also be considered in many of the modeling attempts that assume that the nebular emission freely escapes into 4π steradians.

We can be more quantitative in considering the structure of this ionization front. Recall that the separation of the front and the ionizing star is 0.3 pc. One can derive a formal value of the thickness of the emitting layer from $S_{H\alpha}$ and a knowledge of the electron density, as done by Baldwin *et al.* (1991). We have done this for the central region of Orion, using the [OII] electron density and find values near 0.1 pc. This means that the emitting layer of gas is relatively thin compared with the separation of the star and ionization front.

TABLE III
Reflection Layer Velocities of Different Zones.

Region of Emission being Reflected	Heliocentric Velocity(km s^{-1})
Ionization Front ([SII])	28.1±2:
Low Ionization Region ([NII]+[OII]+[SIII])	20.8±3
Moderate Ionization Region ([OIII])	20.6±2:
HII (All, Low Temperatures)	18.7±2:

The actual case must be even more extreme. As a rule of thumb, we would expect that a substantial fraction of the emission must occur within lengths that are no greater than the detail of the Hα image; otherwise there would have to be a requirement for very ordered flow away from the ionization front. Certainly we see structure in Orion down to at least 10", which corresponds to 0.025 pc., which argues that the emitting thickness is about this same size, i.e. one fourth the formal value. Having the density decrease away from the ionization front will certainly help in producing concentrations of emission very near the front and hence fine structure smaller than 0.1 pc. Therefore, we should probably view the Orion Nebula as a thin shell of emission along the face of the ionization front.

The density of the ionized material as determined from the [SII] (Pogge *et al.* 1992) and [OII] (Jones 1992, Walter 1993, Osterbrock and Flather 1959) doublets drops away a peak near the brightest portion of the nebula, which is southwest from θ^1C Ori. It is not a monotonic decrease for we see in some small regions and certainly on the Trapezium side of the Bright Bar that the density increases significantly above local values. The cause of this general decrease is not understood, but there are some broad considerations that are probably useful. The mass-loss rate through a unit area of the ionization front must depend upon the flux of ionizing photons, because each ionized hydrogen atom is eventually lost through flow into the emitting zone. The density would depend upon both the ionizing flux and the velocity of the material flowing through the front. The flow velocity is primarily determined by the sonic speed in the ionized gas and hence is relatively constant due to the small amount of the changes in the electron temperature (Walter 1993). If this is the case, then we would expect local increases in the density to represent areas getting more flux, i.e. are closer to θ^1C Ori and hence represent irregularities in the ionization front. The movement of the ionizing star must also be an important factor as it forces the front into the molecular cloud.

The role of the density of the neutral material on the molecular cloud side of the ionization front is less clear. In the above discussion of the separation of the ionization front and the ionizing star, we saw that the surface brightness was independent of the local density, which means that areas of unusually high surface brightness are probably closer to the star than their surroundings. Examination of the $S_{H\alpha}$ and N_e plots of Pogge et al. (1992) indicate that the fluctuations are correlated. If the arguments of the preceding paragraph apply, then these correlations are due to the fact that the ionization front is moving into an irregular density distribution region of the molecular cloud and the front progresses more slowly into the denser regions, leaving them as hills in the general structure of the front.

The motion of the ionizing star with respect to the neutral material is very important. The best astrometric data (van Altena et al. 1988) indicates a proper motion of 2.3 mas/yr towards PA=142°, corresponding to a tangential velocity of 5.4 km s^{-1}. The heliocentric velocity of the star (+33 km s^{-1}, Hoffleit 1964) would mean that the star is moving into the molecular cloud at 5 km s^{-1}. At this closure velocity and the star-front separation of 0.3 pc, then the time for the star to move to the present location of the front is 60,000 years and the time to cross the effective emission thickness of 0.1 pc is only 20,000 years. Since the recombination time for the hydrogen is much shorter than this, we can expect that the nebula must continuously change in appearance as the ionization front is pushed forward of the star into the molecular cloud.

3 Foreground Material

Thus far I have addressed only the structure of the principal ionization front, but there is more to Orion than that. There is a partial cover of neutral hydrogen and dust which is more distant from the dominant ionizing star θ^1C Ori. This layer reveals itself in the HI absorption line at 21 cm (van der Werf and Goss 1989, 1990) formed in the thermal continuum from the nebula and in numerous optical absorption lines in the spectra of all four stars of the Trapezium (O'Dell et al. 1993a). This incomplete lid seems to be of highest column density to the east of the Trapezium, in the so called Dark Bay, with the column density decreasing towards the southwest. A more complete description is found in O'Dell et al.(1992) and so only a brief summary need be given here. The inner side is photoionized by θ^1C Ori and the outer (nearer the observer) by 42 Ori. The neutral portions are detected in the absorption lines of HI, CaII, and NaI and the ionized surfaces from velocity shifted components of [OII], [OIII], and [SIII] found in the velocity maps of the main ionization front. A correlation study of column densities and reddening of the nebula determined (O'Dell et al. 1992) that most of the reddening of the nebular emission occurs in the lid. The separation of the lid from the ionizing star is sufficiently large that it does not play an important role in the dynamical flow of the principal ionization front material.

4 A 3-D Model of the Orion Nebula

The recent paper of Baldwin *et al.* (1991) initially explained the arguments used above for determining the distance from θ^1C Ori to the substellar point on the main ionization front. The approach can be generalized to say that the surface brightness of the ionization front in Hα as viewed from the ionizing star should be proportional to the incident flux, which would vary inversely with the square of the distance. They then go on to interpret their recombination line surface brightness relation S $\propto \phi^{-2}$ (where ϕ is the angular distance from θ^1C Ori) as an argument that the nebula is flat and lying in the plane of the sky. They acknowledge that this interpretation ignores some geometric affects. Actually the geometric affects can be very important, since ionizing flux striking a tilted flat wall will have a different length over which the ionizing flux is absorbed as compared with the emitting length. The method they described works, but only for the substellar point. Knowing the geometry of other points requires knowing more physical information. The tilt and distance of the front cannot be determined uniquely without solving the problem beginning with the substellar point.

Zheng Wen (1993) has recently treated this general problem by combining the line of sight density data of Pogge *et al.* (1992) and the extinction corrected $S_{H\alpha}$ of Hester *et al.* (1991a). In this calculation he assumes that there is an exponential drop in density away from the ionization front, although models with an assumed .constant density give very similar results. His model is shown in Figure 1. The surface of the Orion Nebula is not at all flat, rather, it is basically concave with lots of local peaks and depressions. The brightest part of the nebula is a peak displaced to the SW from θ^1C Ori and the NE-SW low ionization Bright Bar that runs just north of θ^2A Ori lies along a long steep wall, thus agreeing with earlier, less quantitative, arguments that the Bright Bar was an ionization front being viewed edge-on(Dopita *et al.* 1974). The high and low regions probably reflect differences in the success of penetration of the advancing ionization front.

5 Turbulence in the Orion Nebula

The existence of random fine-scale motion superimposed on largescale variations of radial velocity have been known for decades (Wilson et al. 1959) and was immediately interpreted by Münch (1958) as turbulence. Similar velocity variations are also seen in many other nebulae and probably have the same interpretation (Joncas and Roy 1986, Roy *et al.* 1986, O'Dell et al. 1986). We know that the Reynolds number for this gas must be very high and hence the flow away from the ionization front would be turbulent rather than smooth. Testing a model for this turbulence is complex because the motion must occur on a hierarchy of scales, with size dependent velocities in three dimensions, and we can only determine the line profile, width, and radial velocity (V) along each line-of-sight.

Fortunately, the theory for a model very similar to Orion was developed very

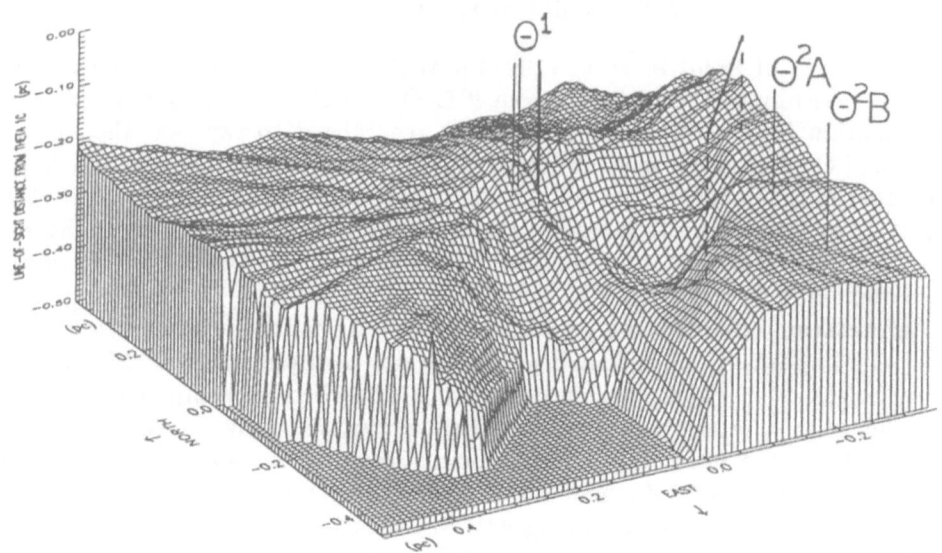

Fig. 1.   A three dimensional model for the principal ionization front of the Orion Nebula as
calculated by Zheng Wen. The ionizing star θ^1C Ori is located at a height of zero. The region
of the Dark Bay ENE of the Trapezium is flat where the required density data is not available,
and is arbitrarily set to a height of -0.48 pc. The 1/e thickness of the HII emission zone is about
0.05 pc near the middle and varies from 0.07 pc to 0.16 pc in the outer regions. The straight line
represents the range of the Bright Bar.

early when von Hoerner (1951) considered Kolmogorov turbulence in a a thin,
plane-parallel nebula.   One can actually generalize his solutions to treat other
power dependences (O'Dell and Castañeda 1987), the spatial velocity varying as
$v^2 \propto r^{2/3}$ for Kolmogorov turbulence.   The most useful way of characterizing
the turbulence is through the structure function $B(\Delta\phi) = <| V(\phi) - V(\phi') |^2>$
where the average of the radial velocity differences is taken over all values of the
separation distance of the two angles ϕ and ϕ'. If there is a power law dependence
of the spatial velocity $v^2 \propto r^n$, then two clear regimes will be defined. When the
distance corresponding to $\Delta\phi$ is much less than the thickness of the emitting layer,
then von Hoerner's theory predicts $B(\Delta\phi) \propto (\Delta\phi)^{n+1}$ and when the separation
distance is much larger than the thickness, then $B(\Delta\phi) \propto (\Delta\phi)^n$. The transition
between these two regimes occur at about the thickness of the emitting layer.
    Sufficient data exists to test this model for four ions O^+, O^{+2}, O^{+3}, and S^{+2}.
Recall that the [OI] emission must come from the ionization zone proper, which
has a thickness of no more than 10^{-4} pc (O'Dell 1993), and the [OII] and [SIII]
come from the very narrow (0.002 pc, O'Dell 1993) low ionization region which
contains ionized hydrogen but neutral helium, and only [OIII] comes from the
main emitting part of the nebula, which the earlier parts of this paper showed to

have a scale about 0.025-0.1 pc. The [OII] (Jones 1992), [SIII] (Wen and O'Dell 1993), and [OIII] (Castañeda 1988) emission all behave in a similar fashion. The power law dependence of the structure function is about 1.2 at small separations and about 0.0 at large. The transition occurs at about 0.07 pc for [OII] and [SIII] and 0.1 pc for [OIII]. [OI] behaves very differently, being a constant power law of 0.6 over the entire range of observation. We can understand the [OI] structure function as indicating turbulence with a $v^2 \propto r^{0.6}$ since even the closest radial velocity samples would be large compared with the ionization front thickness. The power index for the regions away from the ionization front are lower, with the index at large separations being about zero (that is, there is no statistical correlation of the radial velocities). The transition length for [OIII] is in reasonable agreement with the thickness inferred from the surface brightness but the transition length for [OII] and [SIII] is much larger than that inferred for the low ionization region from the line ratios (O'Dell 1993). Perhaps what we are seeing is a breakdown from Kolmogorov type turbulence in the ionization front, through an intermediate stage (the low ionization zone) and failure of the model (the high ionization zone at large separations where the thicknesses can be quite different).

The one remaining problem with interpretation of the turbulence in [OI] is the line width. The FWHM corrected for instrumental and thermal broadening is 10.3 km s^{-1}, while that predicted by von Hoerner's theory is vastly smaller. This discrepancy may be due to fine scale structure in the ionization front, which is only a small fraction of an arc second in corresponding angle while the data was sampled over bins of 3.7". This interpretation would mean that the ambient molecular gas would have corresponding fluctuations in its density.

6   Unusual Features

It should not be surprising that the Orion Nebula should be home of many exotic features, but it seems that nature's imagination continues to exceed our own. The cluster of stars associated with this region is the richest Galactic Cluster (Herbig 1982) and contains as members not only very massive stars like θ^1C Ori but also low mass stars still in the process of gravitationally collapsing to the main sequence. In a study of the fainter cluster members, Herbig and Terndrup (1986) found that pre-main sequence stars as young as 300,000 years were members, which is a good indication of the star formation activity of this region.

Laques and Vidal (1979) originally discovered six emission line objects very near the Trapezium. Unresolvable at ground resolution, these high ionization objects were called partially ionized globules (PIGs) under the belief that they were small versions of the well known Bok globules and lying within the Orion Nebula. Their optical imaging approach could only see objects where the ionizing flux from θ^1C Ori is high and the nebular radiation not at its maximum, so it was not surprising that radio surveys with the VLA at resolutions approaching 0.1" revealed not only the free-free continuum from these objects but also that there was about four times as many (Felli, *et al.* 1993). Quantitative arguments using the low infrared brightnesses and high surface brightnesses in the radio established that these objects were optically thick to ionizing radiation and were ionized from the outside. Their true nature as PIGs or protoplanetary disks was unclear at that time. Clarification came from recent Hubble Space Telescope(HST) monochromatic images in several emission lines and the continuum (O'Dell *et al.* 1993b). In these images one sees that there are many more such objects and that many of them are resolved. A typical object has a low luminosity star in the middle of a region of obscuring dust, surrounded by a shell of ionized gas. The gas shows a bow-shock form bright rim on the side facing the ionizing star (always θ^1C Ori except for one object pointing towards θ^2A Ori). Some even show tails and since these are of high ionization and near the Trapezium, they must be physically close to θ^1C Ori and being shaped by the intense stellar wind of that star. These forms were all anticipated theoretically by John Dyson (1968a, 1968b, 1975) in his attempts to build models for PIGs that would provide an explanation for the anomalously high velocities seen in some parts of Orion. I prefer to use the new term "Proplyds" for these protoplanetary disks found in the Orion Nebula because they are different from similar objects sought elsewhere because these are rendered visible by their presence in an HII region. This visibility is bestowed both by the ionization of part of their gas, but also by detection of their dust component in silhouette against the bright background of the nebula. The argument that they are disks is based on the fact that a large fraction shown their stars in the visible, arguing that they are not spherical shells. Use of the term protoplanetary disks reflects the common wisdom that such disks must precede planet formation.

Orion also contains other objects which may be related to these nascent stars. Monochromatic imaging of Orion has revealed 10 features identified as Herbig-Haro objects. Typically these are bright, low ionization knots, often occurring at the apex of a bow shock feature. Not only are these knots and shocks seen in the optical region (Axon and Taylor 1984, Hester *et al.* 1991b), recent infrared images show many more similar structures in [FeII] and shocked H_2 (Allen and Burton 1993). It is notable that the IR features seem oriented on the source IRc2 buried within the giant molecular cloud while the optical features seem oriented on the Trapezium region. This probably reflects the fact that the optical features are on the surface of the cloud and the IR features within the cloud. The exception to this is the northern Herbig-Haro objects which are seen both optically and in the IR. Perhaps Orion will provide the key information to build an exact model for the Herbig-Haro objects, but at the present it seems impossible to discriminate the two most likely candidates, the first being that the bright knots are "bullets" ejected by mechanisms unknown from young star-disk systems and the second being that these are shocks formed by continuous jets of material coming from similar stars, with the brightest portions being the Mach disk at the head of the shock.

High ionization shocks were also found in the HST images, many at the locations where high radial velocities had been detected. These shocks show a preferential orientation to the SE from a concentration of Proplyds near the brightest part of the nebula. Since ejection of bullets or disks must require shaping by material around a collapsing pre-main sequence star, it seems natural to link the Proplyds to the Herbig-Haro objects and the shocks. The shocks detected in the infrared are all oriented towards the cluster buried within the molecular cloud centered on the BN-IRc2 region and it is reasonable to expect similar star-disk systems forming there. The orientation of the new shocks does argue for a link with the visible Proplyds and it is tantalizing that one of the best defined Proplyds (HST8) falls on the axis of symmetry of the shock containing HH3 and HH4.

A spectroscopic study in [OIII] of the region near the Trapezium that reaches down to levels much fainter than the main body of the nebula (Meaburn, *et al.* 1993) reveals velocities up to 120 km s^{-1} with respect to M42. Even more recent observations (Meaburn and Massey 1993) indicate that there are even more such features and that they are commonly found around the Laques and Vidal sources. This extends and supports the arguments made in the discovery paper (Meaburn 1988) that the objects we call Proplyds are the source of strong, high velocity, outflows. Zheng Wen, Xihai Hu, and I have spectra made in January, 1993 which confirm the reality of Meaburn's observations and show that the high velocity components are seen in a variety of ions. Figure 2 shows a composite drawing of the various types of sources found in M42 by the various methods described above. Examination shows that the Proplyds and compact radio sources are strongly concentrated near the Trapezium, indicating that all of the Proplyds and most of the compact radio sources are photoionized by $\theta^1 C$ Ori and that only a few lie physically near IRc2 inside the molecular cloud. The IR features are oriented towards IRc2, indicating that they too are buried within the cloud and have no optical counterparts except for the previously identified Herbig-Haro objects. The optical features of low ionization are generally aligned on the Trapezium and the shock features are oriented on the region rich in Proplyds.

7   Conclusions

Nothing is static in the Orion Nebula. The ionizing star is moving rapidly into the molecular cloud, pushing the ionization front ahead of it, thus unveiling objects that had been hidden from the intense ultraviolet luminosity of $\theta^1 C$ Ori. Once irradiated, objects such as the Proplyds are rendered very visible through photoionization and recombination and shadowing against the primary ionization front. The interactions taking place between the jets/bullets coming from these same sources which were visible only in the infrared now become optical sources. Of course the gas in the nebula is also changing. Initially trapped in the shock compressed photodissociation region, it is then entrained in the ionization front itself. Once this happens it is rapidly accelerated away from the front, rapidly dropping to a low density and entering the general interstellar medium. The advance of the ionization front is highly irregular, depending upon the local density

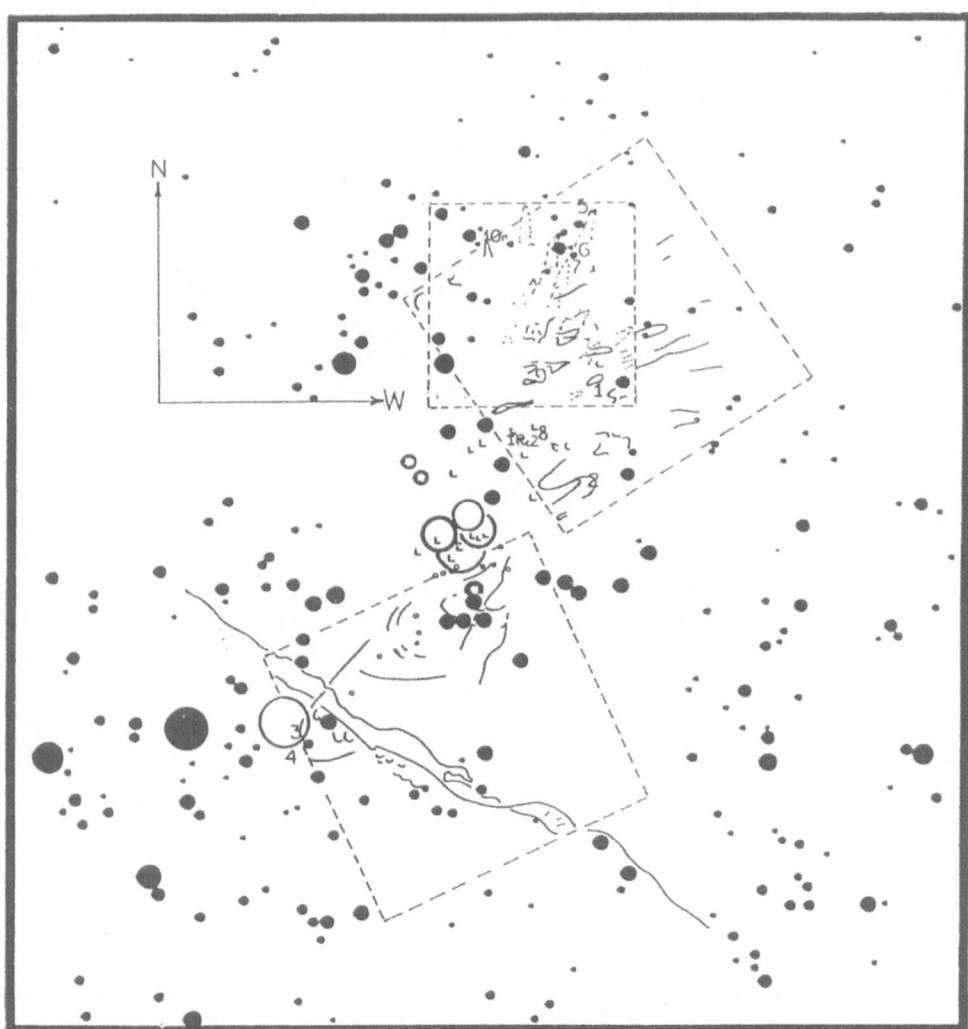

Fig. 2.   Recently discovered interesting objects are shown superimposed on Strand's (1958) star atlas of the region. Proplyds are shown by small open circles and compact radio sources as tic marks. The bright stars of θ^1 Ori and θ^2A Ori are shown as large open circles due to crowding. The boundaries of the HST images are shown as skewed dashed lines and those of Allen and Burton's (1993) IR field as aligned dashed lines. Features delineated by the optical images are shown as solid lines and those seen only in the IR as broken lines. Faint stars with open centers are either Proplyds or compact radio sources. The orientation arrows are 2′ long and centered on $5^h 30^m$, $-5°24′$. IRc2 indicates the position of the bright IR-radio source within the molecular cloud.

of molecular cloud material, so that the front becomes a luminous shroud placed over these irregularities. How fortunate that we live at a time to witness and decipher this.

I am grateful for many conversations with and material from Zheng Wen and Xihai Hu of Rice University, and to Richard Pogge of Ohio State University for the [SII] density data used to construct the 3-D model.

References

Allen, D. A. & M. G. Burton 1993, *Nature*, **363**, 54.
Axon, D. J. & Taylor, K. 1984, *MNRAS*, **207**, 241
Baldwin, J. A., Ferland, G. J., Martin, P. G., Corbin, M. R., Cota, S. A., Peterson, B. M., & Sletteback, A. 1991, *ApJ*, bf 374, 580
Castañeda, H. O. 1988, *ApJS*, **67**,93
Dopita, M. A., Dyson, J., & Meaburn, J. 1974, *ApSS*, **28**, 61
Dyson, J. E. 1968a, *ApSS*, **1**, 388
Dyson, J. E. 1968b, *ApSS*,**2**, 461
Dyson, J. E. 1975, *ApSS*, **35**, 299
Escalante, V., Sternberg, A., & Dalgarno, A. 1991, *ApJ*, **375**, 630
Felli, M., Churchwell, E., Wilson, T., & Taylor, G. B. 1993, *A&AS*, **98**, 137
Goudis, C. 1982, *The Orion Complex: A case Study of Interstellar Matter* (Dordrecht: Reidel)
Herbig, G. H. 1982, *Ann.NY Acad.Sci*, 395, 64
Herbig, G. H. & Terndrup, D. M. 1986, *ApJ*, **307**, 609
Hester, J. J. 1991, *PASP*, bf 103, 853
Hester, J. J., Dufour, R. J., Parker, R. A. R., & Scowen, P. A. 1991a, *BAAS*, **23**, 1364
Hester, J. J. *et al.* 1991b, *ApJ*, **369**, L75
Hoffleit, Dorrit 1964, *Bright Star Catalogue* (New Haven, CN: Yale Observatory)
Joncas, G. & Roy, J. R. 1986, *ApJ* , **307**, 649
Jones, M. R. 1992, *PhD Thesis*, Rice University, Houston, TX
Laques, P. & Vidal, J. L. 1979, *A&A*, **73**, 97
Meaburn, J. 1988, *MNRAS*, **233**, 791
Meaburn, J. & Massey, R. M. 1993, *MNRAS*, in press
Meaburn, J., Massey, R. M., Raga, A. C. & Clayton, C. A. 1993, *MNRAS*, **260**, 625
Münch, G. 1958, *Rev.Mod.Phys.*, **30**, 1035
O'Dell, C. R. 1993, *Revista Mex A&Ap*, in press
O'Dell, C. R. & Castañeda, H. O. 1987, *ApJ*, **315**, L55
O'Dell, C. R., Townsley, L. K., & Castañeda, H. O. 1987, *ApJ*, **317**, 676
O'Dell, C. R., Wen, Zheng, & Hester, J. J. 1991, *PASP*, **103**, 824
O'Dell, C. R., Walter, D. K., & Dufour, R. J. 1992, *ApJ*, **399**, L67
O'Dell, C. R. & Wen, Zheng 1992, *ApJ*, **387**, 229
O'Dell, C. R., Valk, J. H., Wen, Zheng, & Meyer, D. M. 1993a, *ApJ*, **403**, 678
O'Dell, C. R., Wen, Zheng, & Hu, Xihai 1993b, *ApJ*, **411**, 696
Osterbrock, D. E. & Flather, E. 1959, *ApJ*, **129**, 26
Osterbrock, D. E. 1989, *Astrophysics of Gaseous Nebulae and Active Galactic Nuclei* (Mill Valley: University Science Books)
Pogge, R. W., Owen, J. M., & Atwood, B. 1992, *ApJ*, **399**, 147
Roy, J. R., Arsenault, R., & Joncas, G. 1986, *ApJ*, **300**, 624
Rubin, R. H., Simpson, J. P., Haas, M. R., & Erickson, E. F. *ApJ*, **374**, 564
Schiffer, F. H., III & Mathis, J. S. 1974, *ApJ*, **194**, 597
Strand, K. Aa 1958, *Ann.Dearborn Obs.*, **7**, 67
Tielens, A. G. G. M. & Hollenbach, D. 1985, *ApJ*, **291**, 747
van Altena, W. F., Lee, J. T., & Lee, J.-F. 1988, *AJ*, **95**, 1744
van der Werf, P. P. & Goss, W. M. 1989, *A&A*, **224**, 209

van der Werf, P. P. & Goss, W. M. 1990, *ApJ*, **364**, 157
von Hoerner, S. 1951, *Zs.Ap.*, **30**, 17
Walter, D. K. 1993, *PhD Thesis*, Rice University, Houston, TX
Wen, Zheng & O'Dell, C. R. 1993, *ApJ*, **409**, 262
Wen, Zheng 1993, *PhD Thesis*, Rice University, Houston, TX, in preparation
Wilson., O. C., Münch, G., Flather, E. M., & Coffeen, M. F. 1959, *ApJS*, 4, 199
Yorke, H. W 1986, *ARAA*, **24**, 49

An Evolutionary Model For the Wolf Rayet Nebula NGC 2539

J. Meaburn and J. E. Dyson
Department of Astronomy, University of Manchester, UK

and

C. D. Goudis and P. E. Christopoulou
Astronomical Laboratory, University of Patras, Greece

Abstract. The main outer ring of the WR nebula NGC 2539 expands radially with velocity $\approx 26\,\mathrm{kms}^{-1}$. The dynamics are modelled as the interaction of the winds from the O progenitor and the present WR stage with clumpy molecular cloud material and Red Supergiant ejecta. In both stages, the bubbles are driven by wind momentum. The necessary radiative energy loss probably occurs in boundary layers between cool clumps and hot shocked wind.

1   Introduction

The WR nebula NGC 2539 is embedded in the diffuse HII region S298 and its main feature is an outer ring ($\simeq 4\overset{'}{.}2$ diameter). The ionized mass and rms electron density of NGC 2539 are respectively 16 $(D/5\,\mathrm{pc})^{5/2}\,M_\odot$ and 73 $(D/5\,\mathrm{pc})^{-1/2}\,\mathrm{cm}^{-3}$ where D is the distance to the nebula (Schneps *et al* 1981), but the density in the bright filaments is appreciably higher (100–2500 cm^{-3}). The system is associated with three molecular clouds and the nebula contains a substantial amount ($\sim 0.25 - 1.35 M_\odot$) of dust (Marston 1991).

2   Observations

Spectra of NGC 2539 were obtained at 10 Å mm^{-1} with the RGO spectrograph combined with the 3.9m Anglo-Australian telescope. The splitting of the profiles indicates an expansion velocity of 26 km s^{-1}, in good agreement with that determined by Goudis *et al* (1983).

3   Interpretation

The ionized mass suggests a mainly interstellar origin for the nebulosity, although the He/H abundance ratio indicates local He enhancements (Esteban *et al* 1991). The evolution of the nebula must be considered in the context of the evolution of the WR star and its progenitor O phase. We have adopted Maeder's (1990) 25 $M_\odot$ evolutionary track ($Z = 0.02$) as a guide since his calculated mean mass-loss-rate in the WN phase agrees well with that observed. The adopted stellar parameters are given in Table 1. The complex is associated with molecular clouds for which we adopt an interstellar density $n_0 = 10^3 n_3\,\mathrm{cm}^{-3}$ and an effective sound speed $c_n = 2c_2$ km s^{-1}.

Astrophysics and Space Science **216**: 281–283, 1994.
© 1994 *Kluwer Academic Publishers.*

TABLE I
Adopted Stellar Parameters

(a) H-burning Phase		(b) WR Phase
6.7	$< t > /10^6$ yr	0.11
0.22	$< \dot{M}_* > /10^{-6} M_\odot \mathrm{yr}^{-1}$	22
2500	$< V_* > /\mathrm{km\,s}^{-1}$	2500
0.4	$< \dot{E}_* > /10^{36} \mathrm{erg\,s}^{-1}$	40
0.35	$< \dot{\mu}_* > 10^{28} \mathrm{gm\,cm}^{-2}\mathrm{s}^{-1}$	31
2	$S_*(h\nu {\geq} I_H)10^{48} \mathrm{s}^{-1}$	2

3.1 THE H-BURNING PHASE

If an energy driven interaction is assumed, the shell of material traps the ionizing photons early on and the shell expands to unrealistic radii at the end of H-burning ($Rs \approx 16 n_3^{-1/5}$ pc). If the molecular cloud is very clumpy, mixing at interfaces between clumps and hot shocked wind may dramatically increase radiative losses (e.g. Hartquist and Dyson 1993) and a momentum driven flow ensues. The driven shell cannot trap the ionizing photons and it stalls at time $t_{ST} \simeq 3.6 10^3 n_3^{-1/2}$ yr at a radius $R_{ST}(t_{ST}) \approx 0.09 n_3^{-1/2}$ pc. The interstellar gas is ionized out to radius $R_{so} \simeq 0.27 n_3^{-2/3}$ pc.

This HII region expands outwards and the wind shock position is determined by balance between wind ram pressure and the HII region gas pressure. This region reaches pressure equilibrium with the surrounding molecular gas at radius $R_{ST} \simeq 3 n_3^{-2/3} c_2^{-4/3}$ pc when its density is $nI \approx 56 n_3 c_2^2 \mathrm{cm}^{-3}$. The wind evacuates a cavity of radius $R_{WS} \approx 0.7 n_3^{-1/2} c_2^{-1}$ pc.

3.2 THE WR PHASE

We again assume considerable enhancement of radiative losses as the shocked WR wind interacts first with the ejecta and later with interstellar gas. We calculate values of $n_3 c_2^2$ required to fit the observed outer ring radius and velocity using the dynamic age $t_d = 0.5$ Rs/Vs, appropriate to momentum driven solutions. These values are given in Table 2 along with the predicted ionized shell densities ($n_s c_2^2$ - derived assuming an isothermal shock). The predicted mass ($M_1 c_2^2$) and total mass ($M_T c_2^2$) of the ring are given in Table 2. In addition to causing strong cooling, the RSG clumps act as the source for the enhancements noted by Esteban et al (1991).

TABLE II
Derived Parameters of the Outer Ring

D(kpc)	$t_d/10^4$ yr	$n_3 c_2^2$	$n_e c_2^2$(cm^{-3})	$M_i c^2 (M_\odot)$	$M_T c_2^2 (M_\odot)$
4	4.5	0.9	240	43	85
5	5.7	0.6	156	69	111
6	7.3	0.34	88	113	138

4   Discussion

Although the derived ionized nebula masses are high (by factors of about 3–5), agreement with the observations is encouraging. The model could be improved in several ways, e.g. mass addition from the clumps could be taken into account. The bottom line is that clumpiness has major implications for interpreting circumstellar flows.

5   References

Breitschwerdt, D. and Kahn, F. D.: 1993, to appear in proceedings of Manchester Conference on: *Kinematics and Dynamics of Diffuse Astrophysical Media.*
Esteban, C. *et.al.*: 1990, *Astron. and Astrophys.*, **227**, 515.
Goudis, C. D. *et.al.*: 1983, *Astron. and Astrophys.*, **117**, 127.
Hartquist, T. W. and Dyson, J. E.: 1993, *Quart. J. R. astr. Soc.*, **34**, 57.
Maeder, A.: 1990, *Astron. and Astrophys. Suppl.*, **84**, 139.
Marston, A. P.: 1991, *Astrophys. J.*, **366**, 181.

Supersonic Turbulence in Giant Extragalactic HII Regions

Héctor O. Castañeda
Instituto de Astrofísica de Canarias and Isaac Newton Group of Telescopes, La Palma Observatory
La Laguna, E-32000 Tenerife, Spain.

Abstract. I discuss in this paper the more likely physical mechanisms that could provide the energy input for the supersonic motions observed in giant extragalactic HII regions, together with preliminary results of an ongoing observational program that aims to study, with with high spatial and spectral resolution, the kinematics of the ionized gas of the regions.

Key words: HII regions: general — turbulence.

1   Introduction

Giant extragalactic HII regions (GEHR) are sites of active star formation that provide an ideal laboratory to study the interaction between massive stars and the interstellar medium. Clusters of OB stars, generating ionizing photons at a rate of 10^{51}–10^{52} s^{-1}, ionize the low density gas ($N_e \approx 10$–100 cm^{-3}), creating giant complexes with linear dimensions of order 10^2–10^3 pc, varied morphology and an inhomogenous distribution of gas (Shields 1990).

Smith and Weedman (1970) found that the integrated line widths of their emission lines are highy supersonic, which is greatly surprising, as the normal evolution of an HII region would introduce velocity dispersions of order of the sound velocity (≈ 12 km s^{-1}). As supersonic motions are expected to dissipate their energy via shocks in relatively short times compared with the age of the regions, a source that continuously replenishes the lost energy is required. Several mechanisms have been advanced to explain the motions of the gas, the two more likely being *i)* the effect of the stellar winds of the ionizing stars via the collective action of groups of OB stars (Rosa and Solf 1984; Dyson 1979), and *ii)* the virial equilibrium between the gas and the stars proposed by Terlevich and Melnick (1981). These models can be tested comparing their predictions to the observed slope in the correlations established between luminosity, velocity dispersion, linear size, and chemical abundance (O/H). These correlations were originally found by Melnick (1977), and refined by Terlevich and Melnick (1981). Further work by Roy *et al.* (1986), Hippelein (1986) and Melnick *et al.* (1987) has succesfully proved their existence (see Arsenault and Roy 1988).

The main problem with this approach is that the values for the velocity dispersion of the gas are obtained on the line profiles integrated over the entire region. There is lack of evidence that the intrinsic broadening of the region (after intrumental, quantum, and thermal width corrections) that we will call "turbulence" is

Astrophysics and Space Science **216**: 285–289, 1994.
© 1994 *Kluwer Academic Publishers.*

the result of some of the mechanisms previously cited or also the manifestation of other physical processes such as collapse of the gas, a large scale gradient of the radial velocity field or geometric effects. To solve this problem studies with high spatial and spectral resolution over all the surface of the region provide the best option to understand the nature of the velocity field. Together with C. Muñoz-Tuñon and J.M. Vilchez (IAC), R. Terlevich (RGO) and M. Copetti (U. Santa Maria) I am conducting a long term program to study the kinematics of the gas in the main complexes of nearby galaxies. I will discuss in this paper some preliminary results of the project.

2   Observations

Two-dimensional imaging spectroscopy in the Hα and [OIII]λ5007 lines have been obtained for the main regions of M 101, M 33, NGC 2403 and NGC 6822, by means of the TAURUS-II Fabry-Perot imaging spectrograph (Taylor and Atherton 1980). We used the 4.2m William Herschel Telescope at the Observatorio del Roque de los Muchachos, with the Image Photon Counting System (IPCS) as a detector, during two observing runs on July 1990 and January 1991. The 125 microns gap etalon was used, each data cube of 100 frames of 256 × 256 pixels, for a total integration time of 3600 sec; pixel size was 0.26″ × 0.26″. Reduction and analysis of the data was done using TAUCAL, a software package for calibration and analysis of Fabry-Perot spectroscopic data (Lewis and Unger 1992). Single gaussians were fitted to the phase calibrated data cubes, and velocity and line width maps were produced for each region.

3   Discussion

The typical values for the velocity dispersion of GEHR along the line of sight (σ) quoted in the literature are 9–22 km s^{-1} (if the velocity distribution is Maxwellian the r.m.s. spatial velocity is $(3)^{1/3}$ σ). The problem of using integrated line spectra is clearly evident when looking at Figure 1, where we show the value of σ obtained with synthetic aperture spectroscopy from the Hα data cubes for two GEHR, NGC 588 (a single complex in M 33) and NGC 5462 (a patchy structure in emission in M 101, formed by several sub-groups). The velocity dispersion depends on the selected aperture, and extreme care should be taken before assigning a single value to the region.

While synthetic aperture spectra, with large aperture, show gaussian line profiles, spectra obtained with small apertures reveal asymmetric line profiles and multiple components. The study of the velocity maps of the regions uncovers systematic motions in the form of velocity gradients with a well-defined pattern, confirmed by comparison of velocity maps in different emission lines. An example is NGC 588, a giant HII region whose morphology is similar to a planetary nebula's. The velocity map in the [OIII]λ5007 line shows a gradient in the velocity of order

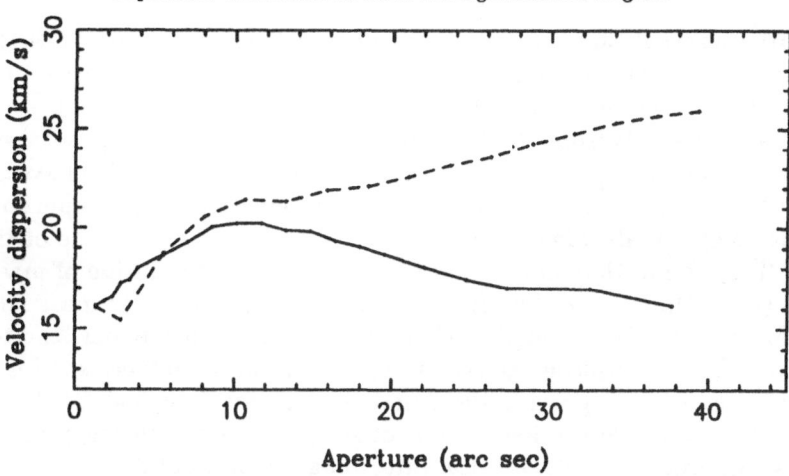

Fig. 1. Variation of the measured velocity dispersion (σ) with aperture for NGC 588 (continuum line) and NGC 5462 (broken line).

of 25 km s^{-1} between the inner and outer zones, the more blueshifted velocities being in the central part of the region. This is also the area where dispersion along the line of sight is larger. This result appears to indicate that random motions are not the only source of line broadening observed in the integrated emission profiles.

It seems obvious that the population of massive stars should play an important role in transfering kinetic energy to the interstellar medium of GEHR, either via stellar winds or as supernova. The effect of stellar winds is evident in the morphology of some of the regions. In Hubble III (in NGC 6822) very peculiar features are observed in the region (Sabalisck et al. 1993): a large fraction of the spectra shows non-symmetric line profiles, in some cases line splittings up to 70 km s^{-1} are measured, and radial velocity gradients of 20 km s^{-1}, between the inner and outer areas of the zones of the filaments that define the region, have been observed.

Typical ages of the regions ($\approx 10^6$ years) indicate the likely presence of Wolf-Rayet (WR) stars, characterized by their strong mass losses ($\approx 10^{-5}$ M$_\odot$ yr^{-1}), and large terminal velocity of their stellar winds ($> 10^3$ km s^{-1}). WR stars have been postulated as candidates for the mechanical energy source of the turbulent motions. To calculate their energy input we need to estimate the number of WR stars in the region; unfortunately, a numerical determination of the ratio WR/O stars is not very accurate. In the case of NGC 604 an analysis of the integrated spectra suggested a ratio of nearly 1 (D'Odorico and Rosa 1981), whereas an actual count of WR stars produces a ratio of 0.1 (Drissen et al. 1990). An examination of the regions in NGC 604 where WR stars have been detected by direct imaging, does not reveal peculiar enhancement in the velocity dispersion where the stars are positioned; the larger turbulent velocities seem to be associated with areas of low surface brightness (Sabalisck 1993). These results, combined with the fact that recent surveys reveal that the average number of WR/O stars in a GEHR is much lower than unity (Drissen 1991), suggest that the effect of WR stars on the overall dynamics of the ionized gas could have been overstimated.

Could stellar winds from all the stellar population provide the total amount of kinetic energy observed in the ionized gas? Leitherer et al. (1992) have calculated the input of mass, momentum, and energy from stellar winds and supernovae to the interstellar medium allowing an estimate of the order of magnitude of the energy transfer. They found that the total wind power from a population of massive stars scales as $(dE/dt)_{winds} \propto Z$. For an instantaneous burst of total mass 10^6 M$_\odot$, (α = 2.35, M$_u$ = 120, M$_l$ = 1 M$_\odot$), and using Fig. 11 of their paper, the total wind power (excluding supernovae contributions) should be $\approx 10^{38}$ erg s^{-1} for Z = 0.01, that in 10^6 years could provide the kinetic energy for a typical GEHR with $\sigma \approx$ 15.0 km s^{-1}, and a mass of ionized gas of 10^5 M$_\odot$, for an efficiency of transfer of kinetic energy from the winds to the interstellar medium of 20%.

Supernovae explosions can also provide the energy requirements, but supernovae rates in giant HII regions are of order 10^{-3} yr^{-1}, hence not providing the continuous energy input to feed the supersonic turbulence (Melnick et al. 1987). The supernova remnants can produce high velocity gas at very low intensity levels, but this phenomenon seems to be located in limited areas of the regions and is unlikely to affect the overall dynamics of the gas (Chu and Kennicutt 1986; Castañeda et al. 1990; Roy et al. 1992).

If giant HII regions are gravitationally-bound systems, observations down to the smallest scales—small compared with the wind driven bubbles—should not produce different line widths. When stellar winds are dominant, we should see radically different line widths, becoming sonic or subsonic (Melnick 1992). Several GEHR of our sample show some areas where the lines are subsonic, but a more careful analysis needs to be done with our data before we can establish the validity of the virial-turbulence model.

Finally I will discuss the implications of the existence of hydrodynamical turbulence within the regions. Gaseous nebula have Reynolds numbers of $\approx 10^6$, and the mass flow is likely to be in a chaotic, random state known as turbulence. In the standard Kolmogorov model, the energy spectrum of turbulence becomes dependent only on ϵ, the mean dissipation of energy per unit time per unit mass of fluid, with the inertial forces transferring kinetic energy from larger to smaller turbulent elements. Let L be the characteristic length of the largest eddies, and Δu the typical velocity, we have then that $\epsilon = (\Delta u)^3/L$ (Landau and Lifshitz 1987). If equilibrium exists in a turbulent state, both rates (energy input and energy dissipation) should be equal. So the classical theory of nebular turbulence should be tested by obtaining the structure function for GEHRs and comparing the typical scale of the correlation function (if it exists) with the determined value of L. Preliminary calculation of the structure functions from the velocity maps for NGC 604 in Hα and [OIII]λ5007 shows a correlation of the radial velocities up to a scale of 70 pc (Fuentes 1993). For NGC 604 the mass of ionized gas is 7.10^5 M$_\odot$,

σ ($\sim (1/3)^{1/3}\Delta u$) = 16.8 km s^{-1}; with $P_{\text{winds}} \sim 10^{38}$ erg s^{-1} (as discussed in a previous paragraph), we have $L \approx 10^2$ pc.. It is tantalizing that very simple physical arguments can actually produce a result very close to the observations, but further work needs to be done in this field. In particular, we expect to obtain the structure function for all the regions of our sample to look for similarities between the functions.

Acknowledgements

I gratefully acknowledge the Spanish Directorate General for Scientific and Technical Research (DGICYT) for a fellowship under which part of this work was carried out, and Monica Murphy for her kindness in correcting the English text. The WHT is operated on the island of La Palma by the Royal Greenwich Observatory at the Spanish Observatorio del Roque de los Muchachos of the Instituto de Astrofísica de Canarias.

References

Arsenault, R., Roy, J.-R.: 1988, *Astron. Astrophys.*, **201**, p. 199.

Castañeda, H.O., Vilchez, J.M., Copetti, M.V.F: 1990, *Astrophys. J.*, **365**, p. 164.

Chu, Y.-H., Kennicutt, R.C.: 1986, *Astrophys. J.*, **269**, p. 202.

D'Odorico, S., Rosa, M.: 1981, *Astrophys. J.*, **248**, p. 1015.

Drissen, L., Moffat, A.F.J., Shara, M.M.: 1990, *Astrophys. J.*, **364**, p. 496.

Drissen, L.: 1991, in *IAU Symposium 143: Wolf-Rayet Stars and Interrelations with Other Massive Stars in Galaxies*, K.A. van der Hucht and B. Hidayat eds., p. 595.

Dyson, J.E.: 1979, *Astron. Astrophys.*, **73**, p. 132.

Fuentes, O.: 1993, private communication.

Hippelein, H.: 1986, *Astron. Astrophys.*, **160**, p. 374.

Landau, L.D., Lifschitz, E.M.: 1987, *Fluid Mechanics*, Pergamon Press, p. 131.

Leitherer, C., Robert, C., Drissen, L.: 1992, *Astrophys. J.*, **401**, p. 596.

Lewis, J.R., Unger, S.W.: 1992, in *Astronomical Data Analysis and Software I*, eds. D.M. Worrall, C. Biesmesderfer, J. Banes, Astronomical Society of the Pacific Ser. Vol. 25, p. 445.

Melnick, J.: 1977, *Astrophys. J.*, **213**, p. 15.

Melnick, J.: 1992, in *Third Canary Islands Winter School: "Star Formation and Stellar Systems"*, eds. G. Tenorio-Tagle, M. Prieto, and F. Sanchez, Cambridge University Press, p. 253.

Melnick, J., Moles, M., Terlevich, R., Garcia-Pelayo, J.-M.: 1987, *Monthly Notices of the RAS*, **193**, p. 219.

Rosa, M., Solf, J.: 1984, *Astron. Astrophys.*, **130**, p. 29.

Roy, J.-R., Arsenault, R., Joncas, G.: 1986, *Astrophys. J.*, **300**, p. 624.

Roy, J.-R., Aubé, M., McCall, M.L., Dufour, R.J.: 1992, *Astrophys. J.*, **386**, p. 498.

Sabalisck, N.: 1993, *Ph. Thesis*, in preparation.

Sabalisck, N., Castañeda, H.O., Muñoz-Tuñon, C., Copetti, M.V.F., Terlevich, R.: 1993, in *Proc. EIPC Workshop: Star Forming Galaxies and Their Interstellar Medium*, F. Ferrini and J. Franco eds., in press.

Shields, G.A.: 1990, *Annual Review of Astron. Astrophys*, **28**, 525.

Smith, M.G., Weedman D.W.: 1970, *Astrophys. J.*, **160**, p. 65.

Taylor, K., Atherton, P.D.: 1980, *Mon. Not. Roy. Astr. Soc.*, **191**, p. 675.

Terlevich, R., Melnick, J.: 1981, *Mon. Not. Roy. Astr. Soc.*, **195**, p. 839.

The Dynamics of the Ring Nebula Surrounding
the LBV Candidate HE 3-519

Linda J. Smith
Department of Physics and Astronomy, University College London
Gower St., London WC1E 6BT, UK

Abstract. High spatial and spectral resolution observations of the ring nebula surrounding the LBV candidate He 3–519 are presented. The data were obtained at the AAT with the UCL echelle spectrograph and cover the Hα and [N II] emission lines for two slit positions. The nebular motions are clearly resolved and have a total velocity spread of -40 to $+100$ km s^{-1}. The shell shows some deviations from spherical symmetry but overall is expanding at 61 km s^{-1} and has an ionized mass of $\sim 2\,M_\odot$. The nebular parameters are found to be similar to those of the AG Car nebula, suggesting that it resulted from a bulk ejection of material $\sim 2 \times 10^4$ yr ago.

Key words: LBVs, nebulae, dynamics, individual: He 3–519

1   Introduction and Observations

Luminous Blue Variables (LBVs) are evolved, luminous, unstable, hot supergiants. They are close to the upper luminosity/stability limit in the HR diagram, and are believed to represent a short but physically important stage in the evolution of a massive O star to a WR star. There are currently six recognised members in the Galaxy (η Car, AG Car, HR Car, P Cyg, WRA 751 and HD 160529), four in the LMC and another twenty (including the Hubble-Sandage Variables) in external galaxies (e.g. Humphreys 1989; Wolf 1992). Observationally, LBVs are characterised by irregular photometric (0.5–2 mag) and spectral variations over timescales of decades. When they brighten visually, they appear to move horizontally across the HR diagram from a hot, minimum phase (T$_{eff} \approx 25\,000$ K) to a cooler (T$_{eff} \approx 10\,000$ K) maximum phase. For example, AG Car has been observed to change over a period of ~ 10 yr from an Ofpe/WN9-type supergiant at minimum (V~ 8) to an A-type supergiant at maximum (V~ 6).

One interesting property of LBVs is that most appear to be associated with circumstellar nebulae, apparently formed of enriched stellar material (e.g. the AG Car ring nebula; Mitra & Dufour 1990), and thus indicative of large stellar eruptions in the past. The subject of this paper, He 3–519, is 20$'$ away from AG Car and is surrounded by a ring nebula 1$'$ in diameter. It has received little attention apart from brief mentions of its spectral similarity to AG Car. A detailed study of He 3–519 and its nebula has just been completed (Smith, Crowther & Prinja 1993). In this paper, I concentrate on the dynamics of the nebula. In summary, for He 3–519 itself, we find that is an evolved massive star close to the upper luminosity/stability boundary in the HR diagram. It has a very dense, slow wind which is unstable, as evidenced by dramatic Balmer line profile variability. The spectrum, stellar parameters and evolutionary state of He 3–519 are very similar

Astrophysics and Space Science **216**: 291–295, 1994.
© 1994 *Kluwer Academic Publishers.*

Linda J. Smith

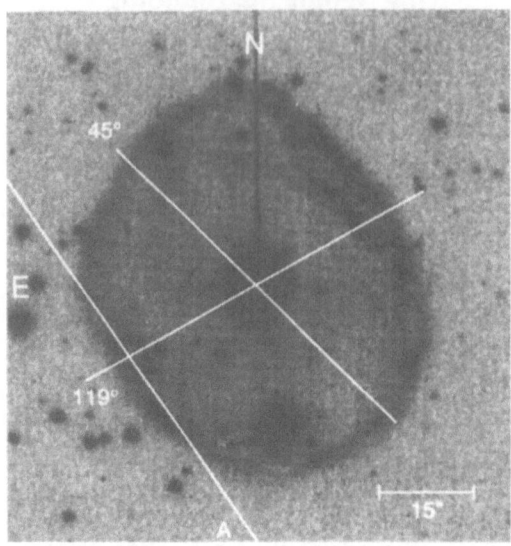

Fig. 1.   A [N II] image of the nebula surrounding He 3–519 reproduced from Stahl (1987). High resolution spectra at Hα and [N II] were obtained for the two slit positions at PA= 119° and PA= 45° and low resolution spectra were obtained at the slit position marked 'A'.

to those of AG Car, leading us to conclude that He 3–519 should be considered as a new LBV.

High spatial and spectral resolution observations of the nebula surrounding He 3–519 were obtained with the UCL echelle spectrograph (UCLES) and a GEC CCD detector at the Anglo-Australian Telescope (AAT) in 1989 April. A slit of dimensions 1″ by 63″ was used with an interference filter to isolate a single order covering Hα and [N II]. The nebula was observed at two position angles of 119° and 45° centred on He 3–519 with exposure times of 2800 and 1800 s respectively. The exact slit positions are shown superimposed on a [N II] image of the nebula from Stahl (1987) in Fig. 1. The data were reduced in the usual way to give 29 individual spectra with spectral and spatial resolutions of 8 km s^{-1} and 2.2″ for each position angle and emission line. Low resolution spectra of the nebula have also been obtained with the RGO spectrograph at the position marked 'A' in Fig. 1. Due to the faintness and heavy reddening of the nebula, the only emission lines detected are Hβ, Hα, [NII] and [S II].

2   Results and Discussion

To study the dynamics of the nebula surrounding He 3–519, the UCLES spectra have been fitted with Gaussian profiles using software available within STARLINK. Velocity maps of Hα and [N II] for each position angle have been constructed from

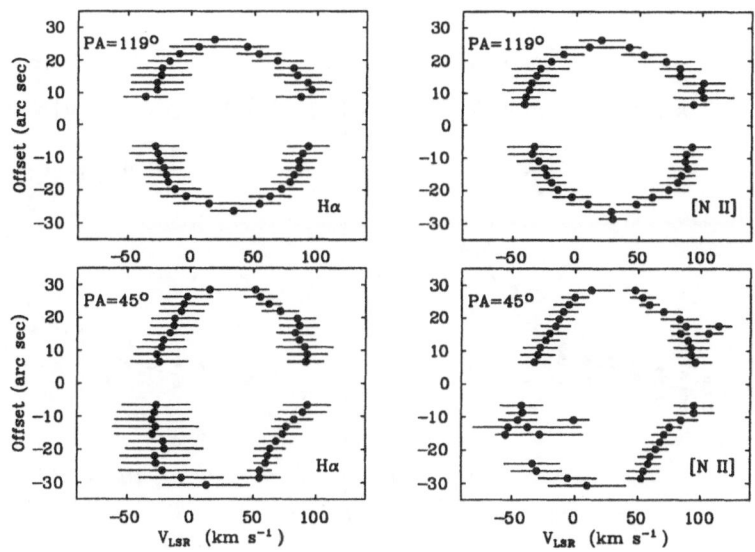

Fig. 2. Velocity maps of Hα and [N II] constructed from the parameters of the Gaussian fits. Filled circles indicate the central velocities and the horizontal bars represent the corresponding FWHMs.

the parameters of the Gaussian fits and are reproduced in Fig. 2. The total velocity spread of the components is from -40 to $+100$ km s^{-1}; the FWHM of each component is represented in Fig. 2 by a horizontal bar. It is apparent that the detailed velocity structure of the nebula differs between the two position angles. For PA$= 119°$, the nebula appears as a simple, symmetric shell expanding away from the central star whereas for PA$= 45°$, the shell is more elongated and has a complicated velocity and FWHM structure. The lower approaching quadrant of the shell appears very different; the Hα emission is twice as broad as elsewhere in the shell (mean FWHM=60 km s^{-1}) and [N II] is very weak or absent. From the Gaussian fits, the mean [N II]/Hα peak intensity ratio is found to be constant throughout the shell and equal to 0.97 ± 0.17 and 1.06 ± 0.19 for PA$= 119°$ and PA$= 45°$ respectively.

In order to derive the dynamical parameters defining the expanding shell (the radius R, expansion velocity v_{exp} and systemic velocity v_{sys}), we have performed a least-squares fit to the observed Hα and [N II] shell velocities v as a function of offset from the central star r assuming spherical expansion centred on the star. The results of the least-squares fits are given in Table I and illustrated in Fig. 3 for the two position angles. Although the two slit positions are clearly different and the shell is slightly elliptical, the details of the least-squares fits given in Table I show that the derived parameters are fairly similar. We will therefore adopt the values given in the final row of Table I which represent the least-squares fit to the

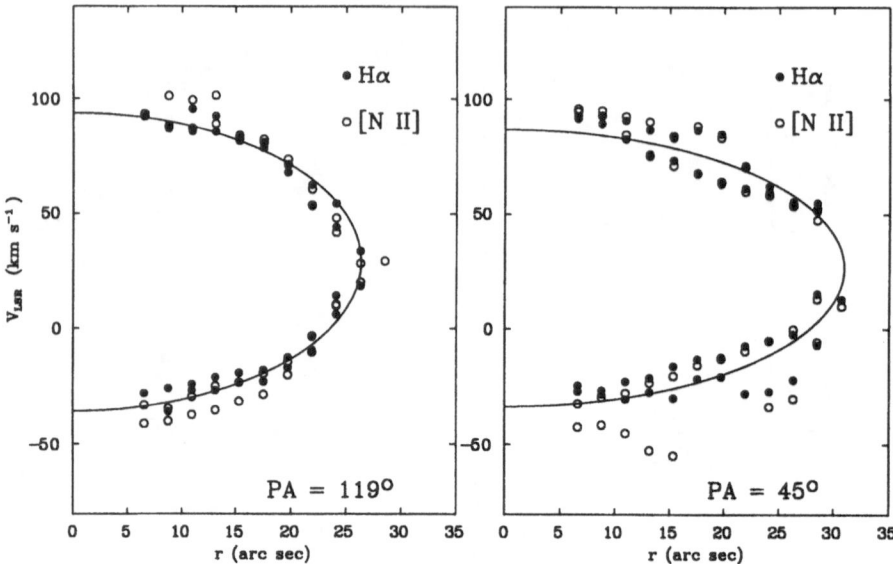

Fig. 3. Least-squares fits to the observed Hα and [N II] velocities as a function of distance from the central star.

entire dataset i.e. both postion angles and emission lines.

TABLE I
Parameters of Least-Squares Fits

Position Angle	Radius (arc sec)	Systemic Velocity (km s^{-1})	Expansion Velocity (km s^{-1})	σ(whole sample) (km s^{-1})
119°	26.3 ± 0.1	28.9 ± 0.7	64.9 ± 0.9	5.8
45°	30.9 ± 0.5	26.5 ± 1.0	60.3 ± 1.4	9.3
119° + 45°	29.3 ± 0.1	27.6 ± 0.8	61.0 ± 0.9	10.0

It is apparent that the expansion of the nebula is not centred on zero, but rather at +27.6 km s^{-1}. A kinematic distance of 8.0 kpc is derived using the Galactic rotation curve of Fich *et al.* (1989). This distance is confirmed by the velocity structure of the interstellar Na I lines present in the high resolution UCLES stellar spectrum. The resulting parameters of the nebula are given in Table II and compared with those of the AG Car nebula from Nota *et al.* (1992). The ionized mass of the He 3–519 nebula has been calculated from the total nebular Hα flux given by Stahl (1987) and the measured electron density (determined from the [S II] doublet present in the low resolution nebular spectra) using the equation given by Barlow (1987), assuming $T_e = 8\,000$ K and a normal helium abundance.

The derived nebular mass of $2.0 \pm 1.0\ M_\odot$ should be considered approximate since the equation used assumes that the electron density is the r.m.s value for the whole nebula whereas we have a measurement for one position only.

TABLE II

Adopted Parameters for He 3–519 Nebula and comparsion with AG Car Nebula

Parameter	He 3–519	AG Car
Shell radius (pc)	1.14 ± 0.14	0.6
Expansion velocity (km s^{-1})	61 ± 1	70
Dynamical age (yr)	$1.8 \pm 0.2 \times 10^4$	8.4×10^3
Electron density (cm^{-3})	300^{+135}_{-105}	~ 500
Ionized nebular mass ($M_\odot$)	2.0 ± 1.0	4.2

Overall, the two nebulae appear to be quite similar, particularly in terms of their expansion velocities and masses. The AG Car nebula has a smaller radius leading to a lower dynamical age. The low mass of the AG Car nebula, together with the observed chemical enrichment (Mitra & Dufour 1990; de Freitas Pacheco et al. 1992), and the unstable nature of the central star, favour the idea that the nebula resulted from a bulk ejection of material from AG Car $\sim 10^4$ yr ago. Given the similarities in the central star and nebular properties, it seems likely that the He 3–519 nebula was formed in a similar manner although the high mass loss of the central star ($1.3 \times 10^{-4}\ M_\odot\ yr^{-1}$) implies a significant interaction between the stellar wind and the ejected material. Over the dynamical age of the nebula, He 3–519 will have lost $\sim$ 2–3 $M_\odot$ of stellar wind material which is directly comparable to the estimated ionized mass of the shell.

References

Barlow, M.J.: 1987, *Mon. Not. Roy. Astr. Soc.* **227**, 161.
de Freitas Pacheco J.A., Neto A.D., Costa R.D.D., Viotti R.: 1992, *Astron. Astrophys.* **266**, 360.
Fich M., Blitz L., Stark A.A.: 1989, *Astrophys. J.* **342**, 272.
Humphreys R.M.: 1989, in: Davidson K., Moffat A.F.J., Lamers H.J.G.L.M. (eds.) Proc. IAU Coll. 113, *Physics of Luminous Blue Variables*, Kluwer, Dordrecht, p. 3.
Mitra, P.M. & Dufour, R.J.: 1990, *Mon. Not. Roy. Astr. Soc.* **242**, 98.
Nota, A., Leitherer, C., Clampin, M., Greenfield, P., Golimowski, D.A.: 1992, *Astrophys. J.* **398**, 621.
Smith, L.J., Crowther, P.A., Prinja, R.K.: 1993, *Astron. Astrophys.*, submitted.
Stahl O.: 1987, *Astron. Astrophys.* **182**, 229.
Wolf B., 1992, in: Drissen L., Leitherer C., Nota A. (eds.) *Nonisotropic and Variable Outflows from Stars*, PASPC **22**, p. 327.

Turbulent Mixing in Wind-Blown HII Regions

The Effect of Local Heating and Cooling in the Bubble

D. Breitschwerdt
Max-Planck-Institut für Kernphysik, D-6900 Heidelberg, FRG

and

F.D. Kahn
Department of Astronomy, Manchester M13 9PL, UK

Abstract. We describe the process of turbulent mixing, occuring at the interface between the stellar wind bubble and the overlying, compressed HII layer. It is shown that hot electrons from the bubble can penetrate into the mixing region, where they deposit some fraction of their thermal energy. This first leads to an expansion of the gas at constant pressure, which soon becomes too inefficient to carry away the additional energy. Therefore pressure can build up in the mixing layer, forcing the gas to expand into the bubble. We also discuss the effect of the over-pressure on the HII shell and find that it is possible to punch a hole at the polar caps, where mixing is most efficient. This may lead to the formation of bipolar outflows. We apply our model to the wind-blown HII region NGC 6334(A), and show that it can explain the observed distortion of the bubble and the bipolar ionization cones.

Key words: stellar winds, HII regions, turbulence, bipolar outflows

1 Introduction

The joint dynamical evolution of a stellar wind bubble (SWB) and a surrounding shell-like HII region, held under pressure by the SWB, has now been studied for 25 years (Pikel'ner, 1968; Dyson and deVries, 1972; Castor *et al.*, 1975). Observations have shown that the classical picture has to be modified in a number of ways. Firstly, the thickness of the HII shell is always larger than the values predicted by the classical model. Secondly, high excitation lines (e.g. SiIV, NV) have been found in the vicinity of O-stars (York, 1974), indicative of temperatures significantly less than 10^7 K. Thirdly, high velocity flows of the order of 100 km/s have been observed in Planetary Nebulae (López *et al.*, 1987), another example of wind-blown HII regions. We conclude that there must be some exchange of mass, momentum and energy between the SWB and the surrounding HII layer. Since the flow pattern resulting from high-speed winds is energy driven (Dyson, 1984), mass addition of dense ionized gas from the shell ultimately leads to cooling and thus reduces the pressure in the bubble. A number of processes have been invoked in the past, which are capable of dissipating some fraction of the energy of the hot shocked stellar wind (HSSW) gas. For example, (Weaver *et al.*, 1977) have considered global heat conduction between the HII/HSSW interface,

Astrophysics and Space Science **216**: 297–301, 1994.
© 1994 *Kluwer Academic Publishers.*

although this might be efficiently suppressed by a tangential magnetic field. It is also well-known that the ambient medium is clumpy on a wide range of length scales. This may lead to mass loading and hence to line emission in an essentially sonic flow (Hartquist and Dyson, 1988). Also large scale density gradients, particularly in groups of massive young stars, can induce Rayleigh-Taylor instability in the HII shell, and so cause mixing with the ambient gas. While all these processes may be operating in their appropriate environments, the existence of compact HII regions like NGC 6334(A) (Rodriguez et al., 1988), which exhibit shell-thickening and distortion of a spherically symmetrical bubble convinces us, that there should be an intrinsic process that promotes mixing at an early stage of evolution. We have found that an acoustic instability should always arise near the HSSW/HII-interface (Breitschwerdt and Kahn, 1988; paper I), and waves may grow there to finite amplitude. There are also observations of interstellar scintillation (Spangler and Gwinn, 1990) that are interpreted as Kolmogorov turbulence associated with extended HII regions. In a later paper (Kahn and Breitschwerdt, 1989; paper II) we described how turbulent mixing can lead to substantial cooling in NGC 6334(A), and deduced a limiting magnetic field strength for the undisturbed ambient medium, before it had been swept up into the shell. The value obtained agrees quite well with observations. In our model we assumed that the field is uniform on scales of parsecs, and therefore that mixing occurs predominantly at the polar caps, where the field strength is lowest. This is consistent with the fact that a bipolar outflow is observed in NGC 6334(A) and other wind-blown HII regions. In the following we discuss the process of local mixing in more detail and the effect of local heating on the evolution of the bubble.

2   Heating of the Turbulent Boundary Layer

To investigate the dynamics of the mixing layer, we assume that the flow is plane parallel, since it is found that the distance from the layer is always smaller than the width of the mixing patch. Further, the recombination and the sound crossing time scales are much smaller than the dynamical time scale in case of NGC 6334(A). From the observations of Rodriguez et al. (1988) one can infer a density of $\rho_0 = 4 \times 10^{-19}$ g/cm^3 of the ambient medium, a wind mechanical luminosity of $L_w = 10^{36}$erg/s and a photon output rate of $S_* = 4 \times 10^{48}$ s^{-1} in the Lyman continuum.

The turbulence in the HII layer on the one hand stretches field lines, thereby increasing the average field strength, and on the other hand brings together magnetic field lines of random orientation, thus promoting fast reconnection (paper II). As a result there is a critical length scale l'_c for which the field strength has a maximum. On the length scales below l'_c the field decreases and hot electrons from the bubble can penetrate the region of reduced magnetic field, depositing some fraction, ΔE, of their energy. We estimate that $\Delta E = k_B T_b l'_c / \lambda_m$, with T_b, λ_m and k_B being the temperature and mean free path of the hot electrons and Boltzmann's constant, respectively. The penetration depth may be somewhat smaller, because at the

HSSW/HII interface the pressure and therefore the magnetic field is larger than at the outer boundary of the shell. Here we ignore such further complications. Thus the rate of energy deposition per unit volume is $W = K_0\rho(1 - (T/T_b))$ with ρ and T being the local density and temperature and $K_0 = \pi^2 e^4 \Lambda(n_b/m_e^2 c_b)$; here Λ, m_e, n_b and c_b denote the Coulomb logarithm, mass of the electron and density and speed of sound in the HSSW, respectively.

3 Time-dependent Expansion of the Mixing Layer

Heating by fast electrons will first cause an expansion of the mixing layer at constant pressure. However, with increasing thickness of the layer due to continuous addition of new material, the rate at which energy can be carried away is too small compared to the energy deposition by the electrons. As a consequence, the pressure in the mixing layer will increase, leading to a dynamical expansion.

3.1 EXPANSION AT CONSTANT PRESSURE

At time τ after injection, the energy absorbed by the mixing layer is given by $\mathcal{E} = \int_0^\tau (dW/dt)\, dt = P_b c_b \sigma_c \tau$, where P_b is the pressure in the HSSW and σ_c is the cross section for Coulomb collisions, and thus $\lambda_m = 1/(n\sigma_c)$. Expansion at constant pressure then requires $(3/2)P_i + nP_b c_b \sigma_c \tau \approx nP_b c_b \sigma_c \tau = (5/2)P_b$. Therefore the density decreases as $n = (2/5)(c_b \sigma_c \tau)^{-1}$. Since expansion occurs subsonically, we have $\rho\, dx/dt = \Phi_m \equiv \epsilon \rho_i c_i$, and therefore $x = (1/5)\epsilon n_0 c_i \sigma \sqrt{k_B T_b/m_e}\ \tau^2$, with Φ_m being the flux of freshly added HII gas, $\rho_i = n_i m_i$ and c_i the density and isothermal speed of sound of the HII layer. In paper I we found the mixing efficiency to be $\epsilon \approx 0.13$. The condition for subsonic expansion is $\epsilon n_i c_i < n c_s = \sqrt{5n P_b/(3m_i)}$ and so $\tau = t_0 = 10^6 (T/T_b)^{1/2}$ s marks the end of this phase.

3.2 EXPANSION DUE TO PRESSURE GRADIENT

The over-pressure in the mixing layer will cause the material to expand at roughly the local speed of sound. Let us denote by $\Sigma = \int_0^x \rho\, dx'$ and l the surface density and extension of the expanding layer, respectively. For $T \ll T_b$, $W \approx W_0 = K_0 \rho$ and so we have the following gas dynamical equations:

$$\frac{d\Sigma}{dt} = \Phi \qquad \longrightarrow \qquad \frac{d\sigma}{d\tau} = \varpi \qquad (1)$$

$$\frac{3}{2}\frac{d}{dt}(P\,l) + \frac{1}{2}\frac{d}{dt}\left[\frac{1}{3}\Sigma\left(\frac{dl}{dt}\right)^2\right] = \Sigma\,W \qquad \longrightarrow \qquad \frac{d\varpi}{d\tau} = \frac{\sigma}{\lambda} - \varpi\sqrt{\frac{\varpi}{\lambda\sigma}} \qquad (2)$$

$$\frac{dl}{dt} = \sqrt{\frac{5P}{3\rho}} \qquad \longrightarrow \qquad \frac{d\lambda}{d\tau} = \sqrt{\frac{\varpi\lambda}{\sigma}}. \qquad (3)$$

The dimensionless form has been obtained by rearrangement and by setting: $\Sigma = A\sigma$, $P = B\varpi$, $l = C\lambda$, $t = D\tau$ and thus $A = (80/27)c_i P_b/(\epsilon W_0)$, $B = P_b$,

$C = (400/81)(c_i/\epsilon)^3 W_0^{-1}$ and $D = (3/5)C(\epsilon/c_i)$. Introducing new variables $g = (\varpi\lambda/\sigma)^2$ and $\xi = \ln\sigma$, enables us to reduce the above equations to $g'(dg'/dg) + (2 - g^{-(1/4)})g' = 2g^{3/4}$, for which an asymptotic solution of the form $g^{1/4} = 1/4(\xi + \text{const.})$ can be found in the case of large surface densities ($g \ll 1$). Thus the dimensionless pressure and distance from the mixing layer are given by:

$$\varpi = \sigma\left[\left(g_1^{\frac{1}{4}} - \frac{1}{4}\ln\frac{\sigma_1}{\sigma}\right)^2 + \frac{1}{2}\left(g_1^{\frac{1}{4}} - \frac{1}{4}\ln\frac{\sigma_1}{\sigma}\right)\right]^{-1}, \qquad (4)$$

$$\lambda = \frac{\sigma}{\varpi}\left(g_1^{\frac{1}{4}} - \frac{1}{4}\ln\frac{\sigma_1}{\sigma}\right)^2, \qquad (5)$$

with g_1 and σ_1 being integration constants. Equations (4) and (5) show that for increasing surface densities the thickness λ of the layer increases relatively slowly, i.e. logarithmically, whereas the pressure ϖ increases almost linearly with σ.

4   Application to NGC 6334(A)

We shall study the effect of local heating in case of the compact wind-blown HII region NGC 6334(A), and so have to determine the transformation constants for the appropriate parameters of this object. These are given by $A = 1.7 \times 10^{-7}\text{g/cm}^2$, $B = P_b = 3.7 \times 10^{-7}\text{dyne/cm}^2$, $C = 5.8 \times 10^{13}\text{cm}$ and $D = 4.5 \times 10^6\text{s}$. The dynamical expansion starts at $t = t_0 = 2.1 \times 10^6$ s, at the end of the expansion at constant pressure. Therefore the boundary conditions of equations (1)-(3) are fixed by $\tau_0 = 0.5$, $\lambda_0 = 2.0$, $\sigma_0 = 0.6$ and, by definition, $\varpi_0 = 1$. The increase in pressure is limited by radiative cooling of the layer, which is most important for the gas that has been mixed in for some time. Another limit is set by the requirement that the electrons shall not lose most of their energy before they penetrate the bulk of the layer, and it turns out that this condition is the more restrictive here. The collision cross section is given by $\sigma_c = 2 \times 10^{-18}(T/10^7[K])^{-2}\text{ cm}^2$ and therefore the critical surface density is $\sigma_{\text{crit}} = (\bar{m}/A\sigma_c) = 1.1 \times 10^2$, with the mean mass $\bar{m} = 2 \times 10^{-24}$ g. Numerical solution of equations (1)-(3) shows that $\sigma = \sigma_{\text{crit}}$ at $\tau = 25$ or $t = 5 \times 10^7$ s, and the maximum over-pressure that is reached is $\varpi = 25$.

The VLA-map of NGC 6334(A) shows two bipolar ionization cones extending about 0.5 pc from the outer shell (Rodriguez et al., 1988). The age of the bubble is about 2.2×10^{11} s, and the estimated electron density in the cones is $\sim 2\times10^3$ cm^{-3}, implying a recombination time scale of $\tau_{\text{rec}} \sim 2 \times 10^9$ s, and therefore it is clear that the lobes have to be kept continuously ionized by the central star. It has been demonstrated (Breitschwerdt and Drury, 1991) that this can be done by a conical ionization front. The unsolved problem, however, is how the ionizing photons can escape from the dense shell, in which they are trapped. In the following we estimate the possiblity of punching a hole into the shell due to the over-pressure; the mixing regions are located at the polar caps and therefore naturally agree with the observed geometry.

The outer shell is largely neutral, and the disturbance will propagate through its thickness Δ in time $\tau_p \sim \Delta/v = \Sigma_0 c_0/\sqrt{P_b P}$, with Σ_0 being the surface density, c_0 the speed of sound, v and P the velocity and pressure of the disturbance. Using similarity solutions (cf. Dyson, 1984) to estimate the dynamical time scale τ_d, we obtain $\tau_p/\tau_d \approx 0.2 t c_0/R_b$, where $R_b = 3.7 \times 10^{17}$ cm is the bubble radius at $t = \tau_d = 2.2 \times 10^{11}$ s and for $c_0 = 0.2$ km/s, $\tau_p/\tau_d \approx 0.01 \ll 1$. The over-pressure can thus be communicated sufficiently fast to the mixing layer to induce fragmentation of the shell at an early stage. This may be a lower limit in the sense that expansion will occur mainly into the HSSW, because the surface density is smaller there. However, continuous injection of ionized gas due to turbulence could maintain an over-pressure. It is also unlikely that the holes will be sealed off by shell material, because the typical time scale is $\tau_s \sim D/c_0 \approx 5 \times 10^{12}$ s $\gg \tau_d$, with $D \approx 1 \times 10^{17}$ cm being the width of the mixing patch at $t = \tau_d$ (paper II).

5 Conclusions

The existence of compact wind-blown HII regions with distorted bubbles and bipolar ionization cones, strongly argues for processes that allow exchange of mass, momentum and energy between the bubble and the HII layer at an early stage of dynamical evolution. We believe that turbulent mixing, even in the presence of magnetic fields, provides such a mechanism. Global cooling of mixed-in, dense HII gas can explain the distortion of the bubble and the thickness of the HII region, as well as the existence of high excitation lines. We have shown, that in case of NGC 6334(A), local heating by hot bubble electrons leads to an over-pressure and hence to a gas-dynamical expansion of the mixing layer into the bubble, which may be observed as high velocity gas of $O(100\mathrm{km/s})$ in emission lines. Moreover, the over-pressure can induce fragmentation of the shell and thus allow the escape of ionizing photons that produce the observed bipolar ionization cones.

References

Breitschwerdt, D, Kahn, F.D.: 1988, *Mon. Not. R. Astr. Soc.*, **235**, 1011 (paper I).
Breitschwerdt, D., Drury, L.O'C.: 1991, *Astron. and Astrophys.*, **245**, 257.
Castor, J., McCray, R., Weaver, R.: 1975, *Astrophys. J. (Letters)*, **200**, L107.
Dyson, J.E.: 1984, *Astrophys. Sp. Sci.*, **106**, 181.
Dyson, J.E., deVries, J.: 1972, *Astron. and Astrophys.*, **20**, 223.
Hartquist, T.W., Dyson, J.E.: 1988, *Astrophys. Sp. Sci*, **144**, 615.
Kahn, F.D., Breitschwerdt, D.: 1989, *Mon. Not. R. Astr. Soc.*, **242**, 209 (paper II).
López, J.A., Falcón, L.H., Ruiz, M.T., Roth, M.: 1987, in: *Planetary Nebulae*, IAU-Symposium No. 131, ed. Torres-Peimbert, S., Kluwer Acad. Publ. Comp., Dordrecht, p. 179.
Pikel'ner, S.B.: 1968, *Astrophys. Letters*, **2**, 97.
Spangler, S., Gwinn, C.R.: 1990, *Astrophys. J. (Letters)*, **353**, L29.
Rodriguez, L.F., Cantó, J., Moran, J.M.: 1988, *Astrophys. J.*, **333**, 801.
Weaver, R., McCray, R., Castor, J., Shapiro, P. Moore, R.: 1977, *Astrophys. J. (Letters)*, **218**, L377.
York, D.G.: 1974, *Astrophys. J. (Letters)*, **193**, L127.

Shock Wave Structure in the Cygnus Loop

J. C. Raymond and S. Curiel
Harvard-Smithsonian Center for Astrophysics,
Cambridge, MA 02138, U.S.A.

Abstract. The interaction of an SNR blastwave with clouds of various sizes determines the appearance of the SNR in all wavelength bands, and it may determine the overall evolution of the remnant. The Cygnus Loop provides excellent examples of the interaction with both large and small clouds at various stages, and its brightness, large size and low reddening make it a natural target for study. We consider X-ray, optical and UV observations of features at the eastern edge of the Cygnus Loop to look for evidence of cloud evaporation, turbulent stripping from a cloud and pressure enhancement associated with the blastwave–cloud interaction. We consider the effects of the sputtering of dust grains on the temperature derived from ROSAT spectra and we briefly consider the clumpiness of Hα emission to be expected from compression of a turbulent magnetic field.

1   Introduction

Theoretical models of supernova remnants in homogeneous media are now quite well developed, but it has become clear that density inhomogeneities in the ISM determine the appearance of most SNRs at all wavelengths. The interaction between a blast wave in a low density intercloud medium and higher density clouds may also govern the evolution of an SNR. Model calculations are simplified by assuming either very large or very small clouds.

At the large cloud extreme, one can consider a blastwave striking a cloud which happened to be in the neighborhood, or one can imagine a shell created by the SN progenitor (McCray and Snow, 1979; Shull *et al.*, 1985). The complexes of bright optical filaments on the eastern and western limbs of the Cygnus Loop are large, slightly rippled sheets of glowing gas, with each tangency of the sheet to the line of sight appearing as a filament (Hester, 1987). Roughly 1/7 of the surface of the Cygnus Loop is covered by these sheets, each of which is a radiative shock wave moving into gas about an order of magnitude more dense than the ambient gas (Hester and Cox, 1986). The optically bright regions are also bright in X-ray, UV, radio and infrared maps (e.g. Green, 1990; Braun and Strom, 1986; Arendt *et al.*, 1992; Blair *et al.*, 1991; Vancura *et al.*, 1993; Seward, 1990; Cornett *et al.*, 1992). The bright radio emsision can probably be attributed to the compression of non-thermal particles and magnetic fields in the cooling region behind a radiative shock. The X-ray and IR enhancements probably result from the overpressure where the non-radiative blast wave encounters a density jump, forcing a secondary shock back into the X-ray emitting gas to further heat and compress it (e.g. Hester, Raymond and Blair, 1993).

At the other extreme, one can concentrate on the cumulative effects of many small clouds overrun by the blastwave. Some of the consequences may be ther-

Astrophysics and Space Science **216**: 303–309, 1994.

mal evaporation (McKee and Ostriker, 1977), enhanced radiative losses ("cloud-crushing"—Cox, 1981), enhanced X-ray brightness in a bow shock enveloping each cloud (Tucker, 1971; McKee and Cowie, 1975), and turbulent mixing as the cloud is disrupted by Rayleigh-Taylor or Kelvin-Helmholtz instabilities (e.g. Stone and Norman, 1992; Klein, McKee and Colella, 1991) which might lead to mass loading (e.g. Dyson and Hartquist, 1987) or efficient radiation from a mixing layer (Slavin, Shull and Begelman, 1993).

The Cygnus Loop is an excellent target for SNR observations because its brightness, low reddening and large size make it accessible at all wavelengths and easily resolvable. It also presents excellent examples of both large- and small-scale interactions.

Figure 1 shows a comparison between HRI ROSAT X-ray and optical narrowband [O III] λ5007 emission of the eastern Cygnus Loop region. This figure is centered on the XZ bow shock, and shows the Spur filament (top right corner), and the filament ASTRO-1 observed by the Hopkins Ultraviolet Telescope (about 2.5′ to the NE of XA). The main X-ray emission seems to be divided into northern and southern sections by a somewhat fainter wedge which intrudes from the west. The northern X-ray emitting shell appears to envelop the XA bow-shock as well as part of the triangular optical structure in this region. This X-ray enhanced emission has been interpreted by Hester and Cox (1986) as an overpressured region resulting from reflection of the blastwave shock by a cloud surface. On the other hand, the southern X-ray emission peaks several arc-secs inside the optical edge and extends about 2′ towards the west. This bright X-ray knot cannot be explained in terms of a shocked cloud (Hester and Cox, 1986), but instead it seems to be the result of shocks reflected from the dense gas along each side of the "right angle" bend of the optical filaments. The edge of the brightest X-ray feature is coincident with the leading edge of the optical emission over a length of several minutes of arc. Hester and Cox (1986) pointed out that this spatial relationship is consistent with a thin (≤ 0.2 pc) zone of enhanced X-ray emission lying immediately behind the much thinner sheet of optical emission.

2   Large Clouds

A simple model for a blastwave encountering a large cloud is a 1-D strong, non-radiative shock encountering a density jump. If the density jump is large enough, a reflected shock moves into the blastwave region, further heating and compressing it, while a transmitted shock moves into the cloud. The high pressure in the region between the shocks enhances the X-ray emission (proportional to n^2 and temperature). It also increases the ram pressure of the cloud shock over that of an undisturbed blastwave. Pressure equilibrium between the shocked cloud and intercloud gasses is expected until radiative cooling sets in.

These predictions can be tested against observations in the northeast Cygnus Loop. Some of the brighter Balmer line filaments trace cloud shocks which are

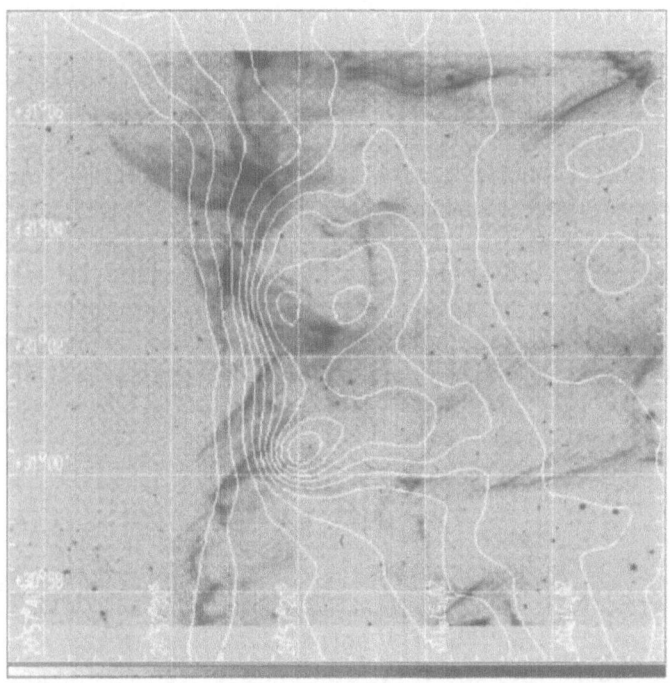

Fig. 1.   X-ray map and optical image of the eastern Cygnus Loop region. Linear X-ray map overlaid onto a linear gray scale image of the [O III] $\lambda 5007$ emission. The X-ray data was smoothed by convolution with a 16″ FWHM Gaussian in order to generate relatively smooth surface brightness contours from the counting statistics of the HRI ROSAT data. The lowest two contour levels are 5 and 10% of the peak emission and the separation between contour levels is 10% of the peak value. The brightest knot has a surface brightness of 0.192 counts arcmin^{-2} s^{-1} and the faintest region on the east side has an average surface brightness of 0.010 arcmin^{-2} s^{-1}. The optical [O III] CCD image was taken with the 48″ telescope at the Whipple Observatory.

just beginning to cool, and a model with $n_c = 1.2$, $n_{ic} = 0.09$, $V_{BW} = 400$ km/s matches the shock speed, X-ray surface brightness enhancement, X-ray scale length and the swept up column in the Balmer filaments, if the shock encountered the density jump 1000 years ago (Hester, Raymond and Blair, 1993). No direct pressure measurement is available, but ROSAT should provide a good test of the predicted temperature enhancement. Comparison of the ram pressure derived from the [O III] surface brightness and Doppler velocity of a 120 km/s radiative shock known as the Spur with the pressure of the adjacent X-ray emitting gas shows a disquieting discrepancy, with the thermal pressure of the X-ray gas too low to drive the radiative shock by a factor of 5 (Raymond et al., 1988). Various geometrical complexities, time dependent effects and non-thermal pressure contributions may play important roles in explaining this feature. Just to the east of the Spur there is a sharp edge to the X-ray shell studied by Hester and Cox (1986). While the blast wave is clearly encountering a high density here, the ROSAT PSPC spectra we are analyzing show a temperature increase from 10^6 to 2×10^6 as one moves *away* from the shock. This is opposite to the predicted temperature enhancement near the shock for the discontinuity case. It is most easily interpreted as a blastwave climbing a factor of two density enhancement as it travels 2–3 pc. The gas appears to be in pressure equilibrium, but caution is needed in interpreting the ROSAT spectra. Models of grain destruction in shocks (Vancura, 1992) show that depletion of Fe and Si (and the liberation of these elements as the grains are destroyed) affects the 1/4 KeV ROSAT band at the factor of two level, while the harder X-rays are dominated by oxygen and neon emission at temperatures below 3×10^6 K, and these are not affected by grain depletion. Thus depletion onto grains increases the ROSAT hardness ratio, making the gas appear hotter than it is for temperatures of 1–2 $\times 10^6$ K. For older SNR such as the Cygnus Loop, this can lead to errors in the inferred temperature as large as 40%.

3   Small Clouds

A basic model for the initial stages of the interaction of a blastwave with a small cloud has been available for some time. McKee and Cowie (1975) , for instance, described the initial formation of a bow shock in the X-ray emitting gas, the propagation of shocks into the front and sides of the cloud, and the eventual dissipation of the bow shock and re-formation of the blastwave behind the cloud. For a blastwave velocity of a few hundred km/s and a density contrast of $\sim$ 10, the cloud shock will be radiative, so one expects a bright optical knot enveloped in an X-ray-emitting sheath. The pressure enhancement in the bow shock is about a factor of 3.

As the interaction progresses, the fundamental issue becomes the mixing of hot and cool gas, both in predicting the optical and X-ray appearance of a shocked cloud, and in evaluating the effect of clouds on SNR evolution. McKee and Os-triker (1977) assumed that hot and cool gas can be efficiently mixed by evaporation while Cox (1981) assumed that magnetic fields completely suppress thermal conduction. In the former model, the enhanced density and reduced temperature of the hot gas lead to rapid cooling and formation of a dense shell, while in the latter model only the radiative losses in the shocked cloud drain energy from the blast wave. Other possible effects, which have been more thoroughly explored in contexts such as stellar wind-driven bubbles, are mass-loading of the flow (e.g. Dyson and Hartquist, 1987) and turbulent mixing of hot and cool gas at a microscopic level on a time scale short compared to the cooling time (Slavin, Shull and Begelman, 1993). The former may decelerate the shell and provide cool gas a velocities higher than expected for radiative shocks (e.g. Meaburn *et al.*, 1989), while the latter could greatly enhance the radiative losses.

Detailed 2-D and 3-D numerical hydrodynamic models show the fragmentation of a cloud as the shear where the hot gas flows past the edge of the shocked cloud generates vorticity and Kelvin-Helmholtz instabilities (Stone and Norman, 1992; Klein, McKee and Colella, 1991), and Rayleigh-Taylor instabilities also play a part. These particular codes do not treat cooling in a way which could investigate thermal instabilities. They do not include thermal conduction or magnetic fields, so they make no predictions regarding the efficiency of evaporation. The codes show fragmentation of the stripped gas to the limits of their resolution.

The Cygnus Loop contains several small, isolated knots which can be compared with models. Hester and Cox (1986) examined a feature called XA, a 2' long triangular feature near the eastern edge of the Cygnus Loop. By comparing optical images with the Einstein HRI image, they showed that evaporation could not explain the observed X-ray morphology, and that a model with a bow shock in the X-ray emitting gas and a converging, roughly conical shock behind the brightest knot could explain the obsevations for a reasonable value of the pressure enhancement in the bow shock.

Our more recent long-slit echelle and ROSAT data are consistent with this picture. The temperature derived from the ROSAT spectra is depressed at XA, implying that the preshock density is higher than normal, suggesting either a centrally condensed cloud or a core-halo cloud structure like that envisioned by McKee and Ostriker (1977). Such a density structure is probably necessary if the optical emission trailing from the bright XA knot is to be attributed to a shock still converging behind the cloud. The waviness of this trailing emission suggests the sorts of instabilities predicted by the numerical models, but the crispness of individual filaments seems at odds with fully developed turbulence, and the models lead one to expect a cylindrical, rather than a conical, geometry. The thickness of the X-ray enhancement seems an order of magnitude larger than could be explained by a mixing layer. However, the interaction which created the XA knot is very recent, as indicated by very high [O III]/Hα ratios, and it is possible that a full turbulent cascade or a strong evaporative flow will develop in the future. Long slit echelle spectra are consistent with a converging 100 km/s concial shock behind the bright optical knot, but they may also be consistent with flow along the apparent lengths of the filaments if the gas was stripped from the edges of the knot.

A cleaner, apparently more recent, example of an encounter between the blast-wave and a cloud has been pointed out by Fesen, Kwitter and Downes (1992). Their Hα images bear a spectacular resemblance to the hydro model of Klein, McKee and Colella at a particular stage in the evolution. The resemblance may be somewhat deceptive, however, in that the feature which looks like the bow shock seems to be an unrelated feature associated only in projection with the cloud (based on proper motion—R. Fesen, private communication—and neutral pre-shock medium—W. Blair, private communication). A more general difficulty in comparing the models to observations is the lack of a detailed treatment of radiative cooling. Instead, the models set $\gamma = 1.1$ to simulate a case in between the adiabatic and isothermal extremes. Spectra of the observed cloud show it to be completely radiative in some places, and completely non-radiative in others, and there is no way to confidently predict the optical appearance from the density structure predicted by the code. Perhaps a first question would be whether or not the fragments stripped from the edge of the shocked cloud lie in an appropriate temperature and ionization range to emit [O III], Hα and [S II]. Planned Fabry-Perot observations should reveal the velocity structure of the emitting material. VLA observations may show whether or not the magnetic field is tangled by turbulence. This may hold the key to the applicability of thermal conduction and mixing layer models.

4   Conclusions

The Cygnus Loop, in general, and the isolated cloud observed by Fesen, Kwitter and Downs in particular, present outstanding opportunities to test theoretical ideas with detailed observations. An ultimate goal is the understanding of the exchange of mass and energy between hot and cool phases of the shocked gas. One major obstacle to this understanding is random chance in the ambient density structure, which could easily introduce morphological features that we will try to interpret as interesting dynamical effects. A second obstacle is the catastrophic thermal instability prdicted to occur in radiative shocks faster than about 150 km/s (e.g. Innes, Giddings and Falle, 1987; Bertschinger, 1986; Gaetz, Edgar and Chevalier, 1988). Most detailed analyses of Cygnus Loop filaments have found them to be either non-radiative (e.g. Hester, Raymond and Blair, 1993) or slower than the unstable velocity range (e.g. Raymond et al., 1988). It could be that unstable shocks are hard to distinguish from steady ones, even if a broad range of spatial, spectral and velocity information is available, but it is also likely that the selection of bright, morphologically simple filaments for detailed study strongly discriminates against unstable shocks. The best candidate for an unstable shock among the regions studied to date is the filament observed with the Hopkins Ultraviolet Telescope by Blair et al. (1991). It lies in an area of depressed X-ray surface brightness (suggesting that T has fallen below 10^6 K), its strong O VI emission implies a shock velocity well within the unstable range and its fuzzy morphology and spectral variations fit the expectation for thermally unstable cooling. Further study of this filament will be an important ingredient in future analyses of large- and small-scale encounters between shocks and clouds.

Finally, it is interesting to ask why most of the filaments in an [O III] image look sharp, while the same filaments appear clumpy and fuzzy in Hα images. Thermal instability seems to be an attractive possibility, but one would expect the instability to clump the cooling gas before it cools to [O III] emitting temperatures. Another possibility is that magnetic pressure supports the gas in the recombination zone, where Hα is formed. If the field is uniform, it will not induce clumping, but the strength of the transverse component of the field varies due to turbulence in the ISM. Diffuse acceleration models for cosmic ray production in shocks postulate that $\delta B/B \sim 1$ in a shock precursor and there is some observational evidence for enhanced turbulence near the Vela SNR (Desai *et al.*, 1992). When this gas is compressed by the shock and allowed to cool, regions of low transverse B will be much more strongly compressed, giving rise to bright knots on various scales and producing a mottled appearance.

Acknowledgements

This work was supported by NASA Grants NAGW-528 and NAG5-1870 to the Smithsonian Astrophysical Observatory.

References

Arendt, R. G., Dwek, E. and Leisawitz, D.: 1992, *Astrophys. J.* **400**, 562.
Bertschinger, E.: 1986, *Astrophys. J.* **304**, 154.
Blair, W. P., *et al.*: 1991, *Astrophys. J. Lett.* **379**, L33.
Blair, W. P., Long, K. S. Vancura, O. and Holberg, J. B.: 1991, *Astrophys. J.* **374**, 202.
Braun, R. and Strom, R. G.: 1986, *Astron. Astrophys.* **164**, 208.
Cornett, R. H. *et al.*: 1992, *Astrophys. J. Lett.* **395**, L9.
Cox, D. P.: 1981, *Astrophys. J.* **245**, 534.
Desai, K. M., Gwinn, C. R., Raynolds, J., King, R. A., Jauncey, D., Flanagan, C., Nicolson, G., Preston, R. A. and Jones, D. L.: 1992, *Astrophys. J. Lett.* **393**, L75.
Dyson, J. E. and Hartquist, T. W.: 1987, Mon. Not. Roy. Astr. Soc. **228**, 453.
Fesen, R. A., Kwitter, K. B. and Downes, R. A.: 1992, *Astron. J.* **104**, 719.
Gaetz, T. J., Edgar, R. and Chevalier, R. A.: 1988, *Astrophys. J.* **329**, 927.
Green, D. A.: 1990, *Astron. J.* **100**, 1927.
Hester, J. J.: 1987, *Astrophys. J.* **314**, 187.
Hester, J. J. and Cox, D. P.: 1986, *Astrophys. J.* **300**, 675.
Hester, J. J., Raymond, J. C. and Blair, W. P.: 1993, *Astrophys. J.* submitted.
Innes, D. E., Giddings, J. R. and Falle, S. A. E. G.: 1987, *Mon. Not. Roy. Astr. Soc.* **226**, 67.
Klein, R. I., McKee, C. F. and Colella, P.: 1991, in S. E. Woosley (ed.) *Supernova*, Springer, New York, p.696.
McCray, R. and Snow, T. P., Jr.: 1979, *Ann. Rev. Astron. Astrophys.* **17**, 213.
McKee, C. F. and Cowie, L. L.: 1975, *Astrophys. J.* **195**, 715.
McKee, C. F. and Ostriker, J. P.: 1977, *Astrophys. J.* **218**, 148.
Meaburn, J., Whitehead, M. J., Raymond, J. C., Clayton, C. A. and Marston, A. P.: 1989, *Astron. Astrophys.* **227**, 191.
Raymond, J. C., Hester, J. J., Cox, D. P., Blair, W. P., Fesen, R. A. and Gull, T. R.: 1988, *Astrophys. J.*, *324, 869*.
Seward, F. D.: 1990, *Astrophys. J. Suppl.* **73**, 781.
Shull, P., Dyson, J. E., Kahn, F. D. and West, K. A.: 1985, *Mon. Not. Roy. Astr. Soc.* **212**, 799.
Slavin, J. D., Shull, J. M. and Begelman, M. C.: 1993, preprint.
Stone, J. M. and Norman, M. L.: 1992, *Astrophys. J.* **390**, L17.
Tucker, W.: 1971, *Science*, **172**, 372.
Vancura, O.: 1992, Ph.D. Thesis, Johns Hopkins University.
Vancura, O., Blair, W. P., Long, K. S., Raymond, J. C. and Holberg, J. B.: 1993, submitted to *Astrophys. J.*.

Catastrophic Cooling Diagnostics

D.E. Innes
Max-Planck-Institut für Aeronomie, Postfach 20,
W-3411 Katlenburg-Lindau, Federal Republic of Germany

Abstract. This paper presents models of optical emission line features that characterise catastrophic cooling in radiative shocks. The computations are based on a 1-D magnetohydrodynamic model. Runaway cooling results in the formation of secondary shocks which travel through the previously shocked cooling layer. Several filaments of emission with specific properties and spectral signatures are produced.

Key words: radiative shocks, spectroscopy

1 Introduction

The dynamics of radiative shocks in the interstellar medium are complicated by the generation of multiple shocks due to catastrophic cooling in the postshock cooling flow. Although this only affects the fast radiative shocks (velocity > 150 km s^{-1}; Innes, Giddings & Falle, 1987a), the optical line intensities from these shocks can be more than an order of magnitude greater than lower velocity steady shocks running into similar ambient material (Innes, 1988). Therefore these shocks may be very visible.

For many years diagnostics for steady shocks have been available (cf. Dopita, 1977; Raymond, 1979; Shull, 1979) and observed spectra have been interpreted within this framework. One important assumption for the interpretation of the spectra has been that the postshock emission region is too narrow to be resolved in either space or velocity. Thus a given shock structure was identified with a single set of line ratios. In contrast the cooling region of an unsteady shock is not only brighter but also significantly broader (Innes, 1992). In this paper we discuss some optical signatures of unsteady shocks.

2 Structural Evolution

We assume that a plane parallel shock with velocity 175 km s^{-1} is travelling into a uniform medium with density, 1 cm^{-3}, and magnetic field, 3 μG, perpendicular to the flow. The ionization is coupled to the flow and the radiation field. Full details of the atomic physics and dynamics are given in Innes (1992).

The evolution of a plane parallel shock is periodic (Gaetz, Edgar & Chevalier, 1988; Innes, 1992). The periodic behaviour of the temperature is shown in Fig. 1. At time zero, in this figure, the shock has just entered the cooling phase. This is followed by the break-up of the cooling layer into three distinct shocks, two

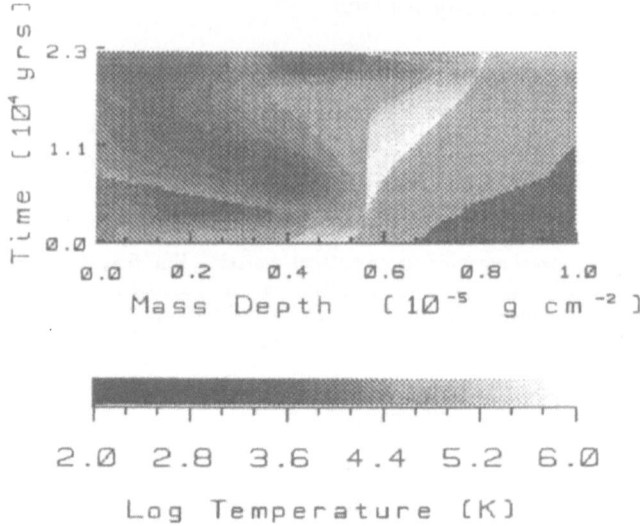

Fig. 1.   Temperature evolution of 175 km s^{-1} shock

moving upstream and one downstream. Later the two upstream shocks merge,
increasing the postshock pressure and, in turn, the shock velocity and postshock
temperature. The main shock then moves out, essentially adiabatically from the
contact. The growth and decline of the photoionization zone downstream of the
contact is visible in this picture.

3   Optical Emission

To obtain emission properties, we follow Innes, Giddings & Falle, (1987b) and
assume that the plane parallel shock structure lies on the periphery of a hollow
sphere, and the emission is computed across the edge. This configuration conveniently illustrates the effects of observing at different orientations to the shock.

3.1   EVOLUTION

Innes (1992) showed that line ratios integrated over space and time are similar
to the steady shock models. The main difference is the relatively stronger [O I].
However, as Fig. 2 shows, if the emission is observed perpendicular to the flow, different emission lines will be strongest at very different places in time and space. In
Fig. 2 the resolution of the simulated image has been degraded to 50 arcsec·kpc in
order to demonstrate that the ionization separation is not due to the low preshock
density used in the model. The cooling length is inversely proportional to preshock
density, so if the length is scaled, these pictures correspond to densities 3–30 cm^{-3},
assuming a spatial resolution of 3 arcsecs and the object is 0.5 to 5 kpc away.

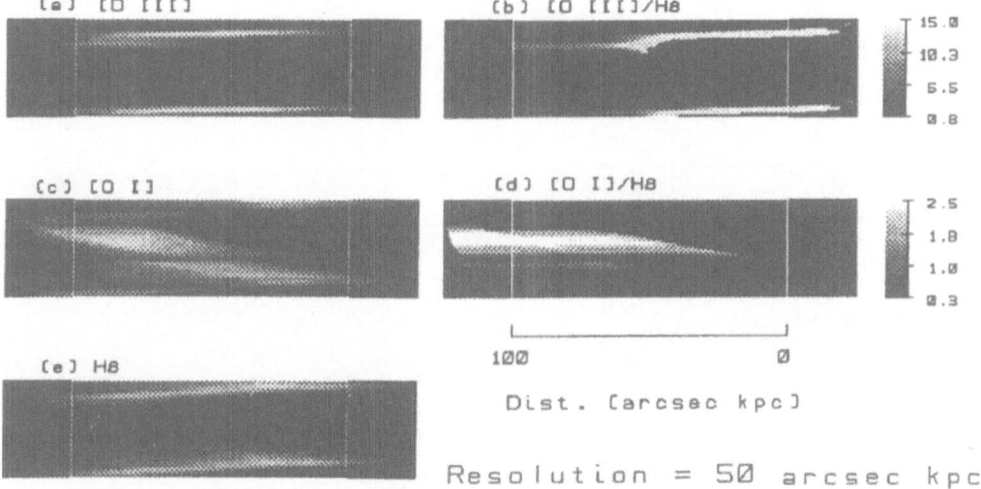

Fig. 2. Evolution of line intensities: a [O III]; b [O III]5007/Hβ; c [O I]; d [O I]6300/Hβ; e Hβ.
The vertical white lines show the width of the unsmoothed region. The RH white line marks the
shock front.

We see that the steady shock upper limit, [O III]/Hβ = 8, is exceeded over
a significant length of the cooling region for a significant fraction of the shock's
evolution. Therefore 'high' [O III]/Hβ ratios result naturally from multiple shock
formation. The highest [O I] is coming from the reverse shock, which does not
become prominent until after the beginning of the adiabatic phase, so the [O I] is
out of phase with the [O III] and the model predicts that [O I] will be generally
downstream of [O III], but there will be no clear relationship between the two.

3.2 REVERSE SHOCK MODEL

Fig. 3 shows the long-slit spectrum of a model with a reverse shock. The reverse
shock, moving into the recently cooled photoionized shell, is separated from the
main shock by a contact discontinuity seen as a sharp drop in temperature across
a constant velocity region. The velocity profile shows that gas behind the reverse
shock is travelling slower, relative to the preshock material, than material in the
contact and in the photoionized shell.

The [O I] spectrum shows three faint filaments: the inner, faintest due to the
uncooled shell; the middle, brightest due to the reverse shock; and the outer from
recombining gas at the contact. In this case, the velocity difference between the
contact and reverse shock is about 25 km s^{-1}. The [O III] emission comes exclu-
sively from the gas near the contact, and is apparent as only one filament.

D.E. Innes

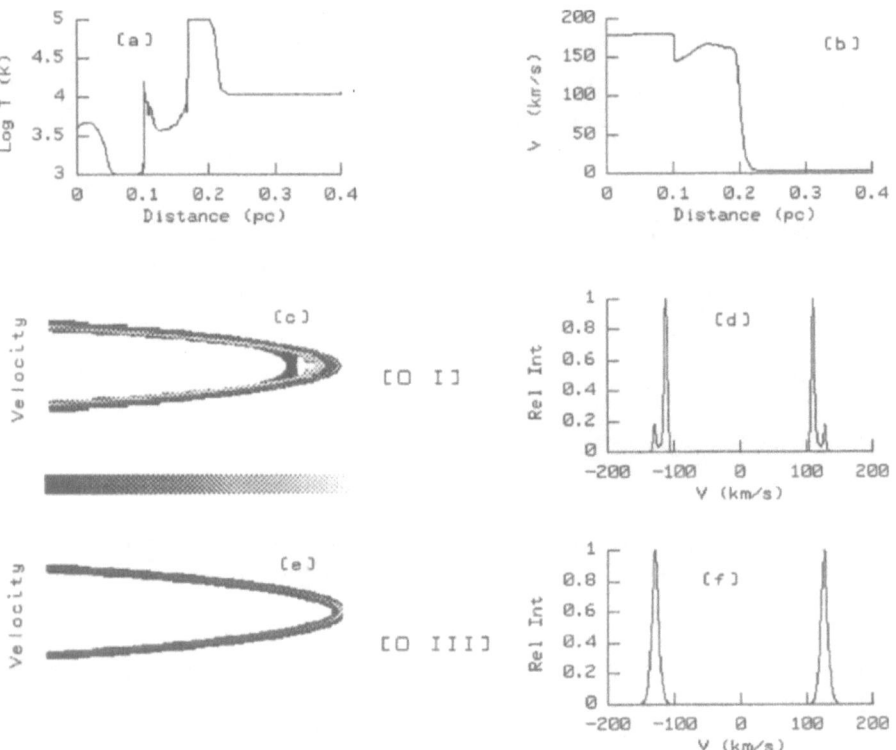

Fig. 3.   Modelled long slit spectra from a structure with a reverse shock: **a** Temperature; **b** Velocity; **c** [O I] long slit spectrum; **d** [O I] line profile perpendicular to the shell; **e** [O III] long slit spectrum; **f** [O III] line profile perpendicular to the shell.

3.3   DOUBLE SHOCK MODEL

The second example is a model with two consecutive shocks travelling upstream. The density behind the second shock is a factor 10 larger than the front shock, therefore the postshock cooling region is 10 times narrower. The velocity profile shows that the front shock is travelling at roughly half the shock velocity.

This phase is characterised by bright high ionization lines showing two components in space and velocity. There is slightly more [O III] produced in the recombination zone of the upstream shock, therefore the lower velocity material is brightest. Most of the [O I] emission is produced in the second shock with faint emission from the photoionized region at higher velocity.

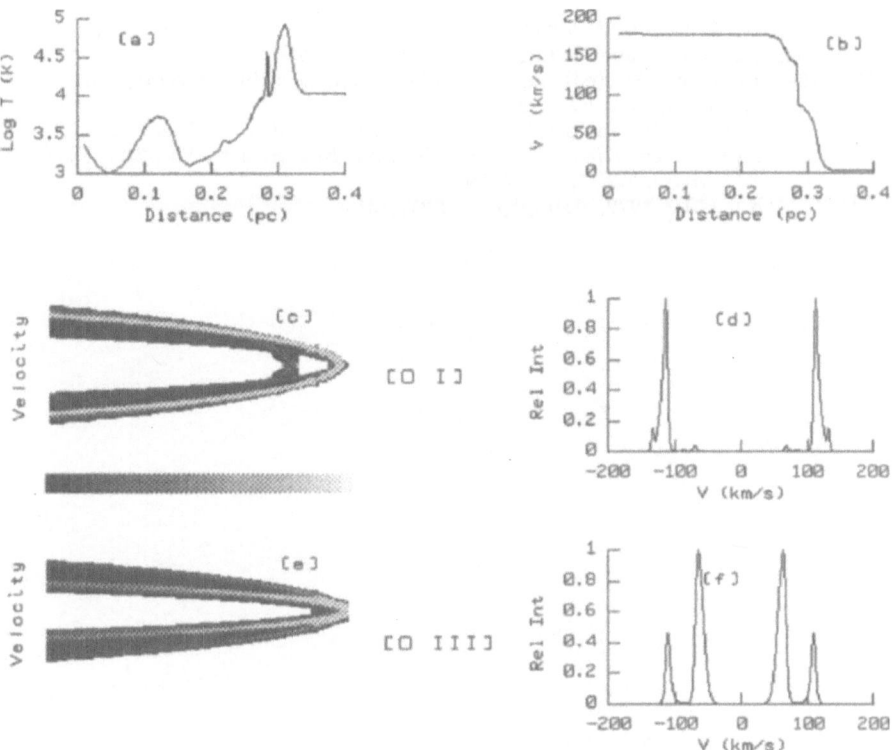

Fig. 4.   Long slit spectra from a structure with two forward shocks: **a - f** as in Fig. 4

4   Conclusions

Computations of the shock emission show that multiple shocks formed in the cooling flow have specific features which can be used to detect the presence of catastrophic cooling. It is shown that during the optically bright phase, multiple velocity and spatial components are expected. The separation between ionization regions is significantly larger than in a steady shock model. This can result in much higher lines ratios with respect to Balmer lines than predicted by the steady shock models. When a reverse shock is present, the low excitation lines (e.g. [O I]) are expected to show multiple filaments. In this case, the velocity of the downstream gas is less than the upstream gas. If the multiple filaments are caused by two upstream travelling shocks, they will be most visible in the high ionization lines and the downsteam gas will have a higher velocity than the upstream gas. Since the reverse shock is most prominent during the adiabatic phase, bright filaments are not expected in [O III] and [O I] simultaneously.

D.E. Innes

References

Dopita M.A.: 1977, *Astrophys. J. Suppl.* **33**, 437.
Gaetz T.J., Edgar J. & Chevalier R.A.:1988, *Astrophys. J.* **329**, 927.
Innes D.E.:1988, In: Kundt W. (ed) Supernova Shells and their Birth Events, p74.
Innes D.E.:1992, *Astron. & Astrophys.* **256**, 660.
Innes D.E., Giddings J.R. & Falle S.A.E.G.:1987a, *Mon. Not. R. astr. Soc.* **226**, 67.
Innes D.E., Giddings J.R. & Falle S.A.E.G.:1987b, *Mon. Not. R. astr. Soc.* **227**, 1021.
Raymond J.C.: 1979, *Astrophys. J. Suppl.* **39**, 1.
Shull J.M. & McKee C.F.: 1979, *Astrophys. J.* **227**, 131.

Star Formation in Shocked Layers

S.J. Chapman, A.S. Bhattal, M.J. Disney, J.A. Turner and A.P. Whitworth
Department of Physics and Astronomy, University of Wales, Cardiff CF2 3YB, U.K.

Abstract. We analyze the gravitational stability of a shocked interstellar gas layer and show how such a layer fragments into protostellar condensations whilst it is still confined mainly by ram pressure. As a consequence, the resulting protostars are massive and well separated. Our analysis is completely general and applies both to layers resulting from collisions between molecular cloud clumps, and to shells swept up by expanding nebulae. We present a numerical simulation of the former scenario, which produces a cluster of 35 massive stars resembling an OB subgroup, with most of the stars in binary systems.

1 Introduction

There are probably many different modes of star formation. At one extreme is what might be called the quasistatic mode, exemplified by the star formation occurring in quiescent GMCs like Taurus-Auriga. This mode appears to operate rather inefficiently to produce sparsely distributed low-mass stars. Shu and his collaborators (*e.g.* Shu, 1991) have suggested that this mode is controlled by ambipolar diffusion in dense cores.

At the other extreme is what might be called the dynamical mode of star formation, exemplified by the star formation in Orion. This mode appears to operate with high efficiency to produce stars with a wide range of masses and concentrated in clusters (*e.g.* Lada, 1991). It is unclear whether regions of dynamic star formation are highly turbulent solely as a consequence of their efficient star formation, or whether it is their dynamic state which engenders a high star formation efficiency. Probably it is both.

In other words, the dynamic star formation mode is self-propagating, at least as far as high-mass stars are concerned (*e.g.* Elmegreen and Lada, 1977). Self-propagating star formation can be most easily understood if massive protostars condense efficiently out of shocked layers. Such layers result either when two clumps of a GMC collide, or when a shell of cool neutral gas is swept up by an expanding nebula (HII region, stellar wind bubble or supernova remnant) around an evolved massive star.

In Section 2 we present the basic theory of the accumulation and gravitational fragmentation of a shocked interstellar gas layer, emphasizing the most general features. In Section 3 we present a numerical simulation of this process performed using an SPH code. We summarize our conclusions briefly in Section 4.

Astrophysics and Space Science **216**: 317–321, 1994.
© 1994 *Kluwer Academic Publishers.*

2   The accumulation and fragmentation of a shocked interstellar gas layer

Consider gas with density ρ_o flowing into a shock front with velocity v_o, measured normal to the front. Suppose that the shocked gas radiates efficiently so that in effect its sound speed relaxes instantaneously to the value a_s. Define the Mach Number of the shock as $\mathcal{M} \equiv v_o/a_s$.

For a collision between two clumps, ρ_o is the mean density in the clumps. The shocked layer forms where the clumps collide. For collisions at finite impact parameter the layer is oblique. v_o is the speed with which the clumps approach the layer, i.e. the velocity component normal to the layer. To first order, the shear in the layer can be ignored.

For a shell swept up by an expanding nebula, ρ_o is the density of the surrounding undisturbed gas. v_o is the speed with which the shell expands, and the shock marks the outer boundary of the shell. To first order, the tangential divergence of the radially expanding shell can be ignored.

The column density of the layer builds up according to

$$\Sigma \sim \rho_o v_o t \sim \rho_o a_s \mathcal{M} t,$$

so that at early times,

$$t \lesssim t_{\text{switch}} \sim (G\rho_o)^{-1/2},$$

confinement of the layer normal to its surface is dominated by ram pressure $P_{ram} \sim \rho_o v_o^2 \sim \rho_o a_s^2 \mathcal{M}^2$, rather than by self-gravity $G\Sigma_{\text{layer}}^2 \sim G\rho_o^2 a_s^2 \mathcal{M}^2 t^2$. Consequently the layer is not strongly centrally condensed normal to its surface. Its mean density is

$$\rho_s \sim \rho_o \mathcal{M}^2,$$

its thickness is

$$Z_{\text{layer}} \sim \Sigma_{\text{layer}}/\rho_s \sim \mathcal{M}^{-1} a_s t,$$

and as long as it remains one dimensional (*i.e.* plane-parallel or spherically symmetric) it has plenty of time to relax to hydrostatic balance.

However, a shock-compressed layer is subject to several instabilities involving motions parallel to its surface. These instabilities have been investigated by Elmegreen and Elmegreen (1978), Vishniac (1983), Bertschinger (1986), Elmegreen (1989) and Lubow and Pringle (1993). As emphasized by Lubow and Pringle, the early short-wavelength dynamical modes are not self gravitating, and are therefore unlikely to condense into protostars, so we shall ignore them. In so doing, we are presuming that they do not disrupt the layer, in the manner suggested by Vishniac (1983), but simply engender weak turbulence within it which then acts as seed perturbations for subsequent gravitational amplification, in the manner suggested by the simulations of Elmegreen (1989). Additional seed perturbations will derive from the pre-existing density structures which the shock overruns.

The fastest-growing self-gravitating mode has a wavelength L_{fastest} and a growth time t_{fastest} given by (*e.g.* Larson, 1985)

$$\frac{L_{\text{fastest}}}{a_s} \sim t_{\text{fastest}} \sim \frac{a_s}{G\Sigma_{\text{layer}}} \sim \frac{t_{\text{switch}}^2}{\mathcal{M}t}.$$

The layer will fragment into protostellar condensations when $t_{\text{fastest}} \lesssim t$, *i.e.* for $t \gtrsim t_{\text{fragment}} \sim \mathcal{M}^{-1/2} t_{\text{switch}}$.

In terms of physical variables, the onset of non-linear fragmentation, and the separations and masses of the resulting fragments, are given by

$$t \gtrsim t_{\text{fragment}} \sim (G\rho_o \mathcal{M})^{-1/2} \sim 3 Myr \left(\frac{n_o}{100cm^{-3}}\right)^{-1/2} \left(\frac{\mathcal{M}}{10}\right)^{-1/2}$$

$$L \gtrsim L_{\text{fragment}} \sim a_s (G\rho_o \mathcal{M})^{-1/2} \sim 0.6pc \left(\frac{T_s}{10K}\right)^{1/2} \left(\frac{n_o}{100cm^{-3}}\right)^{-1/2} \left(\frac{\mathcal{M}}{10}\right)^{-1/2}$$

$$M \gtrsim M_{\text{fragment}} \sim a_s^3 (G^3 \rho_o \mathcal{M})^{-1/2} \sim 6 M_\odot \left(\frac{T_s}{10K}\right)^{3/2} \left(\frac{n_o}{100cm^{-3}}\right)^{-1/2} \left(\frac{\mathcal{M}}{10}\right)^{-1/2}$$

Here n_o is the number density of hydrogen nuclei in all forms, so that with Population I composition $n_o \simeq \rho_o/[2.4 \times 10^{-24}g]$, and T_s is the temperature of the shocked gas (so that if the hydrogen is predominantly molecular, $a_s \simeq 0.2kms^{-1}[T_s/10K]^{1/2}$). We conclude that fragmentation will occur after the layer has been accumulating for a few megayears, and will produce massive protostellar condensations with separations of order a parsec.

3   Numerical simulation of a clump/clump collision

We have simulated collisions between clumps in GMCs for a variety of clump masses M_o, collision speeds v_o, and impact parameters b_o, using our Smoothed Particle Hydrodynamics code, which is designed to treat problems in self-gravitating gas dynamics involving large density contrasts (see Chapman *et al.*, 1993).

We present here the results of a head-on collision ($b_o = 0$) between two identical clouds (mass $M_o = 750 M_\odot$; diameter $D_o = 12pc$) at speed $v_o = 1.7$ km s^{-1} (Mach-9); simulations at finite impact parameter ($b_o > 0$) produce qualitatively similar results. Before the collision, the clumps are in detailed stable hydrostatic equilibrium. The gas has a barotropic equation of state which corresponds to its having isothermal sound speed $a_o \simeq 0.6$ km s^{-1} for $n_o < 300$ cm^{-3} (*i.e.* in the unshocked clouds) and $a_s \simeq 0.2$ km s^{-1} for $n_s > 10^4$ cm^{-3} (*i.e.* in the shocked layer); in between these densities it cools approximately according to $a \propto n^{-1/3}$.

Figure 1 shows the column-density of the compressed layer face-on, i.e. looking along the collision axis, at various resolutions. The layer has condensed into a network of filaments, and the filaments have condensed into strings of beads. Most

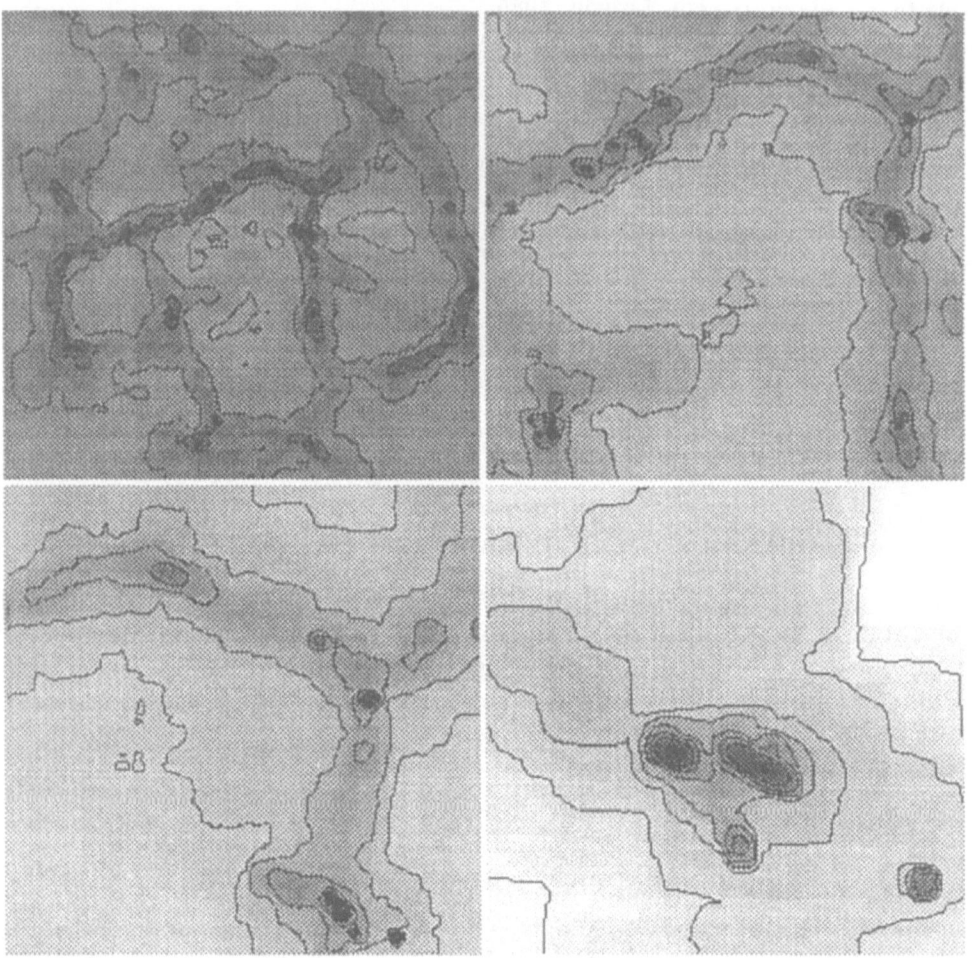

Fig. 1.   Face-on views of a layer formed by a clump/clump collision, showing the whole layer
and then zooming in by stages on one of the binary protostars. The linear size of the frame L,
and the range of column-density N spanned by the grey-scale a nd the contours, are as follows:
(a) $L = 1pc$, $N = 10^{21} - 10^{25}$ cm^{-2}; (b) $L = 0.2$ pc, $N = 10^{21} - 10^{26}$ cm^{-2}; (c) $L = 0.1$ pc,
$N = 10^{21} - 10^{27}$ cm^{-2}; (d) $L = 0.02$ pc $\equiv 5000$AU, $N = 10^{22} - 10^{27}$ cm^{-2}.

of the visible beads contain two or more protostellar discs, usually in binaries but
occassionally in higher multiples. In total there are 35 rotationally supported
and stable protostellar discs in this layer having masses in the range $5 - 40 M_\odot$,
diameters in the range 200-2000 AU, and peak densities in the range $10^{10} - 10^{11}$
cm^{-3}. Figure 1d is a close-up of one binary from near the centre of the frame.

4 Conclusions

By considering the accumulation and gravitational fragmentation of a shocked interstellar gas layer from a very general standpoint, we have been able to show that the protostellar fragments which condense out of it are massive ($M > 6M_\odot$); our numerical simulations indicate that they have no inclination to undergo further fragmentation. This supports the popular paradigm for self-propagating star formation in which the next generation of massive stars condenses out of the dense shell of cool neutral gas swept up by the expanding nebulae around the massive stars of the previous generation.

Our simulations also demonstrate the feasibility of modelling the formation of small clusters of protostellar discs using modern particle methods and supercomputers. With refinements now in hand, it may be possible to follow the evolution of the protostellar discs formed here further, to produce pre-MS stars.

Acknowledgements

This work enjoys the support of SERC grants Nos. GR/H64361 and GR/H86264. JAT is supported by an SERC postgraduate studentship. SJC and ASB are supported by University of Wales postdoctoral and postgraduate studentships respectively. This simulation was performed on the SERC's Cray-YMP at RAL.

References

Bertschinger, E.: 1986, *Astrophys. J.* **304**, 154-177.
Chapman, S. J. *et al.*: 1994, in preparation.
Elmegreen, B. G.: 1989, *Astrophys. J.* **340**, 786-811.
Elmegreen, B. G. and Elmegreen, D. M.: 1978, *Astrophys. J.* **220**, 1051-1062.
Elmegreen, B. G. and Lada, C. J.: 1977, *Astrophys. J.* **214**, 725-741.
Lada, E.: 1991, *Astrophys. J.* **393**, L25-L28.
Larson, R. B.: 1985, *Mon. Not. Roy. Astr. Soc.* **214**, 379-398.
Lubow, S. H. and Pringle, J. E.: 1993, *Mon. Not. Roy. Astr. Soc.* **263**, 701-706.
Shu, F. H.: 1991, in *The Physics of Star Formation and Early Stellar evolution*, eds. C. J. Lada
    and N. D. Kylafis, Kluwer, Dordrecht, pp.365-410.
Vishniac, E. T.: 1983, *Astrophys. J.* **274**, 152-167.

Binary and Multiple Star Formation

J.A. Turner, S.J. Chapman, A.S. Bhattal, M.J. Disney and A.P. Whitworth
Department of Physics and Astronomy, University of Wales, Cardiff CF2 3YB, U.K.

Abstract. From the growing observational evidence that binary/multiple star formation occurs prior to the pre-main sequence (Mathieu, 1992), it is clear that any theory of star formation MUST also explain binary formation. This paper details two formation mechanisms for binary/multiple stars, which occur simultaneously with protostar formation. The protostellar discs we form have masses $5 \to 30 M_\odot$, diameters 200 $\to$ 4000 AU. The binaries/multiples have separations 400 $\to$ 7500 AU. The formation mechanisms were found by conducting numerical simulations of two cloud collisions, using SPH and Treecode gravity with up to 200,000 particles per calculation and a prescribed cooling equation of state.

1 Binary and Multiple Protostar Formation

The collisions proceed with a layer of dense gas forming between the clouds. This becomes Jeans unstable and fragments into many condensations, which collapse and interact with one another to form protostars. All of these collisions form massive protostar(s) and the majority form binary/multiple systems. Those which only produce single protostars have small amounts of bulk angular momentum *i.e.* nearly head-on and slow velocity. Two modes of binary formation occur in our simulations i) accretion induced rotational fragmentation (ARF), and ii) shock induced thermal fragmentation (STF) (Chapman *et al*, 1992).

ARF begins with the layer fragmenting into one condensation (Figure 1a). The gas flowing along the shock layer develops into two accretion streams, increasing the angular momentum of the condensation because the gas flowing along the shock layer has increasing specific angular momentum with time. The condensation goes rotational unstable and loses angular momentum in two ways. It can form a bar which then splits into two, or spiral arms can be excited (Figure 1b). These arms detach and then condense (Figure 1c). The new condensations orbit around the original condensation, and grow by accreting material from the accretion flows. They cause spiral arms to be re-excited in the original condensation, and accrete these spiral arms as they detach, thereby growing at the expense of the original condensation. This leads to the formation of multiple systems (Figure 1d).

STF occurs when the layer fragments into two or more condensations, which fall towards towards each other, and collapse to form individual protostars. They subsequently interact, and form binary/multiple systems by capture.

2 Summary

By conducting realistic simulations of cloud-cloud collisions, we have identified two formation mechanisms for binary/multiple protostar systems. (work supported by SERC (GR/H64361) and Univ. of Wales)

References

Mathieu, R. D., 1992. Disks in the pre-main sequence environment, In: *Evolutionary processes in interacting binary stars*, p.21, eds Kondo, Y., Sistero, R. F. & Polidan, R. S., Kluwer Academic Publishers, Dordrecht

Chapman, S. J. *et al*, 1992 The formation of binary and multiple star systems. *Nature*, **359**, 207

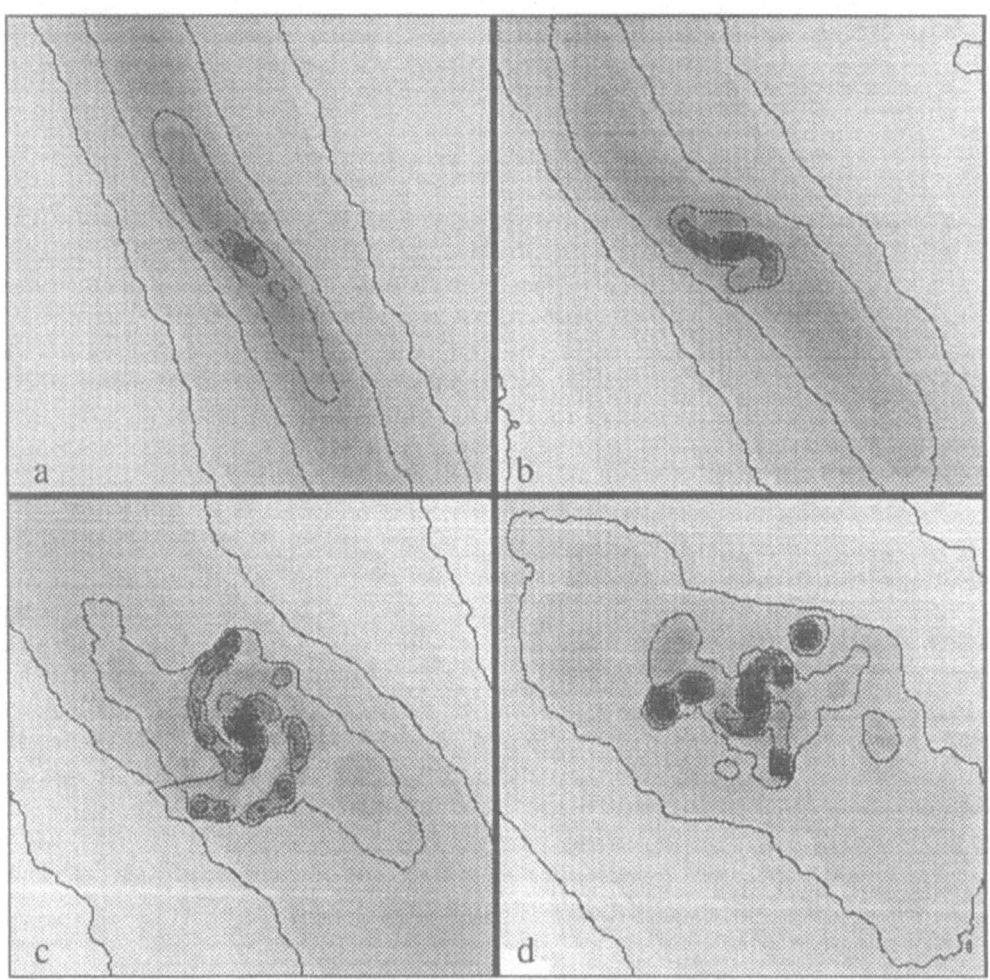

Fig. 1.   An example of ARF. The greyscale represents the column density through the shock layer and are logarithmically scaled, from black representing 6×10^{23} cm^{-2} to white 2.1×10^{18} cm^{-2}.

Galactic Fountains

F. D. Kahn
Department of Astronomy, The University
Manchester M13 9PL, England

Abstract. The hot intercloud medium (ICM) streams away from the Galactic disk into the fountain. It is generally agreed that a flow consisting of thermal gas only reaches a maximum height of a few (3 to 5) kpc. The dynamics is more complicated if cosmic rays are also present. In that case it is possible for some of the gas to be ejected from the Galaxy altogether.

1 Introduction

The notion of a Galactic fountain has existed for quite a long time: an early version, for example, was described by Shapiro and Field (1976). A considerable advance in theoretical understanding was later made by Bregman (1980), whose quasi-static model postulated an upward flow which remains subsonic everywhere. Subsequent developments included a one-dimensional model, driven by the hot intercloud medium (ICM), starting subsonic below and passing through a sonic layer to become supersonic at greater heights (Kahn, 1981). The presence of cosmic rays should be allowed for in any dynamical description. This view has been emphasized especially by Völk and his collaborators; for a recent paper adopting this approach see Breitschwerdt, McKenzie and Völk (1993). The resulting model is more complex in that it allows for some parts of the flow which escape from the Galaxy altogether, while in others the gas returns to the Galactic plane, as it must in a fountain.

It is easy to advance compelling arguments for the existence of a return flow. The hot ICM has too much thermal energy ($> 10^{12}$ erg gm $^{-1}$) to be held down by gravity in the Galactic disk, but not enough ($< 10^{15}$ erg gm^{-1}) to escape altogether from the Galaxy. Further Φ, the rate of flow of mass per unit area from the disk must, on general grounds, be of order P_{IC}/a_{IC}, where P_{IC} and a_{IC} are respectively the pressure and the sound speed in the ICM. Using parameters that are generally accepted ($P_{IC} = 10^{-12}$ dyne cm^{-2}, $a_{IC} = 10^7$ cm s^{-1}) one finds that $\Phi \sim 10^{-19}$ gm cm^{-2} s^{-1}, and spread over the area $\sim 10^{46}$ cm^2 of the Galactic disk, above and below, the inferred rate of loss of mass would be of order 10^{27} gm s^{-1}, or 15 $M_\odot$/year. So high a value is clearly very much larger than the Galaxy can sustain. All such descriptions of an outflow have to incorporate provision for a return flow.

There has long been evidence, principally from 21cm line observations, that such a return flow is actually present (Muller, Oort and Raimond, 1963). The descending gas is observed in the 21cm line and is generally classified either as belonging to an IVC (intermediate velocity cloud) or an HVC (high velocity cloud). All this work has been well summarized by Wakker (1991) in his contribution to IAU Symposium 144 "The Interstellar Disk Halo Connecton in Galaxies" (ed. H.

Astrophysics and Space Science **216**: 325–332, 1994.
© 1994 *Kluwer Academic Publishers.*

Bloemen, 1991); many further references can be found there. Wakker concludes that a fountain model can probably account for the existence of IVC's, but that some other physical mechanism is needed to deal with the HVC's.

The volume edited by Bloemen contains many other relevant papers, as do the Proceedings of IAU Colloquium 120 "Structure and Dynamics of the Interstellar Medium" (eds. Tenorio-Tagle, Moles and Melnick, 1989). In particular there is an important class of models in which the fountains are driven by the combined input of energy from many supernovae exploding in rapid succession and in a limited volume of space (see, for example, Heiles, 1989; Norman 1991). Such clustered sets of supernovae combine to blow a superbubble which then forces its way out of the Galactic disk. Structures of this kind certainly exist in other galaxies (Meaburn, 1980), but their dynamics is quite different and will not be discussed any further here.

2   A Model for a Fountain, Driven by Hot Gas

The theory can be kept tractable by judicious use of simplifications. Here it is assumed that the flow is steady and plane parallel, that centrifugal acceleration balances gravity in the direction parallel to the Galactic plane, and that there is a uniform gravitational acceleration g perpendicular to the Galactic plane. This last assumption becomes reasonable at levels above the bulk of Population I, say at heights z greater than 300 pc. The one-dimensional model also fails at much greater heights, where z is comparable with the distance to the Galactic centre. However the interesting dynamics is associated with the flow through the sonic level, which lies relatively close to the Galactic plane, and so the limitation does not constitute a particular drawback here.

Two equations are needed to describe the flow, the conservation of momentum

$$\Phi u + \kappa(\Phi/u)^{\frac{5}{3}} = \Pi_0 - \Phi g\tau \tag{1}$$

and the simple cooling law (Kahn 1976)

$$\frac{d}{d\tau}\kappa^{\frac{3}{2}} = -q. \tag{2}$$

Here Φ = mass flux, κ = adiabatic parameter, u = flow velocity, positive upwards, $d\tau = dz/u$ = increment of time measured following a fluid element, Π_0 = momentum flux at a reference level, $z = z_0$. The cooling law can usefully be applied in the interesting range of temperatures between 10^5 K and 3×10^7 K. The momentum equation takes the form

$$\Phi u + \{q(\tau_* - \tau)\}^{\frac{2}{3}}(\Phi/u)^{\frac{5}{3}} = \Pi_0 - \Phi g\tau \tag{3}$$

with the help of (2), and here τ_* is the (comoving) time at which the gas has lost all its thermal energy.

To set up a model for the fountain entails solving a critical flow problem. On differentiation of relation (3) it is found that

$$\frac{du}{d\tau}[1 - \frac{5}{3}\{q\Phi(\tau_* - \tau)\}^{\frac{2}{3}}u^{-\frac{8}{3}}] = \frac{2}{3}\frac{q^{\frac{2}{3}}\Phi^{\frac{2}{3}}}{(\tau_* - \tau)^{\frac{1}{3}}u^{\frac{5}{3}}} - g,\tag{4}$$

with a little rearrangement. The critical point in the flow is reached when both the term in square brackets and the right hand side vanish, at the same time τ. This condition leads to the equations

$$u^2 = u_c^2 = \frac{5}{3}\{q(\tau_* - \tau_c)\}^{\frac{2}{3}}\left(\frac{\Phi}{u_c}\right)^{\frac{2}{3}} \equiv \frac{5}{3}\kappa\rho_c^{\frac{2}{3}}\tag{5}$$

and so the flow is sonic at the critical level, and

$$\frac{2}{3}\frac{q^{\frac{2}{3}}}{(\tau_* - \tau_c)^{\frac{1}{3}}}\left(\frac{\Phi}{u_c}\right)^{\frac{5}{3}} = \Phi g.\tag{6}$$

On elimination of $\tau_* - \tau_c$ it follows that the speed at the sonic level is

$$u_c = \left(\frac{20}{27}\right)^{\frac{1}{6}}\left(\frac{q\Phi}{g}\right)^{\frac{1}{3}}.\tag{7}$$

The momentum flux at the sonic level is

$$\Pi = \Pi_c = \frac{8}{5}\Phi u_c \equiv \frac{2^{\frac{1}{3}}}{3^{\frac{1}{3}}5^{\frac{5}{6}}}\left(\frac{q\Phi^4}{g}\right)^{\frac{1}{3}};\tag{8}$$

the coefficient on the right here is 1.52, to three significant figures. The highest point in the fountain is reached at time τ_e and conservation of momentum shows that

$$\tau_e - \tau_c = \frac{\Pi_c}{\Phi g} = 1.52\left(\frac{q\Phi}{g^4}\right)^{\frac{1}{3}}.\tag{9}$$

On the other hand the flow will cool completely by time τ_*, and from (6) and (7) it follows that

$$\tau_* - \tau_c = 0.38\left(\frac{q\Phi}{g^4}\right)^{\frac{1}{3}} \equiv \frac{1}{4}(\tau_e - \tau_c).\tag{10}$$

So starting from the sonic level, the gas will cool off completely in a quarter of the time that it takes to reach maximum height, at $z = z_e$. The gas is therefore completely cold in the higher reaches of the upward flow. The flow speed at the cooling level can be found from the momentum equation

$$\Phi u_* = \Pi_c - \Phi g(\tau_* - \tau)$$

$$= (1.52 - 0.38)\left(\frac{q\Phi^4}{g}\right)^{\frac{1}{3}} = 1.14\left(\frac{q\Phi^4}{g}\right)^{\frac{1}{3}}\tag{11}$$

and so $u_* = 1.20u_c$.

The flow accelerates upwards from z_c until it reaches level z_*. When the thermal pressure has run out, the gas rises further ballistically to level z_e. An elementary calculation leads to

$$z_e - z_* = \frac{1}{2}\frac{u_*^2}{g} = 0.65\left(\frac{q^2\Phi^2}{g^5}\right)^{\frac{1}{3}}$$ (12)

and a reasonable approximation shows that

$$z_* - z_c = \frac{1}{2}(u_* + u_c)(\tau_* - \tau_c)$$

$$= 0.40\left(\frac{q^2\Phi^2}{g^5}\right)^{\frac{1}{3}}.$$ (13)

It seems a reasonable assumption that the critical level in the fountain is not far above the galactic plane, say a few hundred parsecs, and that the gas spends much more time in the supersonic region. Then the momentum flux at the sonic level is little different from the pressure in the ICM, and it follows that

$$\Phi = 0.73\frac{g^{\frac{1}{4}}\Pi_c^{\frac{3}{4}}}{q^{\frac{1}{4}}} = 5.2 \times 10^{-20}\text{gm cm}^{-2}\text{s}^{-1},\ \Pi_c \sim P_{IC} = 10^{-12}\text{dyne cm}^{-2}.$$

Further

$$u_c = 120\,\text{km s}^{-1},\ u_* = 144\,\text{km s}^{-1},$$

$$\tau_* - \tau_c = 16\ \text{million years},$$

and

$$\tau_e - \tau_c = 64\ \text{million years}.$$

Earlier descriptions of the fountains (e.g. Kahn, 1991) contain a discussion of the return flow, in which the descending gas forms a cool layer, headed by a shock. The speed of descent, at any level, is one half the upward speed of the unimpeded fountain at the same level. This conclusion still holds above level z_*, but becomes somewhat inaccurate lower down. An approximate treatment can be devised for the region below z_*, but will not be described here. For the moment it will be enough to know that, in the present model, 200 million years is the longest period that any parcel of gas spends in the fountain, and that the average excursion time is 100 million years. With an area of 10^{46} cm^2 ($\equiv 10^3$ kpc^2) for the two-sided Galactic disk, it follows that a mass of about 8 $M_\odot$ of the interstellar medium enters the fountain flow every year and that some $8 \times 10^8\,M_\odot$ of gas is contained in the fountain any one time. Finally the highest point z_e in the flow lies 3.5 kpc above the cool point, z_*, which itself is 2 kpc above the sonic level, z_c. Clearly this one-dimensional model is close to the limits of validity.

3   A Fountain Model with Cosmic Rays

The dynamics of the fountain flow is changed by the presence of cosmic rays. In general the thermal gas and the cosmic ray gas can diffuse with respect to one another. In a highly ionized gas, such as that in the fountain, the relative speed of the diffusion is held down to about the Alfvén speed $v_A \equiv \sqrt{B^2/4\pi\rho}$; here B is the field strength and ρ the density of the thermal plasma. In the upward flow the magnetic energy density is small in comparison with the kinetic energy density (Kahn and Brett, 1993), and so the diffusion speed will also be much less than the flow speed. Accordingly diffusion effects will be ignored here. The cosmic ray gas can therefore be treated as though it were locked into the thermal gas. The cosmic ray pressure is then given by

$$P_{CR} = \kappa_{CR}\, \rho^{\frac{4}{3}} \, ; \tag{14}$$

the relative importance of the cosmic ray gas in the dynamics is conveniently expressed in terms of the dimensionless parameter

$$\lambda = \kappa_{CR}\, g^{\frac{7}{9}} / \Phi^{\frac{4}{9}} q^{\frac{7}{9}} \, . \tag{15}$$

The equation of momentum balance is now changed to

$$\Pi = \Phi\left[u + \frac{\{q(\tau_* - \tau)\Phi\}^{\frac{2}{3}}}{u^{\frac{5}{3}}} + \lambda\, \frac{q^{\frac{7}{9}}\Phi^{\frac{7}{9}}}{g^{\frac{7}{9}} u^{\frac{4}{3}}} \right] = \Pi_c - g\Phi(\tau - \tau_c); \tag{16}$$

on differentiation of this equation it is found that

$$\frac{du}{d\tau}\left[1 - \frac{5}{3} \frac{\{q(\tau_* - \tau)\Phi\}^{\frac{2}{3}}}{u^{\frac{8}{3}}} - \frac{4}{3}\lambda\, \frac{q^{\frac{7}{9}}\Phi^{\frac{7}{9}}}{g^{\frac{7}{9}} u^{\frac{7}{3}}} \right] = \frac{2}{3} \frac{(q\Phi)^{\frac{2}{3}}}{u^{\frac{2}{3}}(\tau_* - \tau)^{\frac{1}{3}}} - g. \tag{17}$$

One critical point condition is

$$\tau_* - \tau = \frac{8}{27} \frac{q^2\Phi^2}{g^3 u^5} \, , \tag{18}$$

very similar to equation (6); the other condition at the critical point can then be written

$$1 - \frac{20}{27}\left(\frac{q\Phi}{gu^3} \right)^2 - \frac{4}{3}\lambda\left(\frac{q\Phi}{gu^3} \right)^{\frac{7}{9}} = 0 \, . \tag{19}$$

In terms of the dimensionless velocity U_c, defined by $u_c = U_c\left(\dfrac{q\Phi}{g} \right)^{\frac{1}{3}}$ relation (19) becomes

$$1 - \frac{20}{27} U_c^{-6} - \frac{4}{3}\lambda\, U_c^{-\frac{7}{3}} = 0. \tag{20}$$

The table below gives values of U_c in terms of λ, as well as those of β, the fraction of the pressure taken up by cosmic rays at the critical level.

λ	0	0.1	0.2	0.3	0.4
U_c	0.95	0.97	1.00	1.03	1.05
β	0	0.17	0.31	0.43	0.52

The momentum flux Π_c, at the critical level, and the mass flux Φ depend on U_c through the relation

$$\Pi_c = \left(\frac{q\Phi^4}{g}\right)^{\frac{1}{3}} U_c \left(\frac{7}{4} - \frac{1}{9U_c^6}\right). \tag{21}$$

The coefficient on the right changes from 1.52 for $U_c = 0.95$ at $\beta = 0$, to 1.75, for $U_c = 1.05$ at $\beta = 0.52$. The mass flux Φ can then also be expressed in terms of Π_c in the form

$$\Phi = \left(\frac{g\Pi_c^3}{q}\right)^{\frac{1}{4}} \left\{\frac{4U_c^5}{7}\frac{1}{(U_c^6 - \frac{4}{63})}\right\}^{\frac{3}{4}}. \tag{22}$$

The critical level is expected to be not far above the galactic disk and so $\Pi_c \sim P_{IC}$ as before. The parameter λ is defined in terms of the relative importance of the cosmic ray pressure, and in the interval $0 < \lambda < 0.4$, which spans the likely range of values, the coefficient on the right in (22) changes from 0.73 to 0.66. The predicted mass flux is quite insensitive to the value of λ.

The cooling time, starting from the critical level, is

$$\tau_* - \tau_c = 0.26 \left(\frac{q\Pi_c}{g^5}\right)^{\frac{1}{4}} U_c^{-\frac{15}{4}} \left(U_c^6 - \frac{4}{63}\right)^{-\frac{1}{4}}, \tag{23}$$

and the coefficient of $(q\Pi_c/g^5)^{\frac{1}{4}}$ drops from 0.35, for $\lambda = 0$, to 0.20 for $\lambda = 0.40$. The cool level z_* is therefore closer to z_c in fountains where the cosmic ray pressure is more important.

At levels above z_* the flow is driven by cosmic ray pressure, and the momentum flux is

$$\Pi = \Phi u + \kappa_{CR}(\Phi/u)^{\frac{4}{3}} = \Pi_* - \Phi g(\tau - \tau_*). \tag{24}$$

The right-hand side here decreases with τ, and the momentum flux, as a function of u, reaches its minimum value when

$$u = u_{\lim} = \left(\frac{4}{3}\right)^{\frac{3}{7}} (\kappa_{CR}^3 \Phi)^{\frac{1}{7}} = \left(\frac{4\lambda}{3}\right)^{\frac{3}{7}} \left(\frac{\Phi q}{g}\right)^{\frac{1}{3}} \tag{25}$$

for $\lambda = 0.4$. With $\Phi = 5.3 \times 10^{-20}$ gm cm^{-2} s^{-1}, it is found that $u_{\lim} = 98$ km s^{-1}, and the steady flow can ascend no further. Instead an expansion wave has to be fitted on at the level $z_{\lim}$, where the speed $u_{\lim}$ is reached.

At first the head of the expansion wave will be stationary at $z_{\lim}$, since $u_{\lim}$ is the sonic speed. But gravity will change the flow pattern over a time of order

$\sim 10^{15}$ s $\sim$ 30 million years. As a result the return flow downwards will set in, but the descent is hard to describe, and will be dealt with another time.

As long as gravity can be ignored, the flow in the expansion wave satisfies the condition that

$$u + \frac{2}{\gamma - 1}a \equiv u + 6a = 7u_{\text{lim}} \tag{26}$$

is constant; here a is the local sound speed and $\gamma = 4/3$ for the cosmic ray supported gas. The highest speed reached in the expansion fan is $7u_{\text{lim}} = 686$ km s^{-1}, and any gas that is accelerated to this speed will never return to the Galactic disk. However these very high speeds are attained only for very small values of a, and, since the density ρ varies like a^6, by a very small fraction of the gas. Now $a = \frac{1}{6}(7u_{\text{lim}} - u)$, and the fraction of the gas that attains a speed exceeding u is

$$f(> u) = \left(\frac{7}{6}\right)^7 \left(1 - \frac{u}{7u_{\text{lim}}}\right)^7 ; \tag{27}$$

the table below shows the run of f with u

u/u_{lim}	1	2	3	4	5	6	7
f	1	0.28	0.059	0.008	5×10^{-4}	4×10^{-6}	0

There is a problem now in going further with this discussion in that little is known about the run of gravitational potential at large distances from the Galactic centre. However, it seems safe to assume that a parcel of gas will escape if it reaches a speed of 250 km s^{-1}, or above, and the corresponding value of f is 0.13. This window of esape is open for about 30 Myr in every 200 Myr, so about 15 per cent of the time. The overall result is that, averaged over a long period, about two per cent of the gas that ascends into the fountain, or 0.15 $M_{\odot}$ yr^{-1}, will be lost from the Galaxy. A rather larger mass will be raised so far that on return the gas becomes part of the high velocity stream. All these estimates apply to a model of the ISM in which the cosmic ray gas supplies half the interstellar pressure.

4   Conclusion

This model of the Galactic fountain is very robust. It can be adapted to flows in which cosmic ray pressure is important; in that case it reveals new phenomena which are qualitatively different. Very little has been mentioned here about magnetic field effects. The fountain flow drags the field lines out of the disk and allows reconnections to change their topology. The presence of the magnetic field is also important in calculations of the pressure support in the descending flow. This problem has been treated elsewhere (Kahn, 1991) for a fountain without cosmic rays. The discussion will need to be revised to allow for their presence.

References

Bloemen, H. (ed): 1991 *The Interstellar Disk-Halo Connection in Galaxies*, IAU Sympo-
      sium 144, Kluwer Acad. Publ., Dordrecht.
Bregman, J. N.: 1980, *Astrophys. J.* **236**, 577.
Breitschwerdt, D., McKenzie, J. F. and Völk, H. J.: 1993, *Astron. Astrophys.* **269**, 54.
Heiles, C.: 1989, IAU Colloquium 120, p.484.
Kahn, F. D.: 1976, *Astron. Astrophys.* **50**, 145.
Kahn, F. D.: 1981, in F. D. Kahn (ed.), *Investigating the Universe*, D. Reidel Publ. Co.,
      Dordrecht, p.1.
Kahn, F. D.: 1991, IAU Symposium 144, p.1.
Kahn, F. D. and Brett, L.: 1993, *Mon. Not. Roy. Astr. Soc.* **263**, 37.
Meaburn, J.: 1980, *Mon. Not. Roy. Astr. Soc.* **92**, 365.
Muller, C. A., Oort, J. H. and Raimond, E.: 1963, *CR Acad. Sci., Paris*, **257**, 1661.
Norman, C. A.: 1991, IAU Symposium 144, p.337.
Shapiro, P. R. and Field, G.B.: 1976, *Astrophys. J.* **205**, 762.
Tenorio-Tagle, G., Moles, M. and Melnick, J. (eds.): 1989, *Structure and Dynamics of the
      Interstellar Medium*, IAU Colloquium 120, Springer-Verlag.
Wakker, B. J.: 1991, IAU Symposium 144, p.27.

The solution topology of galactic winds

Andreas Poll
Institut für Astrophysik und extraterrestrische Forschung
Auf dem Hügel 71, 53121 Bonn, Germany

Abstract. The system of ordinary differential equations derived from the hydrodynamical equations of a radially symmetric flow has a one dimensional manifold of critical points if there is mass loading in the flow or if the heating and cooling rates are taken to be independent of the thermodynamic variables as is usual.

1 Galactic Wind from an AGN

We want to investigate the galactic wind from an AGN. This wind is thought to be generated by the mass and energy sources of the stellar winds belonging to the central star cluster and possibly by a wind from the center. The steady motion of gas in spherical symmetry is determined by the conservation laws for mass

$$\frac{1}{r^2}\frac{d}{dr}(r^2\rho v) = \dot{\rho} \tag{1}$$

momentum

$$v\frac{dv}{dr} = -\frac{1}{\rho}\frac{dp}{dr} - g - \frac{\dot{\rho}}{\rho}v \tag{2}$$

and energy

$$\frac{1}{r^2}\frac{d}{dr}\left[\rho v r^2\left(\frac{1}{2}v^2 + \frac{\gamma}{\gamma-1}\frac{p}{\rho}\right)\right] + \rho v g = h \tag{3}$$

and the equation of state of an ideal gas

$$p = \rho\frac{kT}{\langle m\rangle} \tag{4}$$

Here r denotes the radial variable, ρ the density, v the flow speed, $\dot{\rho}$ the injection of mass to the flow per unit volume and unit time, p the thermal pressure, $g = \sum_i g_i$ is the sum of outer forces per unit mass, γ the polytropic exponent and h is the net heating rate of the gas including the energy input due to stellar winds. Assuming that the rate of mass injection is independent of the flow, we can solve equation (1) separately and find the mass flux out of a sphere with radius r to be

$$\rho v = \frac{-\dot{M}_c}{4\pi r^2} + \frac{1}{r^2}\int_0^r \dot{\rho}r^2 dr \tag{5}$$

where we define $-\dot{M}_c$ as the wind induced mass loss from the central object. Therefore ρv can be regarded as a known function. Since we are interested in the

Astrophysics and Space Science **216**: 333–335, 1994.
© 1994 *Kluwer Academic Publishers.*

question if the flow is transsonic, it is convenient to rewrite equations (2) and (3) in terms of the speed of sound c and the Mach number M. This yields a system of two linear ordinary differential equations. Both expressions show the expected singularity for $M = 1$. To remove formally this singularity we introduce a new independent variable s by the first equation of the following system.

$$\frac{dr}{ds} = M^2 - 1 \tag{6}$$

$$\frac{dc}{ds} = \frac{1}{2c}(I_1(\gamma M^2 - 1) - I_2(\gamma - 1)) \tag{7}$$

$$\frac{dM}{ds} = \frac{M}{2c^2}(-I_1(\gamma M^2 + 1) + I_2(\gamma + 1)) \tag{8}$$

Here we use

$$I_1 := (\gamma - 1)\left[\frac{h}{\rho v} + \frac{\dot{\rho}c^2}{\rho v}\left(\frac{M^2}{2} - \frac{1}{\gamma(\gamma - 1)}\right) - \frac{2c^2}{\gamma r}\right] \tag{9}$$

$$I_2 := -\frac{\dot{\rho}c^2}{\rho v}\left(\frac{1}{\gamma} + M^2\right) + \frac{2c^2}{\gamma r} - g. \tag{10}$$

Equation (6) shows that the solution curves of this dynamical system change their radial sense of direction if they pass through the plane $M = 1$ in the (r,c,M)–space, thus leading to multiple values of c or M as long as $\frac{dc}{ds} \neq 0$ or $\frac{dM}{ds} \neq 0$. An exception may occur if the right hand sides of equations (7) and (8) both vanish. Putting $M = 1$ into (7) and (8) it is obvious that their right hand sides both vanish under the same condition that $I_1 = I_2$. Therefore we have not only one or several isolated points where a transition through sonic points may happen but a one dimensional manifold: $\mathcal{CL} = \{(r_c, c_c, 1) \mid I_1 = I_2\}$ which we will call critical the line. However, before we can assume that a transition is possible in $(r_c, c_c, 1) \in \mathcal{CL}$ we have to explore the solutions in the neighbourhood of the critical points. It can be shown that the solutions are lying in planes so that they exhibit the well known behaviour of isolated critical points in two dimensions. With AGN parameters none of the nodes ever investigated has allowed a transition from sub- to supersonic motion. So only saddles lead to desired solutions in our case.

In a certain region around the center the flow time scale becomes large compared to the heating time scale. Thus, the flow is in an equilibrium between the net heating and the inflows of stellar winds which determines the wind's temperature and flow speed. In the cases investigated only one solution for each set of parameters followed the equilibrium for small radii and met the sonic plane in a saddle type critical point.

Acknowledgements

I thank Reinhold Schaaf who taught me how to solve essential parts of my topic and Robin Williams for several hints on misprints in the formulae in the poster. In addition I am grateful to the DFG for sponsoring my work in the frame of the project Bi 191/9-2

Galactic Scale Gas Flows in Colliding Galaxies:

3-Dimensional, N-body/Hydrodynamics Experiments

Susan A. Lamb*
NORDITA and Neils Bohr Institute,
Blegdamsvej 17, DK-2100, Københaven Ø, Danmark.

Richard A. Gerber
University of Illinois at Urbana-Champaign, Departments of Physics and Astronomy, 1110 W. Green Street, Urbana, IL 61801, U.S.A.

and

Dinshaw S. Balsara[†]
Johns Hopkins University, Department of Physics and Astronomy, Homewood Campus, Baltimore, MD 21218, U.S.A.

Abstract.
   We present some results from three dimensional computer simulations of collisions between models of equal mass galaxies, one of which is a rotating, disk galaxy containing both gas and stars and the other is an elliptical containing stars only. We use fully self consistent models in which the halo mass is 2.5 times that of the disk. In the experiments we have varied the impact parameter between zero (head on) and $0.9R$ (where R is the radius of the disk), for impacts perpendicular to the disk plane. The calculations were performed on a Cray 2 computer using a combined N-body/SPH program. The results show the development of complicated flows and shock structures in the direction perpendicular to the plane of the disk and the propagation outwards of a density wave in both the stars and the gas. The collisional nature of the gas results in a sharper ring than obtained for the star particles, and the development of high volume densities and shocks.

1 Introduction

Collisions between galaxies can produce large scale flows of both the stellar components and any gas present in the system. All close encounters and interpenetrating collisions result in the overall contraction of each galaxy as it passes close to, or through the other, and a subsequent expansion as the two move apart. At early times in the collision the gas and stars react in similar fashions to the changing gravitational potential, but the collisional nature of the gas soon becomes important (see Gerber and Lamb, 1993), leading to the development of shocks and regions of very high gas density, even if only one of the two galaxies contains any appreciable quantity of gas, and even if the two galaxies do not actually overlap at any time. The resulting distribution of the high density, shocked gas regions, as distinct from the underlying, somewhat different distribution of stars, is more likely to predict the actual optical morphology of interacting disk galaxies, because

---

* On leave from Departments of Physics and Astronomy, University of Illinois at Urbana-Champaign, Urbana, IL 61801, U.S.A.
   [†] Current address: National Center for Supercomputer Applications, Beckman Institute, University of Illinois at Urbana-Champaign, Urbana, IL 61801, U.S.A.

Astrophysics and Space Science **216**: 337–346, 1994.
© 1994 *Kluwer Academic Publishers.*

it is in the regions of high gas density and shocks that we expect large amounts of star formation to take place. In contrast, near-infrared observations are more likely to display the underlying old stellar populations in the galaxies (see Bushouse and Stanford, 1992).

The relationship between the distribution of the star forming regions in these galaxies and the physical properties of the natal gas is of particular importance to an understanding of global star formation in disturbed systems. At present, there is no comprehensive, physical theory of star formation which can be applied to predict what mass range of stars will be produced in different physical circumstances of the interstellar gas, or, for example, what morphological distribution of star forming regions might be expected, given the flow patterns and clumping in the gas. We are attempting to get some information relevant to this overall situation by using an empirical approach in which detailed numerical models are constructed and then compared to real interacting galaxies. The more realistic our models and the more detailed the observations, the more likely we will be able to learn something of true significance to this area of study. As a first step in this endeavor, we have constructed 3-D models of a gas-free elliptical and a gas-rich spiral using a combined N- body/hydrodynamics code, and performed experiments on a Cray 2 computer in which they collide at a variety of angles and impact parameters (see Gerber, Lamb and Balsara, 1991 for a preliminary discussion of these experiments and Gerber, 1993; Gerber, Lamb and Balsara, 1993, 1994, for a current discussion). These experimental collisions can then be compared to observations of real interacting systems, such as the Arp 147 system. In this particular case, a detailed comparison of one of our evolved models and the real system, which contains a galaxy with an incomplete ring in its disk formed by the collision, (Gerber, Lamb and Balsara, 1992) indicates good agreement between model and observed properties, and allows an examination of some of the generic features of this type of collision.

It has been suggested that there is an evolutionary connection between the AGN phenomenon and collisions and mergers of galaxies (see Norman and Scoville, 1988), and between starbursts and AGN (see Perry and Dyson, 1985). Some observational evidence for this has been provided by Simkin, Su and Schwartz (1980). A simplified summary of the suggested scheme follows: the interaction causes gas in the disk of a galaxy to flow towards the central regions. There, a change in physical conditions could lead to star formation, which might then feed an already existing black hole or cause the formation of one. Or, alternatively, the gas flow might fuel an existing black hole directly in some way. The overriding problem to be understood in these scenarios is how angular momentum and energy can be shed by the infalling gas so that a central reservoir of gas or stars can be formed around the black hole and be available to feed it at an appropriate rate. In this connection, a detailed study of the effects of radiating shocks in the gas will be very important. Another part of the puzzle is the possible star formation episode. Under what circumstances would the inflowing gas form stars, where in the flow would they form, and what might the properties of those stars be? For example, would they be supermassive, all very low mass, or follow a Salpeter mass function? A study of the energy budget in starburst galaxies suggests that the IMF is tipped towards the high mass end. At what radius might they form? Observations of nearby Seyferts suggests that in these objects star formation is concentrated in a ring of radius 1 Kpc centered on the nucleus. N- body calculations that merely assume a star formation rate as a function of collision rate between calculational 'gas' particles merely beg the question of star formation in these systems The central region of interest here in these active galaxies is only of the order of a few parsecs in dimension and thus beyond the limits of resolution by current instruments, so any information that we can obtain about the types and clustering of stars formed in other strongly disturbed environments is of possible interest. (We note here that galactic bars can also channel gas towards the central regions and that the same considerations concerning star formation in dense, shocked gas would also apply (see Shloshman, Begelman and Frank, 1989).

The overall rate of star formation in disk galaxies experiencing moderately strong interactions is enhanced by only a factor of approximately two, as observed optically by Bushouse (1986, 1987) and Kennicutt *et al* (1987), and as observed in the far- infrared using IRAS data by Bushouse, Lamb and Werner, 1988. However, the observed new star formation takes place preferentially in the nuclear regions (here of dimension one kpc), rather than in the spiral arms of the disk as in isolated disk galaxies. A moderately intense burst of star formation in the very central regions might go undetected by such relatively low resolution observations. Some galaxies that are classified as merger remnants do appear to have large amounts of star formation in their central regions. These are the extreme IRAS objects which have far-infrared fluxes exceeding 10^{12} $L_{\odot}$ (see Sanders *et al*, 1988, and references therein). Any star formation currently taking place could have been triggered by the first crossing of the two parent galaxies, when a large impulse inward would have been given to the gas and stars, approximately 10^8 or more years ago. Alternately, the final merger of the two galactic nuclei may be implicated in the triggering of a large central star burst. (See Majewski *et al*, 1993, for some observational constraints on high far-IR systems). Much remains to be discovered about possible star formation in active galaxies and its relationship to the AGN phenomenon, however one of several good starting places is an investigation of the gas flows in colliding and merging systems and the changes in physical parameters that result.

The remainder of this paper is divided into three sections. In Section 2 we describe the model galaxies and the numerical methods used and in Section 3 we describe the particular subset of our experiments which we have chosen to discuss here. (These experiments appear to emulate the formation of a class of ring galaxies which have a relatively simple geometry and are thus easier to investigate for our present purposes). Lastly, in Section 4 we present those results of these calculations which are relevant to large scale gas flows and possibly to star formation.

2   Model Galaxies and Numerical Methods

We have chosen to start our investigation of galaxy collisions by modelling systems in which only one of the galaxies contains gas initially. Consequently, we built 3-D models of equal mass galaxies, one of which is a rotating, disk galaxy containing both gas and stars and the other is an elliptical containing stars only. We produced fully self consistent models in which the halo mass is 2.5 times that of the disk. The calculations were performed using a combined N-body/SPH program in which 50,000 N-body (star) particles and 22,000 SPH particles (representing the gas) were utilized. All computations were conducted on the Cray-2 supercomputer at the National Center for Supercomputing Applications, which is situated at the University of Illinois at Urbana-Champaign.

2.1   Numerical Methods

The computer code used to perform the experiments described here is a combined N-body/smooth particle hydrodynamics (SPH) code. That is, it represents both stars and gas using particles, but the evolution of the stars (and any collisionless dark matter) is followed by using N-body techniques, whereas the gas is represented by 'particles' that act as moving interpolation points (see Monaghan, 1985). The gas is considered to form a continuum and the density at any point is obtained by smoothing out the mass of nearby particles, which is done by using an analytic smoothing function. Forces are calculated by taking gradients of a smoothed estimate of the pressure. Shocks can be modelled in this scheme and this code utilizes an artificial viscosity formulation due to Balsara (1990) to accomplish this.

For the experiments described here, we have chosen to use an ideal gas equation of state and to assume that the gas cooling time is less than a typical time step in our calculation. Under these assumptions, each gas particle retains its initial temperature throughout the experimental run.

In the code discussed above the gravitational force is calculated by standard particle-mesh (PM) techniques (see Hockney and Eastwood, 1988). In this method, the gravitational potential is calculated at a restricted number of points on a three-dimensional grid and the force on an individual particle is determined by interpolation between values for surrounding grid points. In determining the gravitational potential on the grid both the stellar and SPH particles contribute. We chose to use a cubic grid with 64 points along each side.

The particular combination of SPH and PM techniques that we use is very suitable for modelling collisions between galaxies in 3- D. The grid method in PM is much faster than the alternative 'tree methods' and can be combined with the SPH, as demonstrated by Balsara (1990). The particle nature of both methods lends itself to the straightforward determination of the gravitational potential. However, one drawback of having a grid tied to the computational space is that the volume of interaction in the experiment must be limited. Thus, the method we have used here is ideally suited to the calculation of the first passage of one galaxy through another but is not suited, in its present formulation, to the calculation of distant encounters and to following energetic collisions to merger. More details of the combined code and its uses can be found in Balsara (1990, 1993), where tests of it are presented. Further discussion of the methods of SPH can be found in, among other places, Lucy (1977), Gingold and Monaghan (1982), Hernquist and Katz (1989), and Balsara (1990, 1993).

2.2 MODELS

The disk galaxy has a radially exponential disk of gas and stars surrounded by a massive, almost spherical halo of gravitating (star) particles, which can be considered to be either stellar or 'dark matter'. The elliptical galaxy consists of a spherical distribution of star particles only. The halo and disk were given 25,000 particles each and the latter were placed in rotation around the center such that Toomre's stability parameter, Q, (Toomre 1964) was 1.5 everywhere in the disk. The three dimensional density distribution, ρ, of the disk is

$$\rho(r,z) = \frac{M_{\exp}}{4\pi H R_d^2} \exp\left(-\frac{r}{R_d}\right) \operatorname{sech}^2\left(\frac{z}{H}\right), \tag{1}$$

where (r,z) are cylindrical coordinates, and $M_{\exp}$ is the total mass of a radially infinite exponential disk with disk radial scale length, R_d. Here, the scale height, H, is set constant with radius, and the disk density distribution is cut off at 4.4 R_d; interior to which radius the integrated mass is $0.93 M_{\exp}$. In the z direction the density distribution of the disk is cut off after two scale heights, where the density is 0.07 of its value at $z = 0$.

The gaseous disk was modelled using approximately 22,000 SPH particles. (This is a factor of seven larger than has been used previously to model such systems). We have set the gas density distribution such that it follows that of the disk stars but has a total mass equal to only one tenth that of the stellar disk. The SPH particles were placed in circular orbits around the center of the disk with no dispersion in their velocities.

We chose to represent the gas-free elliptical galaxy by a spherical King model with 10,000 N-body particles. Its total mass was set equal to that of the disk galaxy (including its halo), and its radius is approximately that of the disk galaxy halo.

No galaxy model is ever perfectly stable, but a model must be stable over a long enough time that changes resulting from the collision can be distinguished from those that would occur in any case. Consequently, we evolved the models in isolation for a period considerably longer than the duration of our collision experiments. We found that the disk exhibited a tendency to produce a low amplitude sheared spiral pattern and expanded slightly in the direction normal to the plane, but no large scale changes in the galaxies occurred on this time scale.

More details of both the numerical methods and the starting models for these numerical experiments can be found in Gerber (1993) and in Gerber, Lamb and

Balsara (1993).

3 The Numerical Experiments

Collisions between galaxies will always lead to large scale gas flows if there is any gas present in either galaxy. The simplest experimental geometry for studying these flows is the case of one galaxy colliding with another along the latter's spin axis and through its center. If two disk galaxies were involved in such a collision then the orientation of the two spin axes would also be a consideration, so it is simpler to start with an investigation of the collision of an elliptical with a disk galaxy. Below we will discuss the results of such an investigation (see Gerber, Lamb and Balsara, 1993) together with the results of an investigation of such collisions in which the impact parameter has been increased in increments up to a value of $0.9R$, where R is the radius of the disk, (see Gerber, Lamb and Balsara, 1994).

The experiment is started with the two galaxies placed at opposite corners of the computational cube, and the two centers of mass are given an initial relative velocity so as to produce a mildly hyperbolic orbit. The initial center of mass separation of the two galaxies is about 7 radial disk scale lengths.

As mentioned above, the particular collision geometry that we have chosen to model leads to the formation of a ring structure in the disk galaxy. Such structures were first investigated computationally for the star particles by Lynds and Toomre (1976). However, this excess density region exists in both the stars and gas, although the detailed structure is somewhat different for the two. The collisional nature of the gas results in a sharper ring than obtained for the star particles, and the development of high volume densities and shocks. The central collision produces a true ring, but off-center collisions result in the formation of only partial rings which have a 'horse-shoe' shape. Real galaxies in interacting pairs are observed to show these types of structure, and it has been natural and informative to associate them with the formation mechanism discussed here (see Lynds and Toomre, 1976; Theys and Spiegel, 1976, 1977; Huang and Stewart, 1988).

As seen in previous studies, our computations show that each galaxy contracts and then expands due to the increase and then decrease in gravitational potential as the other passes through it. The disk participates in that dynamics, initially contracting radially and bowing toward the incoming intruder. Soon after close approach, the inflow has produced a density maximum in time in the central parts of the galaxy. This is the point in time during the first passage that the inward velocities and densities in the center are the highest. The inward motion is stopped when the particles reach their centrifugal barrier. They then move outwards as a result of the declining potential. (Consequently, if one wishes to harness this material as fuel for a hypothetical, central black hole, the problem becomes one of removing large amounts of angular momentum from the gas, or from the stars which may be forming in it, so that it does not re-expand with the overall flow).

The outwardly propagating density wave in the disk is formed because the particles in the inner part of the disk start to move outwards even as Bthe outer parts of the disk are still contracting, creating a density peak where orbits crowd. For a more detailed discussion of this see Gerber, Lamb and Balsara, 1993; Struck-Marcell and Lotan, 1990; and Lynds and Toomre, 1976. As the stellar ring moves outward it expands in the radial direction and a large fraction of the total disk mass is contained within it. The gaseous ring has much sharper, thinner features because the gas is collision-dominated and has no radial velocity dispersion in the initial galaxy model.

When these calculations are done in 3-D with all components contributing to the gravitational potential, as in these experiments, one finds the development of complicated flows and shock structures in the direction perpendicular to the plane of the disk. Thus the outwardly propagating density wave contains a lot of structure which might be expected to lead to star formation, and perhaps considerable radiation directly from heated gas. The dynamical time scale of the collision (typically a few hundred million years) is approximately equal to one revolution time in the outer disk, consequently, the coupling of the particle rotation and density wave disturbance is clearly shown in the numerical results. More discussion of the overall gas flows is contained in Section 4.

We follow the collision as one galaxy passes through the other, only terminating the calculation when particles begin to be lost from the computational box. This is equivalent to following the collision through one dynamical timescale. Both the intruder and the disk galaxy's halo are quite disrupted by the interaction. We do not know the ultimate fate of these components, but enough energy has been pumped into the internal dynamics of these components to make them very diffuse by the end of the experiment. We find that as the density wave reaches the outer disk it has diffused in radius and is becoming somewhat indistinguishable from the background. Thus the timescale for the dynamical disturbance is a few times 10^8 years. However, we note that clumping on scales at or below our experimental resolution, which may occur in the gas in real galaxies, could persist for much longer times, and may lead to star formation over a considerable period.

4 Large Scale Material Flows

On the grossest scales, the material flows in these galaxies consists of a flow inwards towards the center followed by an outflow which takes individual particles to larger radii than they started at. Superimposed on this is the detailed dynamical behavior that we outline below. On even smaller scales (ones below our resolution limit) there is the possibility of flows that may be very important for the formation of stars etc. and other observationally relevant behavior. However, we do not anticipate that the existence of such flows or the formation of stars will have a first order effect upon the flows described here. Rather, they should be considered as possibly producing perturbations on the current results.

In our experiments the two galaxies are of comparable mass and, as a consequence, the disk is bowed considerably out the plane during the collision. Thus, when the material from the inner disk expands radially, it is displaced in the z–direction from the infalling outer material. This results in the stars and gas moving in something close to a toroidal flow in the vicinity of the ring (as Lynds and Toomre speculated would happen). Incoming and outgoing gas flows meet and form a shock on the side of the disk away from the direction of approach (underside). Shocked gas is thus swept up through the disk and carried out with the outflowing material. As a result of this three-dimensional structure the gas ring is broader and less dense than it would have been if the motion had been constrained to only two dimensions. The gas and stars generally flow together throughout the experiments with the gaseous ring lagging the stellar ring only slightly during the outward propagation.

4.1 THE CENTRAL COLLISION

Because of the azimuthal symmetry of this collision geometry, it is useful to average the particle properties over their angular position in the disk for many purposes. For example, we note that a plot of radial position versus radial velocity is multivalued in the ring region because the inner parts of the disk expand 'above' the infalling outer regions. A 3-D plot including the above parameters and the z-direction (see Gerber, 1993) allows one to explore the structure in more detail and to determine the locations of shocks in the flow. For a disk galaxy with dimensions the same as those of the Milky Way we find that the peak inflow velocity (at 22 Myr after the collision) is about 160 km s^{-1} and the velocity difference between the two streams at this time is about 320 km s^{-1}. This relative velocity is equivalent to Mach 10 or greater but, owing to the complex morphology in the region and the multidimensional nature of the shock, it is not really possible to assign a unique shock strength.

As noted above, the full particle motions (where this applies to both the gas and stars) are approximately toroidal in the ring region and this will have an important role when considering possible star formation in these regions.

An analysis of the motion in the azimuthal direction shows a maximum value in the ring region. The collision with the other galaxy gives the particles an inward pull and angular momentum conservation forces azimuthal velocities to increase. Soon after collision the peak in the velocity curve is approximately 250 km s^{-1} but this has increased to about 450 km s^{-1} by 22 Myr after the collision. The value then falls during the continued expansion and the velocity curve behind the ring is relatively flat. We note that rotation velocities in the ring are greater than velocities on either side of it.

Soon after passage of the intruder galaxy the central density peaks at about a factor of 5 above its initial value. After this a ring forms and moves outward. The relative density increases within it are more dramatic and peaks at about

42 Myr after the collision, with a space density of about 10–20 greater than the initial value for the particles involved. At this point the surface density in the ring is about 7–8 times that in the original disk. These high values are maintained beyond 62 Myr after collision, but have declined by 82 Myr.

Further details on this central collision model can be found in Gerber (1993) and in Gerber, Lamb and Balsara (1993).

4.2 The Off-Center Collisions

It is in the results of the off-center collision experiments that the coupling between the rotational motion in the disk and the motion due to the collision can be seen clearly. Features in the 'modified ring' or 'horse-shoe' structure can be seen to spiral outward during the evolution, making approximately one revolution as the ring expands to the edge of the disk.

This coupling of the motions contributes to the detailed morphology that develops in the outwardly propagating density structure in that the differential rotation spreads out the dense region in the disk that is formed when the other galaxy passes through, forming an arc of dense material. Soon after the collision the densest part of the arc is at its apex. However, as the structure evolves, the densest region bifurcates and proceeds to migrate down each arm of the arc. The flows in these arcs are relatively complicated and can only be appreciated by viewing a time sequence of the morphological development of the density enhancements with a depiction of the flow patterns superimposed. In real ring galaxies, such as Arp 147 (see Schultz et al, 1991), there is evidence that star formation proceeds down the arms of the arc. Thus one is drawn to the conclusion that the star formation is related to the increase in density, and to the strong shocks that develop in these regions as a result of the complicated flows. For more discussion of this see Gerber (1993) and Gerber, Lamb and Balsara (1992, 1994).

4.3 Unequal Mass Galaxies in Collision

How dependent are these results on the mass ratio of the two galaxies involved? We have experimented with intruders one- fourth as massive as the target galaxy and find that much of the displacement in the z-direction between outflowing and infalling material disappears. The ring expansion velocity decreases and two rings can appear (for the head on collisions), as predicted by other N-body investigations and epicycle/impulse approximation studies (see Appleton and Struck-Marcell, 1987; Struck-Marcell and Higdon, 1993; Hernquist and Wiel, 1993; Gerber and Lamb, 1993). Thus, many of the strong three-dimensional effects we see in our model are due to the fact that we are colliding together two equal-mass galaxies. However, the equal-mass collisions produce the densest rings and if this translates into the most prominent (or observable) rings, then these experiments have a good chance of being applicable to real systems, such as VII Zw 466, Arp 146, and Arp 147.

346 Susan A. Lamb et al.

Acknowledgements

SAL would like to thank NORDITA and the Neils Bohr Institute for hospitality and support during the completion of this work. SAL and RAG gratefully acknowledged the financial support of the University of Illinois Research Board and of NASA grants NAG 5- 1241 and NGT-70041.

References

Appleton, P. N. and Struck-Marcell, C.: 1987, *Astrophys. J.* **318**, 103.
Balsara, D. S.: 1990. Ph.D. thesis. University of Illinois, at Urbana- Champaign.
Balsara, D. S.: 1993, submitted for publication.
Bushouse, H. A.: 1986, *Astron.J.* **91**, 255.
Bushouse, H. A.: 1987, *Astrophys. J.* **320**, 49.
Bushouse, H. A., Lamb, S. A. and Werner, M. W.: 1988, *Astrophys. J.* **325**, 74.
Bushouse, H. A., and Stanford, A.: 1992, *Astrophys. J. Suppl.* **79**, 213.
Gerber, R. A.: 1993, Ph.D. thesis. University of Illinois, at Urbana- Champaign.
Gerber, R. A. and Lamb, S. A.: 1993, submitted to *Astrophys. J.*
Gerber, R. A., Lamb, S. A. and Balsara, D. S.: 1992, *Astrophys. J. Lett.* **399**, L51-54.
Gerber, R. A., Lamb, S. A. and Balsara, D. S.: 1993, in preparation.
Gerber, R. A., Lamb, S. A. and Balsara, D. S.: 1994, in preparation.
Gingold, R. A. and Monaghan, J. J.: 1982, *J. Comp. Phys.* **46**, 429.
Hernquist, L. and Katz, N.: 1989, *Astrophys. J. Suppl.* **70**, 419.
Hernquist, L. and Weil, M. L.: 1993, submitted for publication.
Hockney, R. W. and Eastwood, J. W.: 1988, *Computer Simulations Using Particles*, Adam Hilger, Bristol.
Huang, S. and Stewart, P.: 1988 *Astron. Astrophys.* **197**, 14.
Kennicutt, R. C., Keel, W. C., van der Hulst, J. M., Hummel, E. and Roettiger, K.: 1987, *Astron. J.* **93**, 1011.
Lucy, L. B.: 1977, *Astron. J.* **82**, 1013.
Lynds, R. and Toomre, A.: 1976, *Astrophys. J.* **209**, 382.
Majewski, S. R., Herald, M., Koo, D. C., Illingworth, G. D. and Heckman, T. M.: 1993, *Astrophys .J.* **402**, 125.
Monaghan, J. J.: 1985, *Comp. Phys. Rep.* **3**, 71.
Norman, C. and Scoville, N.: 1988, *Astrophys. J.* **332**,124.
Perry, J. J. and Dyson, J.: 1985, *Mon. Not. R. Astr. Soc.* **213**, 665.
Sanders, D. B., Soifer, B. T., Elias, J. H., Madore, B. F., Matthews, K., Neugebauer, G. and Scoville, N. Z.: 1988, *Astrophys. J.* **325**, 74.
Schultz, A. B., Spight, L. D., Rodrigue, M., Colegrove, P. T. and DiSanti, M. A.: 1991, *BAAS*, **23**, 953.
Shlosman, I., Frank, J. and Begelman, M. C.: 1989, *Nature* **338**, 45 .
Simkin, S. M., Su, H. J. and Schwartz, M. P.: 1980, *Astrophys .J.* **237**, 404.
Struck-Marcell, C. and Higdon, J. L.: 1993, *Astrophys. J.*, in press.
Struck-Marcell, C. and Lotan, P.: 1990, *Astrophys. J.* **358**, 99.
Theys, J. C. and Spiegel, E. A.: 1976, *Astrophys. J.* **208**, 650.
Theys, J. C. and Spiegel, E. A.: 1977, *Astrophys. J.* **212**, 616.
Thompson, L. A. and Theys, J. C.: 1978, *Astrophys. J.* **224**, 796.
Toomre, A.: 1964, *Astrophys. J.* **139**, 1217.

Gas Flow in a Two Component Galactic Disk

M. Noguchi
Astronomical Institute, Tohoku University, Aoba, Sendai 980, Japan

and

I. Shlosman
Department of Physics and Astronomy, University of Kentucky, Lexington, KY 40506-0055, USA

Abstract. Numerical simulations of two-component (stars + gas) self-gravitating galactic disks show that the interstellar gas can significantly affect the dynamical evolution of the disk even if its mass fraction (relative to the total galaxy mass) is as low as several percent. Aided by efficient energy dissipation, the gas becomes gravitationally unstable on *local* scale and forms massive clumps. Gravitational scattering of stars by these clumps leads to suppression of bar instability usually seen in heavy stellar disks. In this case, gas inflow towards the galactic center is driven by dynamical friction which gas clumps suffer instead of bar forcing.

Key words: gas dynamics, galactic disks, bar instability, nuclear activity

1 Introduction

Many numerical studies carried out so far have confirmed that large scale stellar bars arising in dynamically unstable galactic disks are one of the most efficient tools to supply galactic nuclear regions with interstellar gas and possibly initiate various activities there (*e.g.*, Schwarz 1984). However, most theoretical studies on galactic gas dynamics in the past treated the interstellar gas as a massless component whose dynamics is governed by the stellar gravitational field, neglecting possible back-reaction of gas to the stellar component.

In order to check the validity of such a simplified treatment, we study numerically the effect of gas on the dynamical behavior of a two-component (stars + gas) self-gravitating galactic disk embedded in a live halo. The stars are modelled as a system of collisionless particles and the interstellar gas is represented by an ensemble of finite size particles which individually correspond to giant molecular clouds. The gravitational interactions between all the particles are calculated using a TREE method. Collisions between gas particles are treated by means of the 'sticky particle' method (*e.g.*, Roberts and Hausman 1984). Namely, the radial component of the relative velocity of the colliding gas clouds is multiplied by $f_{coll}(< 1)$ and their direction of motion is reversed.

We consider three model families A, B, and C, for which the total disk mass fraction, f_d, is 0.2, 0.5, and 1.0, respectively. In each family, the gas mass fraction, f_g, is varied between 0 (*i.e.* purely stellar disk) and f_d with the total disk mass fixed (*i.e.*, the stellar disk has a mass fraction $f_d - f_g$ in general).

M. Noguchi and I. Shlosman

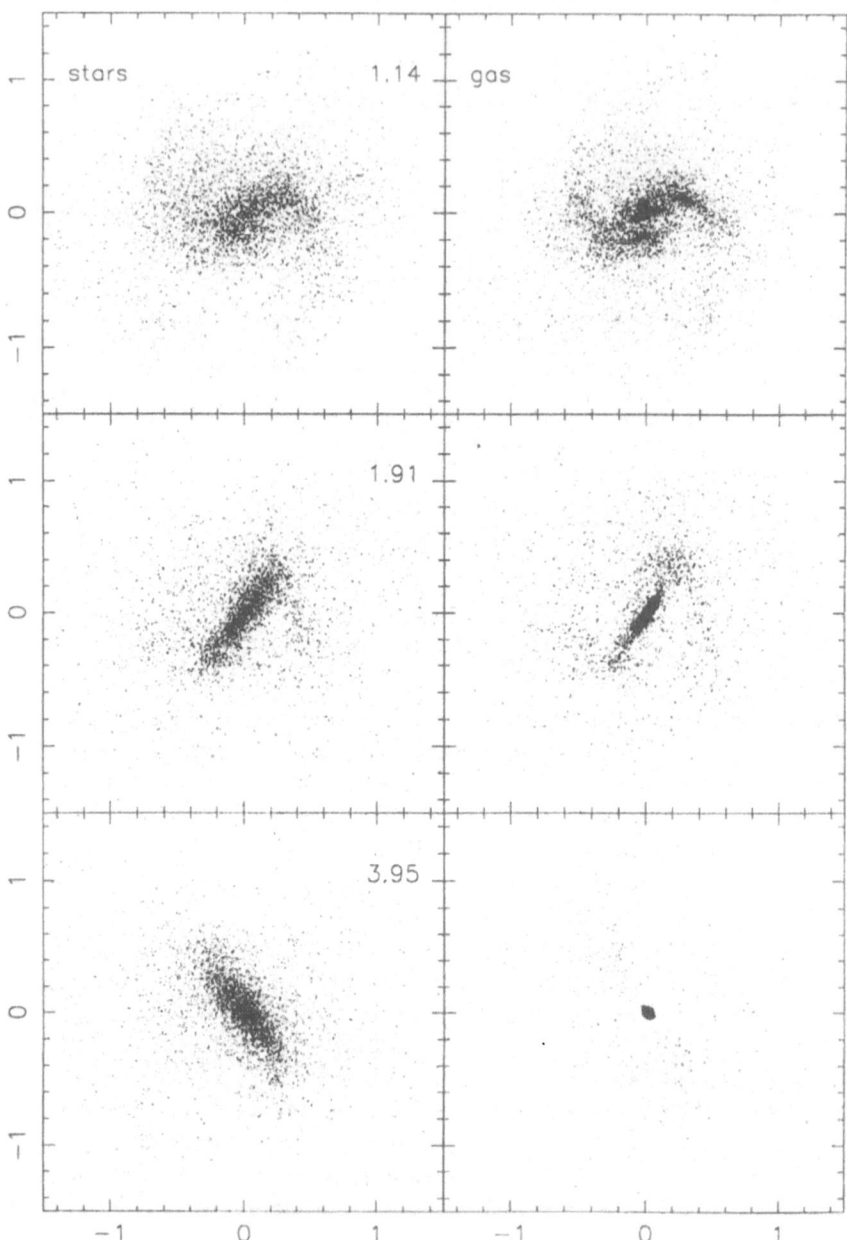

Fig. 1.   One model from B-family with a small gas fraction (f_d=0.5, f_g=0.01, f_{coll}=0.25). Only disk particles are plotted. The time in units of the disk rotation period is given in the upper right corners.

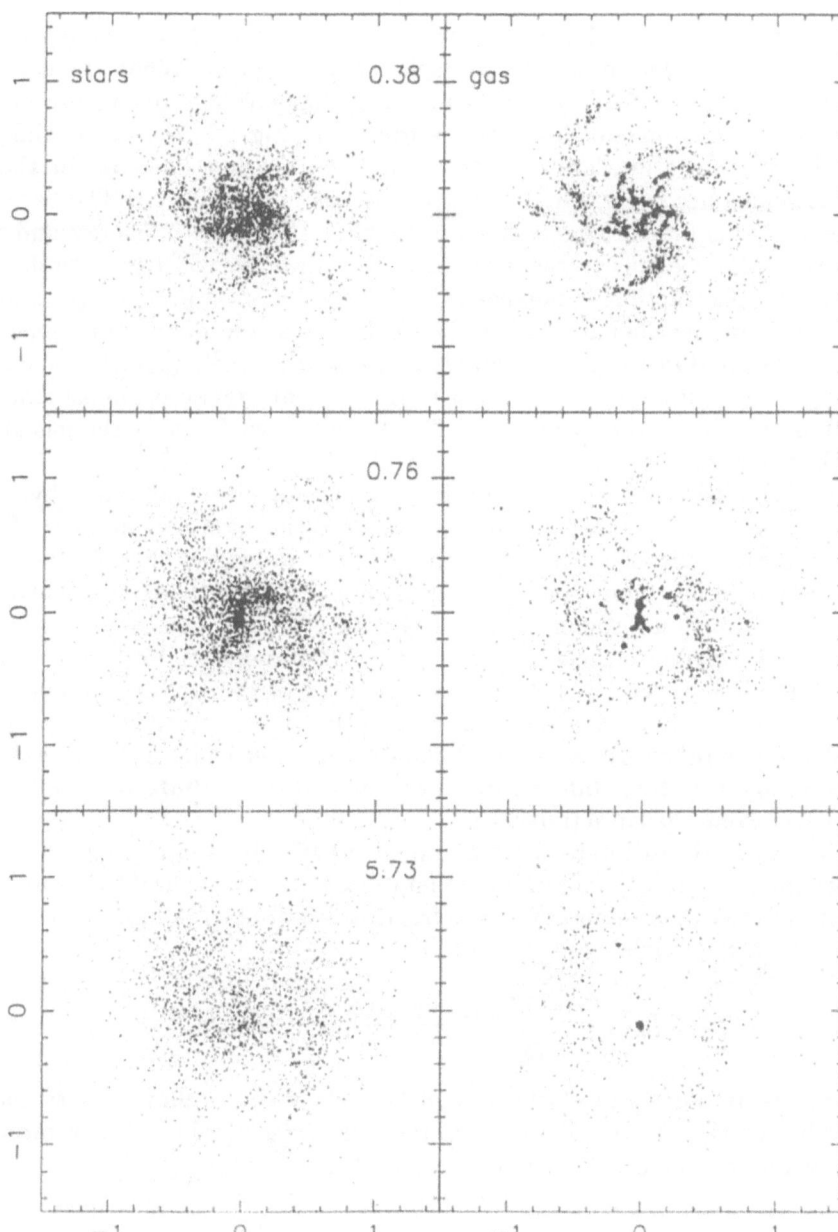

Fig. 2.   One model from B-family with a large gas fraction (f_d=0.5, f_g=0.1, f_{coll}=0.25). Only disk particles are plotted. The time in units of the disk rotation period is given in the upper right corners. Note distinct clumps in the gas disk.

2   Results

We concentrate on B-family, whose purely stellar model exhibits violent bar instability and see the effect of addition of self-gravitating gas. When the gas fraction
is small ($f_g \approx 0.01 - 0.025$), the inclusion of gas has no effect on the bar instability
and the stellar component develops a strong bar (Fig.1). The gas initially forms
many clumps, but these are only transient. As the bar develops in the stellar
disk, gas clouds become gradually trapped by bar potential well. The bar deprives
the gas of its angular momentum. Due to the loss of kinetic energy and angular
momentum, the gas falls to the disk center within a few rotation periods.

When the gas fraction in the disk is increased, *i.e.* $f_g > 0.05$, we see a dramatic
change of dynamical behavior (Fig.2). In this case, the gas clumps arising from
local gravitational instability are stable and maintain their identity for long time.
The stellar disk does not develop a bar in this case but remains almost axisymmetric still after a few rotation periods. The efficiency with which the gas stabilizes
the disk is remarkable.

The mechanism of bar formation in disks is not clear yet (*e.g.* Lynden-Bell
1979). Analysis of dominant particle orbits in strong bars reveals their alignment
with the major axis of a bar. We infer that a process of bar formation must
be accompanied by increasing correlation between increasingly eccentric orbits
within the corotation resonance. The massive gas clumps efficiently scatter stellar
particles and randomise their motion. A correlation between long axes of different
orbits will be destroyed . Net effect is the heating of the stellar component.

In these gas rich models, the gas inflow due to bar forcing is not realized.
Instead the gas inflow is driven by dynamical friction which massive gas clumps
experience against stars. Inflow rate can be as high as $\sim 10 M_\odot yr^{-1}$ and sufficient
to maintain even quasar activity.

We can roughly estimate the amount of heating by gas clumps and derive a
criterion for bar stability in two component disks utilizing the criterion by Ostriker
and Peebles (1973). The minimum gas fraction required to suppress bar formation,
$f_{g,crit}$, is then given by,

$$f_{g,crit} = \frac{0.14}{\tau_{bar}^{1/4}}[(\frac{f_d - f_{g,crit}}{0.56} - \frac{1}{2})^2 - \alpha_0^4]^{1/4}, \tag{1}$$

where τ_{bar} is timescale of bar growth and α_0 is the initial stellar velocity dispersion
divided by typical rotational velocity of the disk. Numerical results for models A-C
show a good agreement with this criterion (Fig.3).

3   Discussion

Both a numerical study and a simple analytical consideration have shown that the
gas fraction required to prevent bar instability is as low as several percent of the
total galaxy mass. This fraction is well within the range observed for nearby disk

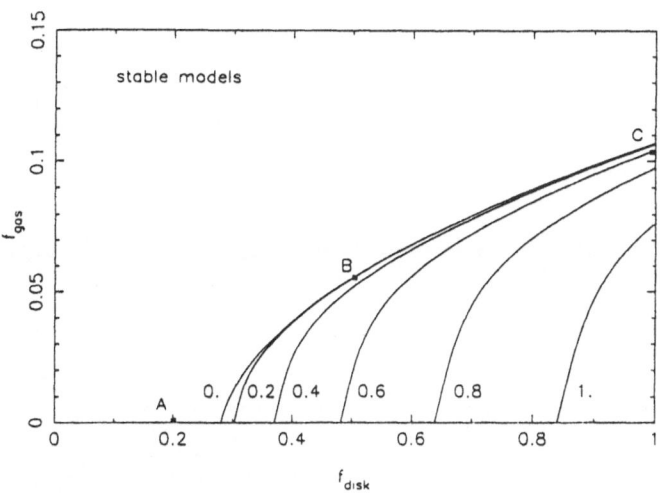

Fig. 3. The critical gas mass, $f_{g,crit}(f_{gas})$, plotted against $f_d(f_{disk})$ for various α_0 from equation (1). Filled squares indicate numerical results.

galaxies of late morphological types. The gas hence may serve as other stabilizing agent than massive haloes against bar formation in galactic disks. Interestingly it is observationally suggested that late type disk galaxies tend to have a bar less frequently than early type ones (Sellwood 1983). The present result seems to have even larger implications when we consider the dynamical evolution of galaxies in the early cosmological epoch, when gas mass fraction in the disk was much larger than in the present epoch.

Acknowledgements

We are grateful to Lars Hernquist for providing us with version of his collisionless TREE-code and to the University of Kentucky Center for Computational Studies for a generous allocation of the computing time. This work was supported in part by NASA grant NAGW-767 to the Joint Institute for Laboratory Astrophysics at the University of Colorado, Boulder.

References

Lynden-Bell, D., *Mon.Not.R.astr.Soc.*, **187**, 101 (1979).
Ostriker, J.P., and Peebles, P.J.E., *Astrophys.J*, **186**, 467 (1973).
Roberts, W.W., and Hausman, M.A., *Astrophys.J*, **277**, 744 (1984).
Schwarz, M.P., *Mon.Not.R.astr.Soc.*, **209**, 93 (1984).
Sellwood, J.A., in *Internal Kinematics and Dynamics of Galaxies*, IAU Symp. 100, ed. E. Athanassoula, p 197 (1983).

How Faithful Are N-Body Simulations of Disc Galaxies?—Artificial Suppression of Gaseous Dynamical Instabilities

Alessandro B. Romeo
Onsala Space Observatory, Chalmers University of Technology, S-43992 Onsala, Sweden

Abstract. High-softening two-dimensional models, frequently employed in N-body experiments, do not provide faithful simulations of real galactic discs. A prescription (♠) is given for choosing meaningful values of the softening length. In addition, a local stability criterion (♣) is given for choosing meaningful input values of the Toomre parameter for a given softening length. Such a criterion should also provide a key to a correct interpretation of computational results in terms of real phenomena.

1 Introduction

N-body simulations employing particle-mesh codes have nowadays become a very powerful tool for investigating the dynamics of disc galaxies. In particular, two-dimensional N-body models in which the stars and the cold interstellar gas are treated as two different components have successfully been applied in studies of spiral structure (e.g., Salo, 1991; Thomasson, 1991). A correct interpretation of computational results in terms of real phenomena poses serious problems, also because there are quantities introduced for numerical reasons which do not have clear physical counterparts. One of such artificial quantities is the softening length of the modified (non-Newtonian) gravitational interaction between the computer particles, and its value can critically affect the results of N-body experiments. It is thus of fundamental importance to have a prescription for choosing meaningful values of the softening length. From the stability point of view, it has been suggested that softening introduces a quite reasonable thickness correction for a two-dimensional model and that, even where softening is not introduced directly, a grid has a similar effect (e.g., Sellwood, 1986, 1987; see also Sellwood, 1983 for an extensive discussion of the additional softening introduced by a finite grid size; Byrd *et al.*, 1986). In this paper the analogy between numerical softening and finite-thickness effects is investigated in detail on the basis of a local linear stability analysis, and in particular the question "How faithfully does the softening mimic the thickness of galactic discs?" is addressed. It is found that high-softening two-dimensional models, frequently employed in N-body experiments, do *not* provide faithful simulations of real galactic discs. A prescription (♠) is given for choosing meaningful values of the softening length. Grid effects are also estimated.

Strictly connected with that problem is the choice of meaningful input values of the local stability parameter for a given softening length. In contrast to the softening length, the Toomre parameter is directly related to observable quantities, has a clear physical meaning, and its output values in N-body experiments can be compared to those predicted by theories of spiral structure and secular heating.

Astrophysics and Space Science **216**: 353–357, 1994.
© 1994 *Kluwer Academic Publishers.*

In this paper a local stability criterion (♣) is found in virtue of the descriptive similarity between numerical softening and finite-thickness effects. Such a criterion should indeed provide a key to this problem.

A more thorough discussion is given by Romeo (1993). In this short paper we just focus on a few points.

2 Local Stability

It is convenient to adopt the following scaling and parametrization:

$$\bar{\lambda} \equiv \frac{k_{\mathrm{H}}}{|k|}, \quad \text{where} \quad k_{\mathrm{H}} \equiv \frac{\kappa^2}{2\pi G \sigma_{\mathrm{H}}}; \tag{1}$$

$$\alpha \equiv \frac{\sigma_{\mathrm{C}}}{\sigma_{\mathrm{H}}}, \quad \beta \equiv \frac{c_{\mathrm{C}}^2}{c_{\mathrm{H}}^2} \quad (0 < \alpha < +\infty, \ 0 < \beta < 1); \tag{2}$$

$$Q_{\mathrm{H}} \equiv \frac{c_{\mathrm{H}} \kappa}{\pi G \sigma_{\mathrm{H}}} \quad (\textit{local stability parameter}); \tag{3}$$

$$\eta \equiv k_{\mathrm{H}} s \quad (0 < \eta < +\infty). \tag{4}$$

In these formulae, k is the local radial wavenumber of the perturbation, κ is the epicyclic frequency, σ_i and c_i $(i = \mathrm{H}, \mathrm{C})$ are the unperturbed surface densities and the equivalent planar acoustic speeds of the stars (H) and the cold interstellar gas (C), respectively, s is the softening length of the modified gravitational interaction. The case $\eta = 0$ represents the limit of an unsoftened gravitational interaction. There exists a critical value of the softening length beyond which the model is locally stable even for vanishing Q_{H}^2:

$$\text{STABILITY OF COLD MODELS}: \quad s > s_{\mathrm{crit}} = \frac{1}{e} \frac{2\pi G \sigma}{\kappa^2}, \tag{5}$$

σ being the total unperturbed surface density. This two-component extension of Miller (1972, 1974) criterion for cold models $(c_i = 0)$ is indeed the limiting case of a more general *local stability criterion* for cool models $(c_i > 0)$, which can be viewed as the softened two-component extension of Toomre (1964) criterion:

♣ ┃ **Local Stability Criterion** ┃ ♣

$$\text{STABILITY OF COOL MODELS}: \quad \boxed{Q_{\mathrm{H}}^2} > \bar{Q}^2, \tag{6}$$

$\bar{Q}^2 = \bar{Q}^2(\alpha, \beta, \eta)$ being the global maximum of the marginal stability curve derived by Romeo (1993). In particular, in low-softening standard star-dominated regimes

$$\bar{Q}^2 \approx 1 + 4(\alpha - \eta) \quad [\alpha \ll 1; \ \beta, \eta = O(\alpha)]. \tag{7}$$

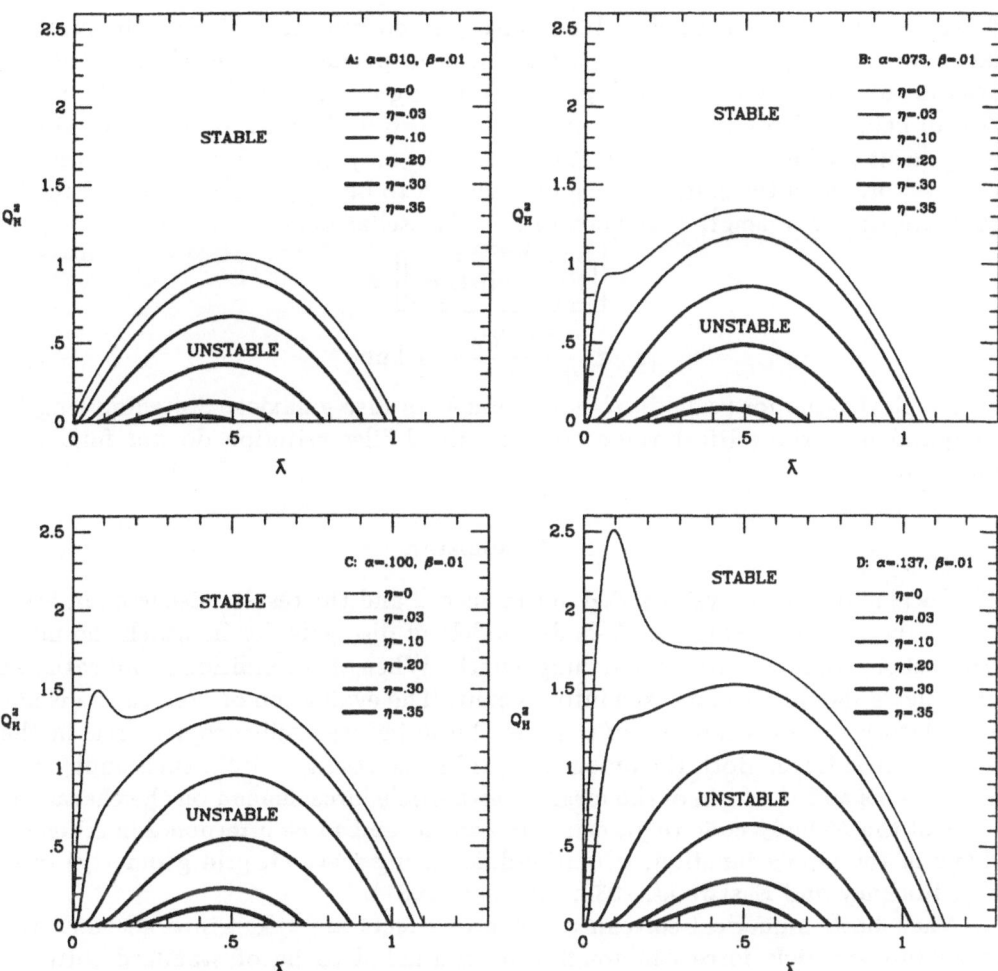

Fig. 1.   Two-fluid marginal stability curves in the $(\bar{\lambda}, Q_H^2)$ plane for some values of the local parameters η, α and fixed $\beta = 0.01$. The case $\eta = 0$ represents the limit of an unsoftened gravitational interaction.

3 Results

In presenting the results of the local stability analysis performed in this paper, we have considered the standard star-dominated and the peculiar gas-dominated regimes already investigated in the context of thick two-component galactic discs (Romeo, 1990, 1992). The marginal stability curves shown in Fig. 1 should qualitatively be compared to those shown in Fig. 4 of Romeo (1992). It is apparent that, because of the highly stabilizing role of numerical softening, the local linear stability properties of two-dimensional N-body models can indeed be considerably

356Alessandro B. Romeo

different from those of thick galactic discs, as derived analytically. In particular, *note* the suppression of the gaseous peak in peculiar gas-dominated regimes
even for exceedingly low softening. The softening can faithfully mimic the thickness of galactic discs or, more precisely, the effective thickness-scale of the stellar
component [defined in Eq. (6) of Romeo (1992)] only in standard star-dominated
regimes, provided the softening length is chosen to be very short compared to the
characteristic wavelength corresponding to the stellar peak:

♠ |Prescription| ♠

$$\boxed{s} \ll \frac{1}{2}\frac{2\pi G\sigma_{\rm H}}{\kappa^2} \sim 1 \text{ kpc}, \qquad (8)$$

as a typical value for realistic N-body models of disc galaxies. Softening lengths
comparable to the critical value given by the Miller criterion do *not* fulfil this
prescription.

4   Discussion

The local stability analysis carried out in Sect. 2 and the results discussed in Sect.
3, strictly speaking, apply to N-body models of disc galaxies in which the mesh
size is short compared to the softening length. When this condition is not satisfied
grid effects should also be taken into account. The evaluation of such effects is not
straightforward for a number of reasons. Mesh-induced anisotropies arise in the
inter-particle force. Both the amount of grid noise resulting from such unphysical
fluctuations and the form of the mean inter-particle force depend on the characteristics of the N-body code (e.g., mass-assignment and force-interpolation schemes,
shape of the Green function, potential-differencing method, grid geometry) (see,
e.g., Hockney and Eastwood, 1988; Sellwood, 1987).

When both numerical softening and grid effects are taken into account, the
mean inter-particle force can roughly be estimated to be of standard softened
type, with an *effective softening length* given by

$$s_{\rm eff}^2 = s^2 + \Delta^2, \qquad (9)$$

Δ being the mesh size. This can be deduced for CIC and TSC schemes from Fig.
1a of Efstathiou *et al.* (1985), once a reasonable extrapolation is made to include
the presence of numerical softening. Thus, the local stability analysis performed
in Sect. 2 and the results presented in Sect. 3 are still expected to apply, provided
the numerical softening length is replaced by the effective softening length of the
modified gravitational interaction.

5   Concluding Remarks

The suggestion that softening introduces a quite reasonable thickness correction
for a two-dimensional model is quantitatively confirmed in low-softening standard

star-dominated regimes. A constant effective softening length would then ideally correspond to a constant scaleheight of the stellar component. On the other hand, a realistic simulation of the vertical structure of disc galaxies would in any case require a proper three-dimensional model.

Although the local stability properties of high-softening two-dimensional N-body models are considerably different from those of thick galactic discs, the propagation properties of the spiral waves are still expected to be physically plausible in standard star-dominated regimes, provided the effective softening length is chosen to be shorter than the critical value given by the Miller criterion [cf. the more restrictive prescription (♠)]. In choosing the input values of the local stability parameter as well as in comparing its output values to those predicted by theories of spiral structure and secular heating, it should then be borne in mind that the stability threshold is *not* unity (Toomre 1964 criterion for unsoftened one-component models), but the value given by the local stability criterion (♣) discussed in Sect. 2.

References

Byrd G.G., Valtonen M.J., Sundelius B., Valtaoja L.: 1986, *Astron. Astrophys.* **166**, 75

Efstathiou G., Davis M., Frenk C.S., White S.D.M.: 1985, *Astrophys. J. Suppl.* **57**, 241

Hockney R.W., Eastwood J.W.: 1988, *Computer Simulation Using Particles*. Hilger, Bristol

Miller R.H.: 1972, In: Lecar M. (ed.) Proc. IAU Colloq. 10, *Gravitational N-Body Problem*, Reidel, Dordrecht, p. 213

Miller R.H.: 1974, *Astrophys. J.* **190**, 539

Romeo A.B.: 1990, PhD thesis, SISSA, Trieste, Italy

Romeo A.B.: 1992, *Mon. Not. Roy. Astr. Soc.* **256**, 307

Romeo A.B.: 1993, *Astron. Astrophys.* (submitted)

Salo H.: 1991, *Astron. Astrophys.* **243**, 118

Sellwood J.A.: 1983, *J. Comput. Phys.* **50**, 337

Sellwood J.A.: 1986, In: Hut P., McMillan S.L.W. (eds.) *The Use of Supercomputers in Stellar Dynamics*, Springer-Verlag, Berlin, p. 5

Sellwood J.A.: 1987, *Ann. Rev. Astron. Astrophys.* **25**, 151

Thomasson M.: 1991, PhD thesis, Chalmers University of Technology, Göteborg, Sweden

Toomre A.: 1964, *Astrophys. J.* **139**, 1217

Long-Lived Spiral Structure in N-Body Simulations:
Work in Progress

Alessandro B. Romeo
Onsala Space Observatory, Chalmers University of Technology, S-43992 Onsala, Sweden

Gustaf Rydbeck
Onsala Space Observatory, Chalmers University of Technology, S-43992 Onsala, Sweden

and

Mauri J. Valtonen
Tuorla Observatory, Turku University, SF-21500 Piikkiö, Finland

Abstract. Work is in progress for constructing physically consistent N-body models capable of supporting a long-lived spiral structure. Such models would represent a missing link between theory and observations of disc galaxies.

1 Summary

As a practical application of the analysis carried out by Romeo (1994) (hereafter Paper I), work is in progress for constructing physically consistent N-body models capable of supporting a long-lived spiral structure, and for evaluating the secular heating induced by it. Such models would represent a missing link between theory and observations of disc galaxies. For this purpose, we are writing auxiliary programs for selecting input values consistent with the prescription (♠) and the local stability criterion (♣) given in Paper I, and corresponding to regimes of spiral structure in which a fruitful comparison between theory and simulations can be made. At the same time, we are modifying an N-body code developed by Thomasson (1989) for setting the disc up more satisfactorily, and we are writing programs for simulation-data reduction.

References

Romeo, A. B.: 1994, "How Faithful Are N-Body Simulations of Disc Galaxies? —Artificial Suppression of Gaseous Dynamical Instabilities". In: J. E. Dyson and E. B. Carling (eds.), *Manchester Conference on Kinematics and Dynamics of Diffuse Astrophysical Media* (Paper I), Kluwer, Dordrecht.

Thomasson, M.: 1989, Research Report No. 162, Chalmers University of Technology, Göteborg, Sweden.

Astrophysics and Space Science **216**: 359, 1994.
© 1994 *Kluwer Academic Publishers*.

The Use of Gravitational Microlensing to Scan the Structure of BAL QSOs

D. Hutsemékers, J. Surdej* and E. Van Drom
Institut d'Astrophysique, Université de Liège, 5, av. de Cointe, B-4000 Liège, Belgium

Abstract. Approximately 10% of the QSOs show broad absorption lines (BAL) in their spectra which, if interpreted in terms of Doppler velocities, reveal the presence of high velocity gas outflows. One of these BAL QSOs is known to be gravitationally lensed. It therefore constitutes a good candidate to search for microlensing effects, i.e. the selective amplification of different line forming regions. Considering current models for the BAL region, we have investigated the effects of moving microlenses on the line profiles, and we conclude that these effects strongly depend on the adopted model. A regular spectroscopic monitoring of lensed BAL QSOs would therefore be highly valuable to distinguish between the various models proposed so far to interpret the origin of broad absorption lines.

Key words: BAL QSOs, Gravitational lensing

1 Introduction

The BAL QSOs constitute a class of approximately 10% of the total number of QSOs. Their spectra are characterized by broad absorption lines (BALs), blue-shifted with respect to the emission, and indicating gas outflows at very high Doppler velocities, up to 0.2 c (see e.g. Turnshek 1988, for a review).

In order to interpret these line profiles, two main classes of kinematical and geometrical models have been proposed for the BAL region (BALR).
1) In the *cloud model* (Junkkarinen 1983; Weymann *et al.* 1985), the absorption is assumed to arise in a large number of small clouds, the total solid angle subtended by these clouds being fairly small as seen from the continuum source. These clouds do not significantly contribute to the emission lines which are formed as in normal QSOs. In this case, BAL QSOs are essentially normal QSOs which just turn out to be adequately oriented, i.e. with their absorbing clouds located along the line-of-sight.
2) For the *P Cygni type model* (Scargle *et al.* 1970; Drew & Boksenberg 1984) the absorption is produced in a relatively smooth flow having roughly the spherical symmetry (but in which discrete clouds may be embedded). In this case, the absorbing material may contribute to the emission lines and, more specifically, it can be at the origin of emission observed at high velocities. BAL QSOs may consist of a physically different class of QSOs, eventually being normal QSOs seen at another evolutionary stage.
It is clear that these two models are rather extreme, and that hybrids are possible.

---

* Also, Maître de Recherches au FNRS

Astrophysics and Space Science **216**: 361–365, 1994.
© 1994 *Kluwer Academic Publishers*.

Also, there seems to exist subclasses of BAL QSOs, like the low-ionization or the PHL5200-type BAL QSOs, which probably necessitate slightly different modeling.

Since one of these BAL QSOs is known to be gravitationally lensed (Magain et al. 1988), the selective magnification of some line forming regions is expected, revealing the structure of the QSO. Considering the two proposed types of BALR models, we have investigated microlensing effects on simulated BAL profiles. In this framework, we discuss the case of the lensed BAL QSO H1413+117, deriving some possible constraints on the flow models.

2   The effects of gravitational microlensing on BALs

Microlensing corresponds to the gravitational lens effect due to a star or a group of stars (which may belong to a galaxy) located along the line-of-sight. While the separation between the micro-images is too small to be detected ($\sim 10^{-6}$ arcsec), microlensing can be at the origin of a strong (de-) magnification of the source, or of parts of it (Chang & Refsdal 1979, 1984; Kayser et al. 1986).

Microlensing is characterized by an effective lensing size, the so-called projected Einstein radius which essentially depends on the lens (star) mass, and by generic magnification patterns made of critical points and lines (the caustics). If a source with a size comparable to the Einstein radius of the lens crosses these caustics, it is (de-) magnified and the flux of radiation received by a distant observer is (de-) amplified.

For typical QSO and lens distances, the projected Einstein radius of a solar mass lens becomes comparable to the size of the continuum emitting region of QSOs. A microlensing event will therefore (de-)magnify the continuum emitting region but not the broad emission line region (BELR) which is generally assumed to be much more extended. The projected BALR, on the other hand, coincides with the continuum emitting region, such that microlenses can magnify the BALR, or parts of it (cf. the individual clouds if any), relative to the BELR. Since microlenses are moving with respect to the source, they can act as scanners of the QSO BALR with typical timescales of months (for the case of H1413+117).

Considering simple cloud and P Cygni type models for the BALR, we have investigated the possible effects of microlenses on the line profiles (cfr. Hutsemékers 1993, and Hutsemékers et al. 1992, 1993), adopting the Schwarzschild and Chang-Refsdal lens models which contain all generic microlensing features. An example of such effects is illustrated in Figure 1. We namely found that the two BALR models have rather different signatures when affected by microlensing effects. Basic differences may be summarized as follows:

• for the *cloud model*:

- clouds with different optical depths and smaller than the continuum region may be selectively magnified, inducing variations in the absorption profiles; for sufficiently small black clouds, the relative profile variation is directly proportional to the radius of the magnified cloud;

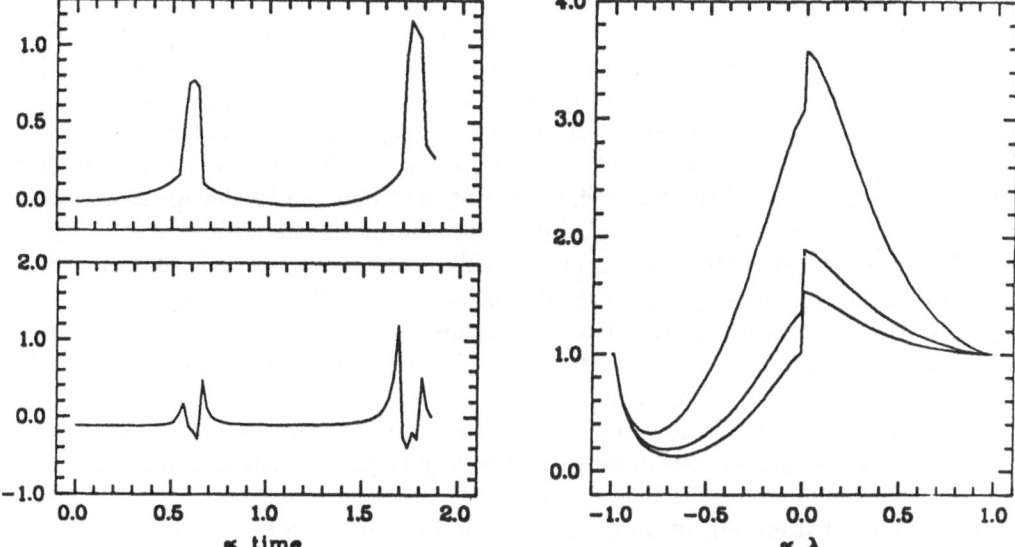

Fig. 1.   Example of microlensing effects when a BAL QSO crosses a caustic pattern in the source plane, the P Cygni type model being assumed for the BALR and the Chang-Refsdal model for the microlens. In the first figure (upper left) the magnitude difference is plotted against a quantity proportional to the time, illustrating the variation of the continuum flux. The second figure (lower left) illustrates the corresponding variation of the total equivalent width of the line profiles. The last figure shows resulting P Cygni type profiles at selected moments.

- it is likely that only some parts (i.e. some velocities) of the profile are affected;
- depending on the optical depth of the magnified cloud, the variations in the absorption line may be uncorrelated with those of the emission line equivalent width;
- since the continuum region is simultaneously magnified, this effect, only differential, remains small or then requires a very precise lensing configuration;
• for the *P Cygni type model*:
- strong modifications of the P Cygni type profiles are induced, including possible enhancements of the emission;
- the behaviour is essentially the same at all velocities;
- the variations in the absorption line are correlated with those of the emission line equivalent width;
- the total equivalent width variation shows a characteristic behaviour (Fig. 1).

   It seems therefore clear that the effects due to microlensing may constitute a powerful tool for distinguishing between the BALR models, provided that a regular spectroscopic monitoring becomes available.

3   Microlensing in H1413+117

The lensed BAL QSO H1413+117 consists of 4 images roughly forming a 1"×1" square. Evidence for microlensing in one of the image (D) was first reported by Kayser *et al.* 1990, on the basis of photometric variations. Further evidence comes from the fact that the spectrum of image D is slightly different from that of the other images (Angonin *et al.* 1990). The CIV and SiIV line profiles are similarly affected, in contrast with the intrinsic variations reported in the spectra of other BAL QSOs (see e.g. Barlow *et al.* 1992, and references therein).

The emission line differences may be easily interpreted by a microlensing effect: if the continuum emitting region is magnified, and the BELR is not affected as expected for a normal lens, the emission line equivalent width decreases as observed, after normalization to the amplified continuum. The difference observed in the absorption lines then arise because of
- the selective magnification of a small cloud optically thicker than the average; but this kind of interpretation necessitates very large continuum amplifications on short time scales,
- the apparent dimming of an emission line located at high velocities and superimposed on the varying part of the absorption line.
This latter interpretation is more viable in terms of microlensing, but requires the intrinsic emission line to be double-peaked. Such an intrinsic emission line profile is quite unusual and not typical of standard P Cygni type profiles. Ad-hoc models with turbulence may possibly account for such line profiles (Hutsemékers 1994). It is also striking to notice that the velocity separation between these two emission peaks exactly corresponds to the velocity separation between the Lyα and NV emission lines, possibly giving support to some line locking effects in driving the flow.

4   Conclusions

The different BALR models have clearly different signatures during a microlensing event. Microlensing events may therefore constitute a powerful tool for distinguishing between the BALR models provided that a regular spectroscopic monitoring of H1413+117 becomes available. In all cases the expected characteristics of the variations induced by microlensing are different from those reported for intrinsic variations.

If the spectral differences observed between the images of the quadruple QSO H1413+117 are actually due to microlensing effects, they are more convincingly interpreted by adopting a non-standard P Cygni type model for the BAL region. However, only the study of the temporal evolution of the observed spectral differences will provide us with more definite conclusions.

Although difficult, a spectroscopic monitoring of BAL QSOs known to be lensed, or located behind a galaxy, is badly needed.

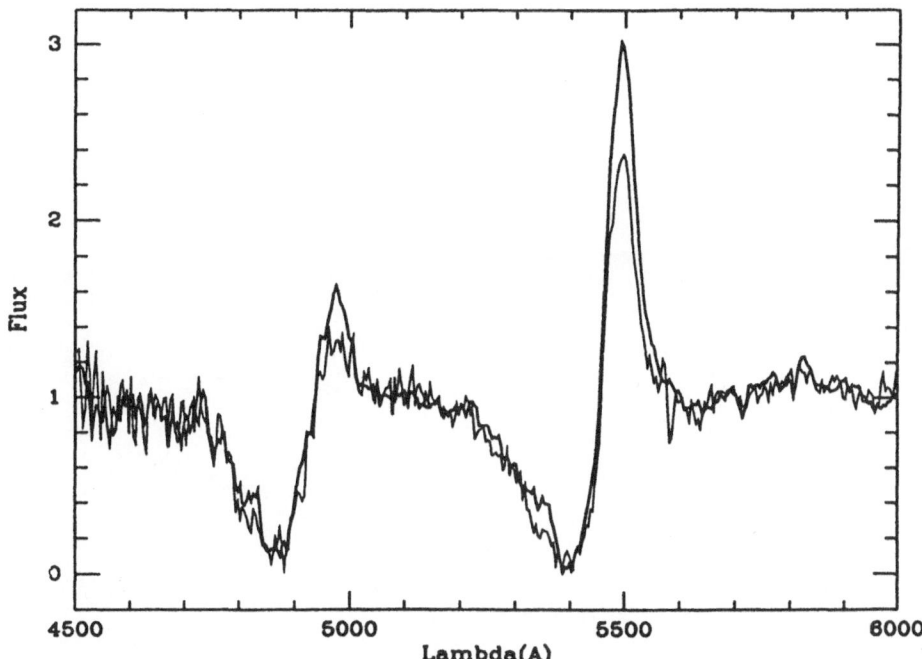

Fig. 2. The differences of the SiIV and CIV line profiles between the normalized spectrum of the component D of the BAL QSO H1413+117 (thin line) and the average spectrum (thick line) of the three other components (A,B,C) (cfr. Angonin et al., 1990, and Hutsemékers, 1993)

References

Angonin, M.C., Remy, M., Surdej, J., Vanderriest, C.: 1990, *A&A* **233**, L5

Barlow, T.A, Junkkarinen V.T., Burbidge, E.M., Weymann, R.J., Morris, S.L., Korista, K.T.: 1992, *Ap.J.* **397**, 81

Chang K., Refsdal, S.: 1979, *Nature* **282**, 561

Chang K., Refsdal, S.: 1984, *A&A* **132**, 168

Drew, J.E., Boksenberg, A.: 1984, *MNRAS* **211**, 813

Hutsemékers, D.: 1993, *A&A*, in press

Hutsemékers, D.: 1994, in preparation

Hutsemékers, D., Surdej, J., Van Drom, E.: 1992, *Gravitational lenses. Lecture Notes in Physics Vol. 406*, Kayser R., Schramm T., Nieser, L. (eds.), Springer-Verlag, p.373

Hutsemékers, D., Surdej, J., Van Drom, E.: 1993, in preparation

Junkkarinen, V.: 1983, *Ap.J.* **265**, 73

Kayser, R., Refsdal, S., Stabell, R.: 1986, *A&A* **166**, 36

Kayser, R., Surdej, J., Condon, J.J., Kellermann, K.I., Magain, P., Remy, M., Smette, A.: 1990, *Ap.J.* **364**, 15

Magain, P., Surdej, J., Swings, J.P., Borgeest, U., Kayser, R., Kühr, H., Refsdal, S., Remy, M.: 1988, *Nature* **334**, 327

Scargle, J.D., Caroff, L.J., Noerdlinger, P.D.: 1970, *Ap.J.* **161**, L115

Turnshek D.A.: 1988, *QSO Absorption Lines: Probing the Universe*, Blades, J.C., Turnshek, D.A., Norman, C. (eds.), Cambridge, p. 17

Weymann, R., Turnshek, D.A., Christiansen, W.A.: 1985, *Astrophysics of Active Galaxies and Quasi Stellar Objects*, Miller (ed.), Oxford, p.333

Anomalous component motion in the mas double radio source, 0646+600

Chidi E. Akujor
*Onsala Space Observatory, Onsala, S 439 92, Sweden**

and

Richard W. Porcas
Max-Planck Institut für Radioastronomie, Auf dem Hugel 69, Bonn, Germany

Abstract. 0646+600, an 'optically quiet quasar' has been imaged with global VLBI at 5GHz in 1989 and 1992 and at 8.4 GHz in 1991. Between 1989 and 1992, the overall separation of the two bright components decreased by 0.25 mas, and a weak middle component or extension seen in 1989 was not detected in the later observations.

1 Introduction

0646+600 is one of the compact double/triple radio sources found in our survey of 'optically quiet quasars' – objects which resemble quasars in radio structure (bright radio cores) and spectrum (flat), but are unidentified optically (Akujor and Porcas, 1992). It is a 0.8 Jy (5GHz) variable radio source and shows no large-scale extended structure (Stanghellini *et al.*, 1990). We had associated it with an 'empty field' on the basis of the radio positions of Patnaik *et al.* (1992) survey, but Meisenheimer and Röser (1983) report an identification with a 19 mag (red only) object.

We observed 0646+600 with global VLBI network at 5 GHz (epoch: 1989.79; Akujor and Porcas, 1992). The resulting map showed two almost equally bright components, A and B. The separation of two components was $3.0mas$ and there was a weak component or extension of A which we label A_e and which contains a flux density of $\sim 50mJy$. Further observations were made to discover whether the source was either a double or triple, and above all to measure the motion, if any, of the components. Although superluminal motion is common in the cores bright flat-spectrum sources, no definite seperation speed has been measured in compact doubles/triples.

2 Observations and Results

0646+600 was observed at 8.4 GHz (1991.17) and reobserved at 5 GHz (1992.22) with transatlantic baselines. At both frequencies and epochs, A_e was not detected, although this feature had been confirmed independently by the Caltech-Jodrell group (A. Polatidis, private communication). More significantly, between the two epochs at 5 GHz the overall size of the source (that is, A, B separation) decreased

---

* Present address: NRAL, Jodrell Bank, UK

Astrophysics and Space Science **216**: 367–368, 1994.

by 0.25*mas*.   0646+600 is unusual because it is a gigahertz-peaked spectrum

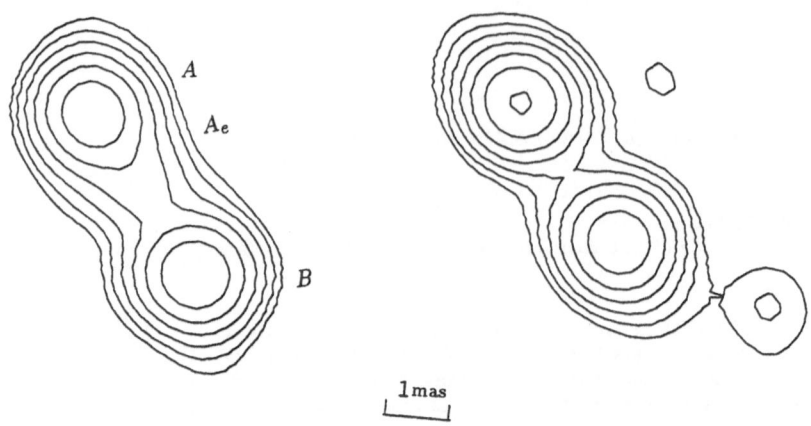

Fig. 1.   Global VLBI maps of 0646+600 at 5 GHz; epochs: 1989.79 (left) and 1992.22

(GPS) source with a structure resembling the compact triples/doubles (Philips and Mutel, 1982), but shows structural changes never seen in these types of sources. We may be witnessing another example of superluminal expansion, that is, of A_e between the two stationary components A and B, somewhat similar to that seen in 3C395 (Simon *et al.*, 1988). The most plausible reason for the apparent decrease in overall size appears to be that a new strong component is ejected westwards from a core within A, thus shifting its centroid while A_e has fadded.

These observations indicate that structural changes in 0646+600 are occuring on time scales shorter than the 2 year interval between the two observations at 5 GHz. More frequent observations with improved resolution are being undertaken.

Acknowledgements

We thank Dr Ian Browne for comments on this paper. CEA acknowledges Swedish NFR fellowship at Onsala.

References

Akujor, C. E. and Porcas, R. W.: 1992, in *Extragalactic Radio Sources–From Beams to Jets*, Roland J., Sol H., Pelletier G., eds., Cambridge Univ. Press, p. 134.
Meisenheimer, K. and Röser, H.-J.: 1983, *Astron. Astrophys. Suppl.* **51**, 41.
Patnaik, A. R., Browne, I. W. A., Wilkinson, P. N. and Wrobel, J. M.: 1992, *Mon. Not. Roy. Astr. Soc.* **254**, 655.
Philips, R. B. and Mutel, R. L.: 1982, *Astron. Astrophys.* **106**, 21.
Simon, R. S., Hall, J., Johnston, K. J., Spencer, J. H., Waak, J. A. and Mutel, R. L.: 1988, *Astrophys. J. (Letts.)* **326**, L5.
Stanghellini, C., Baum, S. A., O'Dea, C. P. and Morris, G. B.: 1990, *Astron. Astrophys.* **233**, 379.

Effects of Dense Medium Surrounding Galactic-sized Radio Sources

Everton Lüdke
N.R.A.L., University of Manchester, Jodrell Bank, Macclesfield SK11 9DL, U.K.

Abstract.
    In this paper, I have analysed the subarcsecond polarization structure of two high-z compact steep-spectrum quasars. Morphology suggests that the jets are interacting strongly with intergalactic medium. Models of bending by ram pressure equilibrium in a cooling flow and alignment of magnetic field lines by jet–IGM shock suggest that the CSS jets are light, supersonic and mildly relativistic. Particle energy index variations along the jet suggests replenishment triggered by such interactions.

Key words: Radio Polarization: Galaxies, Quasars: CSS sources

1 Introduction

Compact Steep–spectrum sources (CSSs) are a sub–class of quasars and radio galaxies with small projected linear sizes (< 15 Kpc), and source plasma is dominated by particle population with steep energy spectrum ($N(E)dE \propto E^{-\gamma}$, $2.3 < \gamma < 3.4$). CSSs galaxies are similar to normal double sources but quasars show very distorted jets from interactions with intergalactic medium. In this paper I present high-resolution polarization maps of two CSS sources extracted from a sample observed with the upgraded MERLIN (Lüdke *et al*, 1993) and simple kinetic models for source–environment interactions, to constrain the jet dynamics.

2 Discussion

High-definition maps of the CSS quasars 3C380 (z=0.692), 3C43 (z=1.459) have been obtained with MERLIN at 4995 MHz with 60 mas resolution. Although these sources show depolarization asymmetry, these quasars cannot be fit into unifying schemes due to the distortion of the extended components. 3C43 has a sharp bend probably due to beam-IGM interaction (Akujor *et al*, 1991). The polarized flux is constant along the jet (m=6 %) as one goes to the south, but is enhanced to 33 % along the sharp bend. There is also a enhancement of m near the eastern boundary where m=12 %. Regions of increased polarized flux are independent on positions of regions of higher radio brightness suggesting that different physical conditions are occurring along the jets. $\vec{H}$ lines are parallel to the large-scale jet direction, which is bent by 110° with the respect to the original direction. In regions where the polarization shows that compression occurs (Laing, 1980) the energy index is flat $1.5 < \gamma < 2.5$, indicating re-acceleration of particles. 3C380 has been discussed by Wilkinson *et al* (1991) and superluminal motion data suggests that the VLBI jet is inclined by 12° to the line-of-sight. The map show little evidence for Faraday rotation. Moreover, the percentage of polarization m increases from 12 % to 25 %

on the jet while the VLBI jet has integrated $m = 42\%$.

Hughes, Aller and Aller (1985) showed that in a current of optically thin plasma driven by a weak shock front, magnetic field lines $\vec{H}$ are aligned into the plane of the compression wave, changing the percentage of polarization of the unperturbed plasma. The expressions derived by the authors enable the compression of an initially randomly-orientated magnetic field to be estimated. They can be applied for a more general case of compression and the compression ratio κ between the shocked and unshocked plasma have been derived for the CSS sources: $\kappa = 0.66$. This value suggests that cold gas entrainment may also be occurring. If the jets are colliding with a cloud in a X-ray halo, a simple kinetic model (Bicknell, 1991) predicts the jet velocity to be $\log v_{jet} \approx -2.76 + \log M - 0.5 \log \kappa$, where M is the jet Mach number with respect to the environment ($M > 6$). An alternative hypothesis to explain 3C43 jet is jet bending by cooling flow winds. If I take typical cooling flow values for IGM temperature and densities and a model for ram-pressure equilibrium of bent mild–relativistic jets (O'Dea, 1985), jet velocities are constrained in the range $0.01 < v_{jet} < 0.3c$.

3 Conclusions

From the observed data, I conclude that increasing of the percentage of polarization in distorted CSS quasars can be explained by magnetic field compression due to strong jet/IGM interactions. At 50 mas resolution, Faraday rotation effects appears to be small. Observations of CSSs radiogalaxies suggest that they have similar compression ratios. However, the derived κ 's are not very high, even for the most heavily-disrupted jets. This suggests that medium may be efficient in confining lobes of galactic-sized radiogalaxies. Simple analytical models for jet/IGM dynamics give similar limits for jet velocity.

I would like to thank Dr. R.G. Conway for useful comments and suggestions and the Brazilian agency CAPES for my PhD scholarship at NRAL.

References

Akujor, C. E., Spencer, R. E. and Saikia, D. J.: 1991, *Astron. Astrophys.* **249**, 337.
Bicknell, G. V.: 1991, *Proc. Astron. Soc. Australia* **9**(1), 93.
Hughes, P. A., Aller, H. D. and Aller, M. F.: 1985, *Astrophys. J.* **298**, 301.
Laing, R. A.: 1980, *Mon. Not. Roy. Astr. Soc.* **193**, 439.
Lüdke, E. *et al*: 1993, in preparation.
O'Dea, C. P.: 1985, in *Physics of Energy Transport in Extragalactic Radio Sources*, ed. A. Bridle and J.A. Eilek, NRAO Workshop no. 9, p.64.
Wilkinson, P. N., Akujor, C. E., Cornwell, T. J. and Saikia, D. J.: 1991, *Mon. Not. Roy. Astr. Soc.* **248**, 86.

8.4 GHz VLA Observations Of The CfA Seyfert Sample

Marek J. Kukula and Alan Pedlar
N.R.A.L., Jodrell Bank, Macclesfield SK11 9DL, U.K.

S.W. Unger
R.G.O., Madingley Road, Cambridge CB3 0EZ, U.K.

and

S. Baum and C. O'Dea
S.T.Sc.I., 3700 San Martin Drive, Baltimore, MD 21218, U.S.A.

Abstract.
We present the results of an 8.4 GHz VLA survey of the CfA Seyferts. We find that the luminosity functions for Seyfert 1s and 2s are essentially identical, but that the type 2 objects are more likely to contain extended radio structures. This seems to be consistent with unified models for Seyferts, in which the differences between the two classes are largely due to orientation effects.

1   Introduction

Increasingly, the evidence from optical and infra–red wavebands suggests that the difference between Seyfert 1 (broad, permitted lines) and Seyfert 2 (narrow, forbidden lines) nuclei is due largely to orientation effects rather than intrinsic differences between the two classes (Osterbrock, 1992, and references therein). This has led to the development of 'unified scheme' models for Seyferts, in which an obscuring torus of material surrounding the active nucleus blocks our view of the central Broad Line Region in Seyfert 2s. Radio plasma is ejected along the axis of this torus either in the form of jets (as in the Seyfert galaxy Markarian 3 (Kukula *et al.*, 1993)) or as a string of discrete plasmons.

Since radio emission should not be affected by the torus, radio observations provide an excellent means for testing for similarities between the two classes of Seyfert; if the unified schemes are correct then we would expect the mean radio luminosity of Seyfert 1s and 2s to be roughly the same.

The results of previous radio surveys of Seyfert galaxies (Ulvestad and Wilson, 1984a,b; 1989) have been ambiguous and it seems that the samples on which they were based suffered from selection biasses and incompleteness. Clearly, a well–defined sample of Seyferts which is both unbiassed and complete is required.

2   The CfA Seyfert Sample

The CfA Seyfert Sample consists of 48 galaxies (26 Seyfert 1s and 22 Seyfert 2s) from the CfA Redshift Survey. The sample is magnitude–limited and, because the objects are selected purely on the basis of a characteristic AGN emission–line spectrum, it avoids the problems of bias and incompleteness which were inherrent in previous samples.

Astrophysics and Space Science **216**: 371–372, 1994.
© 1994 *Kluwer Academic Publishers.*

Edelson (1987) observed the CfA Seyferts at 1.5, 6 and 20 cm with the VLA and reported no statistical difference between the luminosities of Seyfert 1s and 2s. However, at a resolution of 15″ his images contained little information on the radio structure of the Seyfert nuclei and in many cases there was considerable confusion of radio emission from the disc of the host galaxy with that of the active nucleus.

To remedy this situation we have observed the CfA Seyferts at 3 cm (8.4 GHz) using the VLA in A– and C–Configurations. At this frequency our A–array maps have a resolution of 0.25″ and a sensitivity of 70 μJy – much higher than those achieved in previous surveys.

3   Preliminary Results

A comprehensive analysis of the radio data is currently under way, but already several results have emerged. Statistical work has been carried out using Survival Analysis techniques so that upper limits could be taken into account.

Detection Rate: The C–Configuration observation,s with their 3″ resolution, detected 90% of the sample at 3 cm, with similar detection rates for both Seyfert types. Meanwhile, the high–resolution A–Configuration maps have enabled us to resolve the detailed structure of the objects and 55 individual radio components have been identified.

Luminosity Functions: The small number objects with large luminosities makes the the high–luminosity end of the distributions unreliable, but for L$\leq$ $10^{22.5}$ W, the luminosity functions for type 1 and type 2 objects are consistent at the 93% probability level.

Structure: If the difference between Seyfert types is indeed due to orientation, so that type 1s are seen 'head on' and type 2s 'side on', then we would expect the extended radio structures of Seyfert 1s to suffer from a greater degree of foreshortening than those of Seyfert 2s. The A–Configuration maps are consistent with this: extended structure has been resolved in 41% of the Seyfert 2s, compared to only 12% of the Seyfert 1s.

Although some work still remains to be done, all the results so far indicate that Seyfert 1s and 2s belong to the same underlying population and that the observed differences between them are the result of their orientation relative to the line of sight.

References

Edelson, R. A.: 1987, *Astrophys. J.* **313**, 651.
Huchra, J., and Burg, R.: 1992, *Astrophys. J.* **393**, 90.
Kukula, M. J., *et al.*: 1993, *Mon. Not. Roy. Astr. Soc.*, **264**, 893.
Osterbrock, D. E.: 1992, *Astrophys. J.* **404**, 551.
Ulvestad, J. S. and Wilson, A. S.: 1984a, *Astrophys. J.* **278**, 544.
Ulvestad, J. S. and Wilson, A. S.: 1984b, *Astrophys. J.* **285**, 439.
Ulvestad, J. S. and Wilson, A. S.: 1989, *Astrophys. J.* **343**, 659.

Relativistic Jet Simulations

Mark K. Bowman
N.R.A.L., University of Manchester, Jodrell Bank, Macclesfield SK11 9DL, U.K.

Observations on parsec scales suggest that the flow in extra-galactic radio sources has both bulk and thermal Lorentz factors which are significantly greater than one. Using a simple transformation of variable, Chiu (1973) demonstrated that the dynamical equations for a relativistic, nondegenerate fluid (Synge (1957)), can be expressed in a form which is mathematically identical to the Newtonian conservation equations.

The axisymmetric jet presented here was simulated using Godunov's method to solve the transformed equations in the steady state case, (Fig. 1). Initially, the jet is in pressure equilibrium with its surrounding and all the flow is parallel. At this point the fluid has a 4-velocity component in the axial direction of 10.6 and a temperature of $\sqrt{10} \times 10^{13} K$ which or a Mach number of 15. The jet expands into an atmosphere with a pressure that decays as a power law of index -2, but which becomes constant at a distance of 40 intial jet radii from the starting point. Due to the inertia of the expansion, the jet continues to expand and adiabatically cool into the region of constant external pressure. The subsequent over-expansion results in recollimation of sufficient strength for the compression characteristics to cross and a standing oblique shock to be formed. The shock is reflected by the jet axis. As this reflected shock reaches the outer boundary, the jet is once more overpressured and another expansion fan forms. The jet continues to oscillate radially with weakening recollimation shocks.

Figure 2 depicts the Lorentz factor along a streamline for the jet of Fig. 1. A large rise occurs across the initial expansion fan. In the temperature range of this example the sound speed tends towards a maximum of $c/\sqrt{3}$. Consequently, such 'hot' flows can be supersonic but still have thermal Lorentz factors that are much greater than seen in the bulk motion. During adiabatic cooling the fluid tends to reduce the difference in these Lorentz factors leading to large bulk accelerations.

The final wavelength of the radial oscillations is a function of the intial Mach number, but independent of starting temperature and hence adiabatic index. Also, in initially cooler jets the first recollimation shock is weaker than in the simulation discussed here. However, the rate of shock strength decay, from one shock to the next, is slower due to the reduced dissipation.

References

Synge, J. L.: 1957, *The relativistic gas*, North-Holland Publishing Company, Amsterdam.
Chiu, H. H.: 1973, *The Phys. of Fluids* 16, 825.

Astrophysics and Space Science **216**: 373–374, 1994.
© 1994 *Kluwer Academic Publishers*.

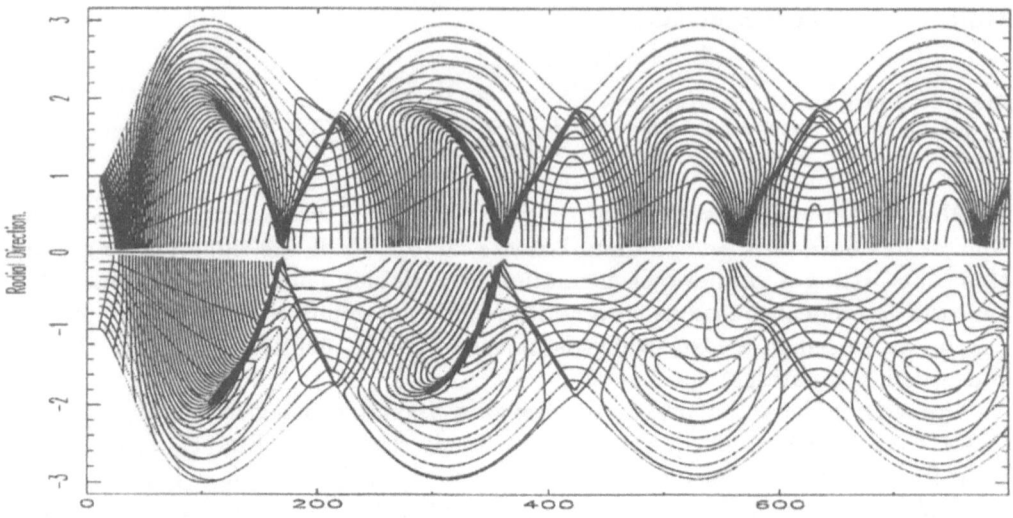

Figure 1. Plot of a jet starting from the left with a Mach number of 15 and with a temperature of $\sqrt{10} \times 10^{13} K$. The plot shows streamlines with pressure contours in the top half and temperature contours in the lower half. Both axis are in units of the initial jet radius although the Z–axis is greatly compressed for ease of display. The maximum pressure is 1.75×10^{-6}. There are 50 contours on the upper half of the plot and the ratio between successive levels is 1.18. The maximum value of γ is 1.993. There are 50 contours on the lower half of the plot and the ratio between successive levels is 1.003.

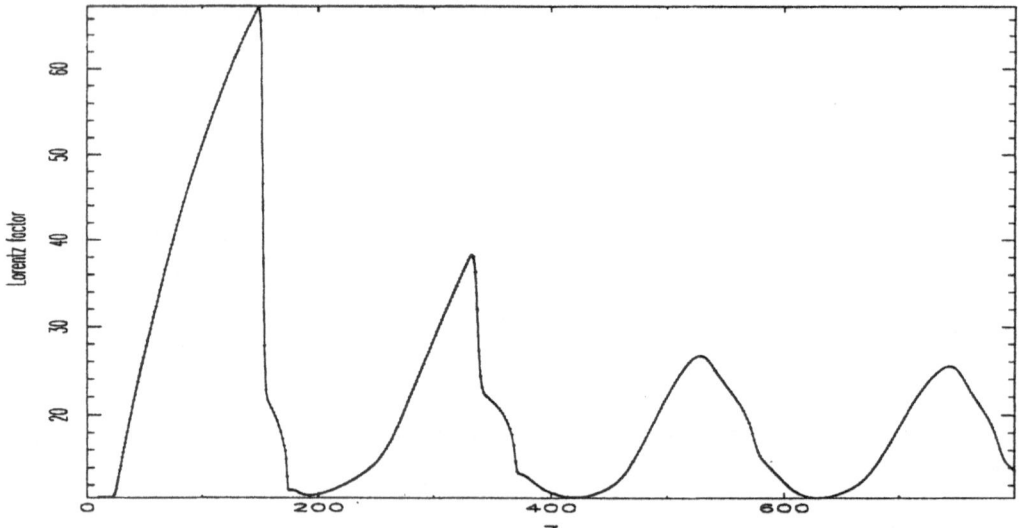

Figure 2. Bulk Lorentz factor along a streamline for a jet with initial Mach number of 15 and temperature of $\sqrt{10} \times 10^{13} K$. The Z-axis is in units of the initial jet radius. The streamline shown started 1/10 of the way out from the centre of the jet.

Active Galactic Nuclei Flow Velocities and the Highest Energy Cosmic Rays

J. J. Quenby and B. Drolias
Astrophysics Group, Blackett Laboratory,
Imperial College, London SW7 2BZ, U.K.

and

R. Lieu
Centre for EUV Astrophysics, University of California,
Berkeley, CA 94720, USA

Abstract.
    The dependence of the maximum energy cosmic rays can reach via diffusive shock acceleration in AGN jets on flow speeds is discussed. It is shown that in highly inclined termination shocks where the speed of the de Hoffman-Teller frame is crucial, a good independent knowledge of the jet speed is required to properly assess the extent of the cosmic ray spectrum.

1   Introduction.

The termination shocks of AGN relativistic jets provide the most plausible location for the acceleration of the highest energy cosmic rays in our local supercluster. Diffusive shock acceleration, involving first order Fermi acceleration across the shock interface is the mechanism at work. To estimate the possible maximum energy which can be arrived at by this process, the flow speeds, magnetic field configuration and extent and the effective particle scattering mean free path must all be known. While synchrotron measurements allow estimates of the field strength, arriving at the flow speeds is more difficult because of relativistic beaming effects and confusion between observation of phase and group speeds. The need for better information on the flow speeds is emphasised in this paper. It is perhaps too optimistic to regard measurements of the upper cutoff of the cosmic ray spectrum as the way to establish the speed of the jet, however.

Standard analytical theory for diffusive shock acceleration for non relativistic, upstream and downstream flow speeds V_1 and V_2 for particle velocity $v \approx c$ yields a spectrum $\frac{dn}{dp} \propto p^{-\alpha}$, where $\alpha = \frac{r+2}{r-1}$ and $r = \frac{V_1}{V_2}$ and an acceleration time constant,

$$\tau(p) = \left(\frac{3}{V_1 - V_2} \right) \left(\frac{K_1}{V_1} + \frac{K_2}{V_2} \right) \tag{1}$$

where the diffusion coefficient, $K = \frac{\lambda v}{3} \cos^2 \psi$, and ψ is the field shock normal angle.

Astrophysics and Space Science **216**: 375–378, 1994.
© 1994 *Kluwer Academic Publishers.*

2 Relativistic 'Speed Up' of Diffusive Shock Acceleration

We describe here results of numerical simulations of diffusive shock acceleration under the test particle approximation. While a full treatment should include the non-linear effects of the back reaction of the cosmic ray gas on the MHD flow, it is believed that the basic results obtained are correct at the top end of the spectrum where escape is important and there is less relativistic gas energy available to modify the low energy plasma. The assumptions in the model, which is a guiding centre approach, are that the scattering mean free path $\lambda \propto r_g$ (gyroradius) and we take $\lambda = 41r_g$ based on interplanetary experience, the scattering is isotropic and occurs with a probability $P_{(z)} \propto \exp\left(-\frac{z}{\lambda(\cos\theta)}\right)$, for distance z along $\underline{B}$ with pitch angle θ, cross field diffusion is neglected which for hard sphere scattering amounts to the condition $\left(\frac{\lambda}{r_g}\right)^2 \cot^2\psi > 1$, or $\psi \leq 88°$ and that the magnetic moment invariant, $\frac{p\sin^2\theta}{B}$ is conserved across the shock in the $\underline{E} \equiv 0$ or de Hoffmann-Teller frame, obtained by adding a velocity along the shock to make $\underline{V} \parallel \underline{B}$.

As shown by Quenby and Lieu (1989), for parallel shocks both a spectral flattening and a relativistic speed-up is obtained (Kirk and Schneider, 1987 first noticed the spectral flattening). For $V_1 = 0.96$ c, $V_2 = 0.32$ c or a compression ratio of 3, the mean momentum increment per cycle of shock scattering, $\frac{\Delta p}{p} = 10.5$ due to a γ^2 enhancement related to the relativistic difference $V_1 - V_2$. The spectral index flattened to $\alpha = 1.2$ from the non-relativistic value of 2.5 and the acceleration time of equation (1) was reduced by a factor 13.5.

Modifying the programme to accommodate inclined shocks now demonstrates that it is the upstream flow speed seen in the de Hoffman-Teller frame which is crucial. For a compression ratio of 4, corresponding to the strong shock, test particle limit, figure 1 shows the ratio of computed to predicted (equation 1) acceleration time constant as a function of the de Hoffman-Teller frame upstream flow speed. Values of $\psi = 70°$ and $\psi = 80°$ were chosen, so as not to get too close to the region where cross-field diffusion is important. λ remains independent of distance in these simulations, the initial particle $\gamma = 10$, particles are injected 100 λ upstream and removed 150 λ downstream after multiple shock crossings. We see that a near factor 10 speed up is still achieved. Figure 2 shows the spectral flattening obtained for the same simulations.

3 The Crucial Parameters Determining The Extent Of Cosmic Ray Acceleration.

We suppose the top end of the cosmic ray spectrum to arise from one or more AGN's in our local supercluster, where propagation is not 'killed' by the microwave background, where the jets terminate in a hot spot ~ 10 kpc in radius with fields

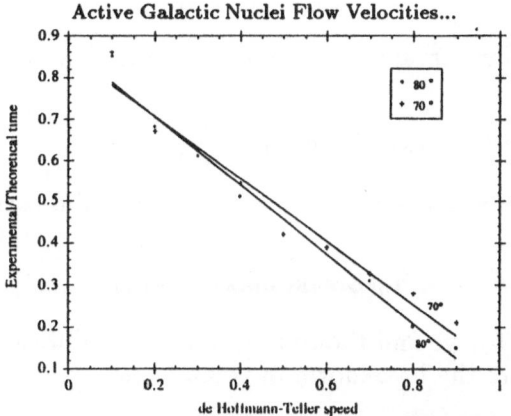

Fig. 1.   Ratio of experimental to theoretical time to accelerate particle ξ versus upstream flow velocity in the de Hoffmann-Teller frame (in units of c).

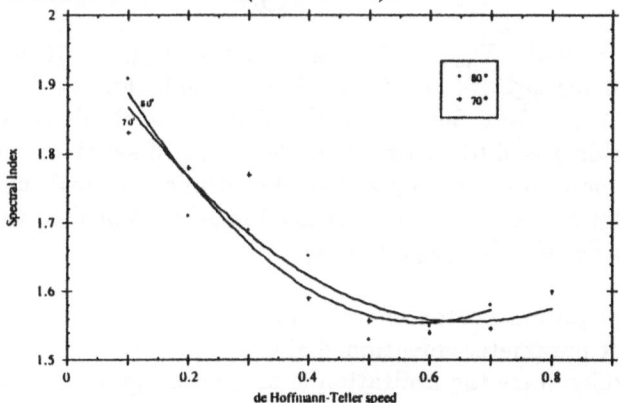

Fig. 2.   Differential number spectral index versus upstream flow velocity in the de Hoffmann-Teller frame (in units of c).

$\sim 4\ 10^{-4}$ gauss in strength, the possibility of highly inclined shocks exists and $V_1 \sim 0.1c$.

To be specific, let R = radius of trapping/acceleration region = escape distance along the shortest flux tube to the outside. Then the escape time is:

$$\tau(\text{esc}) \simeq \frac{R^2}{\lambda c} \tag{2}$$

where $\lambda = N r_c$ and $N \geq 10$. Equation (1) may be written as:

$$\tau(\text{acc}) = \frac{F\lambda}{fc} \tag{3}$$

where $F = g\cos^2\psi$ is a factor less than unity and g represents the speed up factor demonstrated numerically (section 2). For AGN flows a fraction of c, $f \sim 0.01$.

Equating (2) and (3) we obtain a maximum possible r_g, when acceleration time $\equiv$ escape time of:

$$r_g \,(\text{diff, max}) = \frac{R}{N} \left(\frac{f}{F}\right)^{1/2} \qquad (4)$$

Independently, we define a maximum attainable gyroradius corresponding to breakdown in trapping,

$$r_g(\text{Break, max}) = \frac{R}{X} \qquad (5)$$

Comparing (4) and (5), we find the condition that the diffusive shock acceleration limit is reached before the breakdown in trapping is:

$$X_{\min} < N \left(\frac{F}{f}\right)^{1/2} \qquad (6)$$

For $g \sim 0.1$, $f \sim 0.01$, $X_{\min} < 3N \cos \psi$. Since $X_{\min} \leq 10$ is reasonable and $N \geq 10$, we find for $\psi \leq 72°$, the factor F is crucial. Factor F depends linearly on the relativistic speed-up factor. So provided the algebraic combination of flow speeds in (1) yields $f \sim 0.01$ or less, it is the speed of shock acceleration in AGN hot spots which determine the top end of the cosmic ray spectrum. Furthermore, the maximum rigidity or gyroradius attained is proportional to this speed up factor, g, to the power of $g^{-1/2}$ (equation 4).

In practice maximum proton energies of 10^{20} eV are expected to be attained with the hot spot parameters mentioned above, and correspondingly higher heavy nuclei total energies since the limitation is magnetic rigidity dependent.

4   Conclusion.

AGN hot spots corresponding to the termination jet shocks are likely sites of cosmic ray acceleration up to magnetic rigidity of 10^{20} volt, provided flow speeds of about 0.1 c are present and the field configuration corresponds to a near-perpendicular shock. The mechanism for acceleration depends on the relativistic speed-up of the diffusive shock process as seen in the $E \equiv 0$ or de Hoffman-Teller frame.

5   References.

Kirk, J. G. and Schneider, P.: 1987, *Astrophys. J.* **322**,256.
Quenby, J. J. and Lieu, R.: 1989, *Nature* **342**, 654.

Hidden Broad Line Regions and Anisotropy in AGN

David J. Axon*
Space Science Division of ESA, Space Telescope Science Institute, 3700 San Martin Drive, Baltimore MD, USA

and

J. H. Hough, S. Young and M. Inglis
Division of Physical Sciences, University of Hertford, Hatfield, Herts AL10 9AB, UK

Abstract. We discuss the recent advances made in the search for hidden broad line regions in Seyfert and Narrow Line Radio Galaxies (NLRG) using both spectropolarimetry and infrared spectroscopy/spectropolarimetry. Two important results which support the Grand Unification Theory are presented. In the first of these we report that the famous Seyfert 1 NGC 4151 has a scattered component to its broad line region, and is most likely an object where we view obliquely into its occulting torus. In the second, we show high signal to noise observations of scattered broad lines in the NLRG of 3C 234.

Key words: Active Galactic Nuclei; polarimetry.

1 Introduction

While the concept of anisotropy in radio loud objects has been firmly established in our minds for some time (Scheuer and Readhead, 1979), it is only comparatively recently that we have started to understand that the radiation fields of the radio quiet objects, such as Seyferts, are themselves anisotropic. Two key observational discoveries have prompted this awareness. Firstly, the detection of a *"hidden BLR"* (Antonucci and Miller, 1985) in the polarized flux of the Seyfert 2 NGC 1068, showed that in reality it was a Seyfert 1 in which our line-of-sight to the BLR was obstructed by a dusty torus. The BLR can be seen in polarized light because it is reflected to us by particles situated above and below this torus. In at least some cases then, Seyfert 1 and Seyfert 2's are physically the same kind of object viewed at a different orientation. Secondly, optical spectroscopy of Seyfert galaxies (both type 1 and 2) with linear radio sources revealed a new type of large scale $\sim 10-40$ kpc extra-nuclear emission line region, the ENLR, which is closely aligned with the radio structure (Unger *et al.*, 1987) Though the gas in the ENLR appeared kinematically quiescent, its emission line ratios corresponded to those of a diffuse gas photoionized by an AGN continuum (i.e similar to the NLR). Simple order-of-magnitude calculations showed that the number of ionizing photons escaping into the ENLR must be as high as 10 or 15 times that expected from the analysis of the directly observed nuclear emission lines, the so called *Ionizing Photon Deficit*. There could be only one reasonable explanation of this result: the radiation field of the AGN must be collimated along the radio axis. Since no concrete evidence

---

* On Leave from Nuffield Radio Astronomy Laboratory, University of Manchester, Jodrell Bank, Macclesfield, Cheshire, England

Astrophysics and Space Science **216**: 379–398, 1994.
© 1994 *Kluwer Academic Publishers*.

for a relativisticaly beamed continuum could be found, the anisotropy must be due to *selective shadowing*: the photon beam in each of these nuclei was escaping along the pole of a torus, just as Antonucci and Miller (1985) had envisaged to explain their NGC 1068 result. Almost immediately it became clear that the extended emission line regions and UV continua in high redshift radio sources are even more tightly collimated along their radio jets, and require even larger anisotropic radiation fields. (e.g. McCarthy *et al.*, 1987; Chambers *et al.*, 1987).

People started to think of *Grand Unification* (GU) : perhaps not only could the radio properties of the core and lobe dominated sources be explained as a single population by supposing that in the former the radio jets pointed at us, while in the latter they were directed in the plane of the sky (e.g. Orr and Browne, 1982), but perhaps this could be extended to include the balance between the observed numbers of narrow and broad emission line sources (Bartel, 1989).

This general topic of anisotropy and GU in AGN has been covered at length in a number of excellent recent reviews (Antonucci, 1993; Fosbury, 1994; Tadhunter, 1994; McCarthy, 1993). Our intention here is to concentrate on just one aspect of this problem, namely the presence of hidden BLR in narrow line galaxies of all types. We will start by summarizing the current state of the search for these. Then we will move on to discuss the statistical characteristics of the known objects, and describe some of the outstanding issues and how observations can resolve them. Along the way we will present new results which show that the famous Seyfert 1 NGC 4151 has a significant scattered component in its visible BLR; a result which provides support for the GU picture. Finally we will turn our attention to radio galaxies, and discuss their optical polarization structure and provide improved evidence that hidden BLR are also present in NLRG.

2   Narrow Line Galaxies with Hidden BLR

Since the original discovery by Antonucci and Miller (1985), there has been a great deal of spectropolarimetry of other Seyfert 2's, largely by the Lick team and by this collaboration, aimed at establishing the frequency of hidden BLR. A completely independent approach to the problem pursued by many groups, has been to try and observe the BLR directly by looking through the tori of Seyfert 2's using IR spectroscopy. Both these techniques have led to the detection of broad lines in a number of the narrow line galaxies. Table 1 gives a list of the known objects and summarizes the relevant literature. The table is arranged in the following way. The object name is given in Column 1. Column 2 identifies whether a broad line region has been detected using optical spectropolarimetry, with the associated references cited in Column 3. The nomenclature used in Column 2 is that a letter Y implies a confirmed detection, while a ? indicates a claimed but slightly less firm result, and a blank means unknown. A similar scheme is adopted for the IR spectroscopy. Column 4 has an entry if broad lines have been detected using IR spectroscopy/spectropolarimetry , with the reference identified in Column 5.

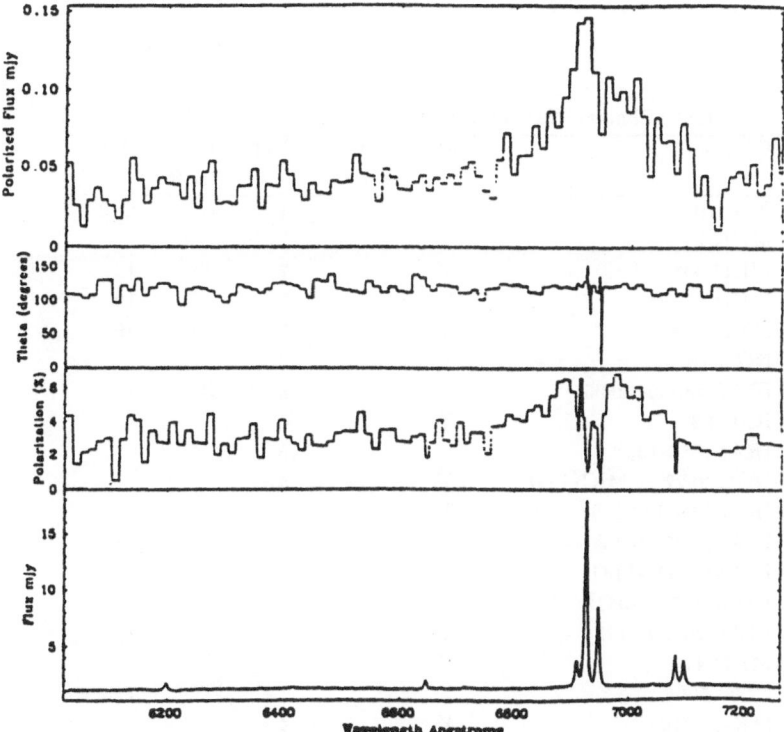

Fig. 1.   The hidden BLR of IRAS 110548-131. One of the most spectacular hidden BLR's .
From top to bottom the panels show the polarized flux, position angle of polarization, degree of
polarization and the total flux. The flux units are mJy.

There are three possible codes in Column 4; an N means that only narrow lines
are seen, while a T followed by a ? indicates that broad line have definitely been
detected but it is not clear if these are reflected or transmitted, and finally an R
indicates that the broad lines are seen in reflection. Again a blank means unknown.

Many of the objects in Table 1 are IRAS Seyferts with hot dust. Figures 1
to 3 show examples from this new crop of IRAS selected hidden BLR galaxies.
The first of these, IRAS 110548-131, is a really dramatic example in which the
scattered broad lines are extremely broad and strong in the polarized flux spectrum. The other two cases, IRAS 22017+0319 and IRAS 20460+1925, have much
narrower and asymmetric broad scattered lines but are still clear cut cases. Other
examples of hidden BLR will be shown in various places in the paper to illustrate
the discussion.

TABLE I
List of known Narrow Line Galaxies with Hidden BLR

Object	Polarized Optical BLR	Ref.	IR BLR	Ref.
NGC1068	Y	1	N	8, 9
MKN 3	Y	2		
MKN 348	Y	2	T?	10
MKN 463E	Y	2	T?	10
MKN 533	Y	2	T?	10
NGC 7674	?	2		
IRAS 23060+0505	Y	3	R	11, 3
IC 5063	Y	4		
IRAS1958-183	Y	5		
NGC 2622 = MKN1218	Y	5		
IRAS13349+2438	?	6		
IRAS1445-1447	?		T?	11
IRAS110548-1131	Y	7		
IRAS 22017+0319	Y	3		
IRAS 20460+1925	Y	3	T?	12
MKN 477	Y	4	T?	13
NGC 5066	?		T?	14
MKN 1210	Y	4		
MKN 7212	Y	4		
Was 49B	Y	4		
MCG -05.23.16	?		T?	14

References

1. Antonucci and Miller (1985)
2. Miller and Goodrich (1990)
3. This paper
4. Inglis et al. (1993)
5. Goodrich (1989)
6. Wills et al. (1992)
7. Young et al. (1993)
8. Ward et al. (1994)
9. Depoy (1987)
10. Ruiz et al. (1994)
11. Hines (1991)
12. Nakajima et al. (1991)
13. Rix et al. (1990)
14. Blanco et al. (1990)

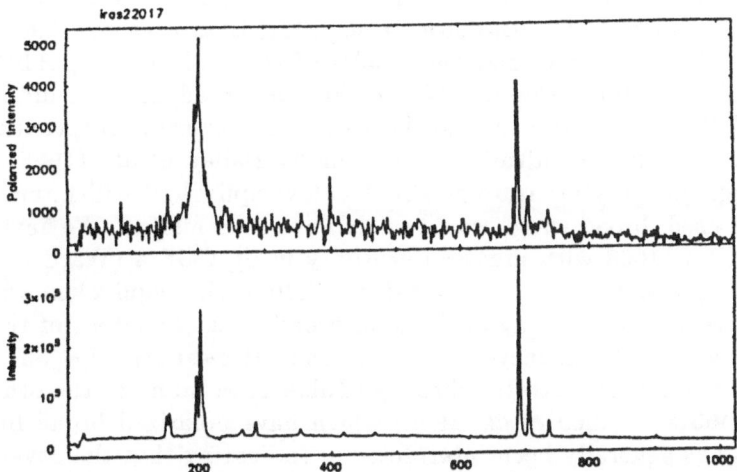

Fig. 2.   Comparison between the polarized and total flux profiles of IRAS 22017+0319. The top panel shows the polarized flux and the bottom panel the total flux. Both spectra are in units of counts, and the spectral region encompasses Hα on the left with [OIII] λλ5007, 4959 and Hβ on the right.

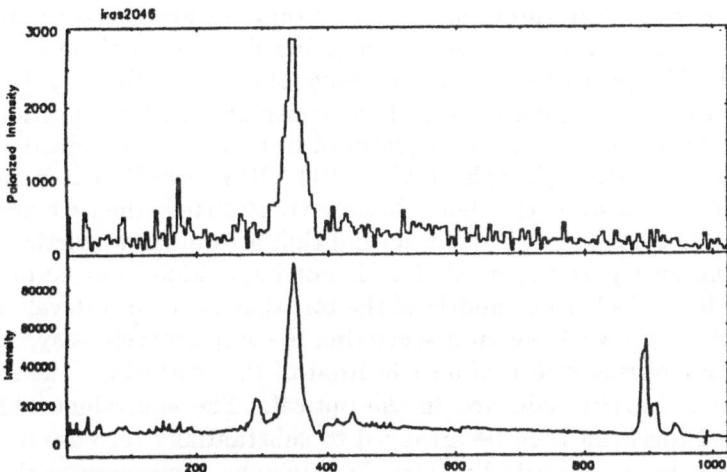

Fig. 3.   Comparison between the polarized and total flux profiles of IRAS 20460+1925.

2.1   Infrared Observations: The covering factor of the obscuring torus

With the advent of sensitive IR array spectrometers, like CGS 4 on UKIRT one of the most active areas of research has been to try and directly detect broad lines in the near IR (since $A_K/A_V \sim 0.1$) (Hines, 1991; Nakajima *et al.*, 1991a, b; Ruiz *et al.*, 1994; Rix *et al.*, 1990; Carleton *et al.*, 1984; Blanco *et al.*, 1990). The last 2 columns of Table 1 summarize the results of the bulk of the published work. In the case of NGC 1068, Depoy (1987) looked for broad Brα but did not find it and deduced that $A_V \sim 100$mags to the BLR. This figure is substantially larger than the A_V of ~ 25 magnitudes argued for by Bailey *et al.*, (1986). Depoy's limit is in fact not very accurate as Brα has low equivalent width, and with the resolution he used, broad wings would easily have been missed. We have therefore relooked at NGC 1068 with greater sensitivity using CGS 4 (Ward *et al.*, 1991) at Paα and Paβ, both of which have substantially higher equivalent widths than Brα. Again we do not find any evidence of broad wings on either of the Paschen lines, implying that the extinction of the torus is at least A_V of ~ 35. Broad IR lines have however been detected directly (Table 1) in many of the other objects (e.g. MKN 463E—Blanco *et al.*, 1990) which have polarized broad lines in the visible. This has generally been interpreted as indicating that the covering factor of the tori in these galaxies is substanially less than that in NGC 1068. Here we want to illustrate a potential flaw with this conclusion using IRAS 23060+0506 as an example. Hough *et al* (1990) showed that this object was heavily polarized and suggested it was a buried QSO. The optical spectropolarimetry shown in Figures 4 and 5 supports their interpretation as it shows that there is a hidden BLR. As in MKN 463E, IR spectroscopy reveals broad Paα (Hines, 1991). Taken at face value, one would interpret this to mean that we are therefore able to see into the torus Unfortunately, this neat conclusion does not look so convincing in the light of the IR spectropolarimetry we have obtained with CGS 4 (Figures 6 and 7). Both the level of polarization of the continuum and the postion angle of polarization (when corrected to the equatorial frame) are in excellent agreement with the broad Paα data (Hough *et al.*, 1991). However, Paα is polarized the same as the continuum. If the line is largely transmitted, then we now have to suppose that this polarization is dichroic, and that a toroidal magnetic field in the torus aligns the dust particles in it! This is not impossible , but certainly starts to extend the limited physical models of the tori that have been developed so far. Explaining this result with electron scattering is comparatively easy, all one has to do is place a covering screen of dust in front of the scattering zone so that the scattered light is heavily reddened in the optical. The equivalent width of the broad scattered lines can then be arranged to substantially increase in the IR as this screen then becomes optically thin. In principle, a measure of the covering factor can be obtained by comparing the IR Pa ratios with the case B values and combining these with the ratios of the scattered Balmer lines and fluxes.

Faced with this new twist, it looks like we need to determine whether the broad IR lines seen in most of the objects in Table 1 are scattered or transmitted by using IR spectropolarimetry.

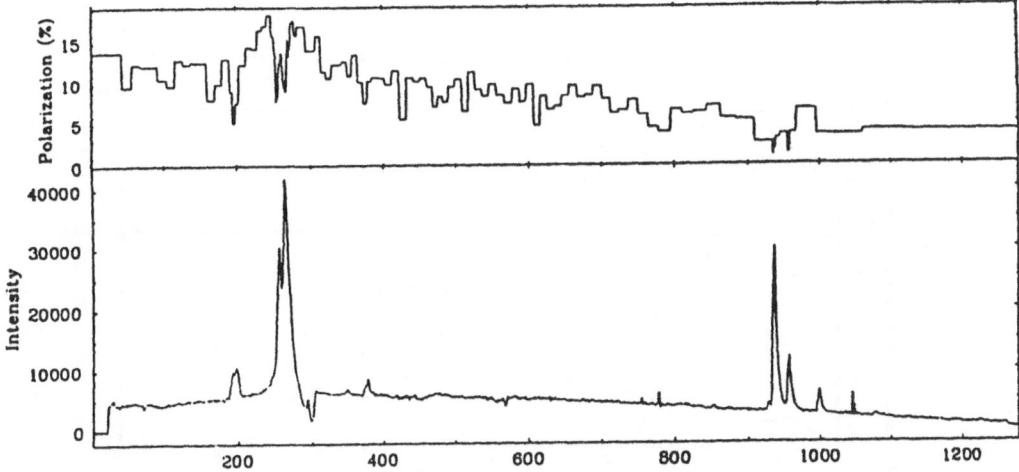

Fig. 4.   Optical spectropolarimetry of IRAS 23060+0506. Details as in Figure 2.

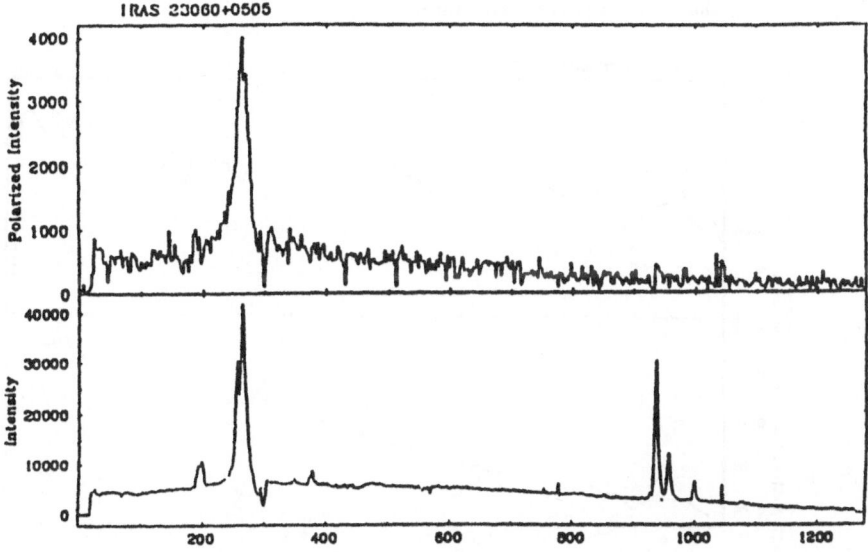

Fig. 5.   A comparison between the total and polarized flux of IRAS 23060+0506 revealing the hidden BLR.

3   General Properties of Hidden BLR AGN

There are now sufficient of these galaxies for us now to recognize some general characteristics:-

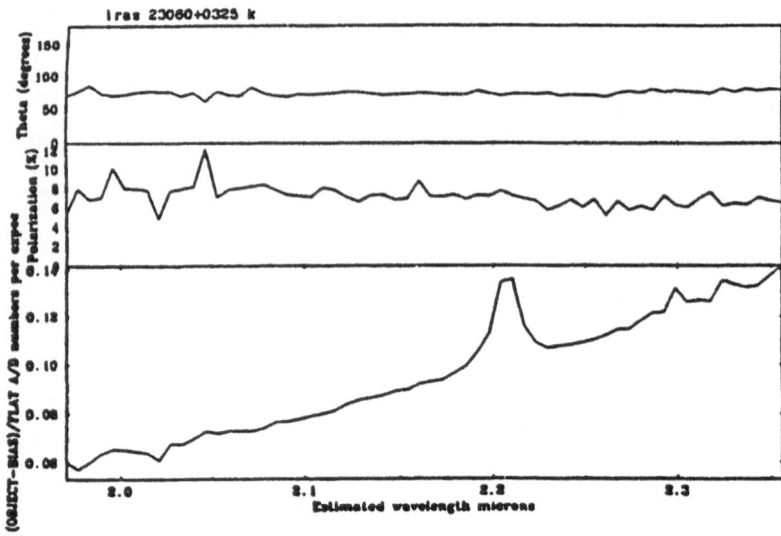

Fig. 6.   Paα spectropolarimetry of IRAS 23060+0506, obtained with CGS 4. Shown from top to bottom are the position angle of polarization, degree of polarization, and total intensity. The position angles shown are relative to the instrumental frame and not equatorial North. Note that the line is polarized the same as is the continuum.

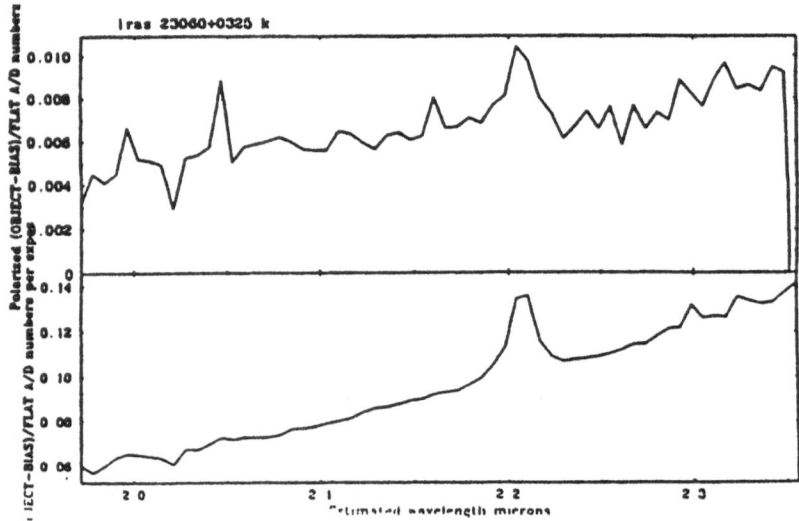

Fig. 7.   A comparison between the total and polarized flux in Paα for IRAS 23060+0506

1. **Warmish IRAS colours.** Figure 8 shows a colour-colour plot of the $25-60\mu$ and $60-100\mu$ spectral indices for a number of Seyfert galaxies (type 1 and type 2). Those galaxies with hidden broad line regions are identified; it can be seen that they have generally flatter spectra. Part of the hot dust is associated with the immediate environment of torus. These galaxies may also have more

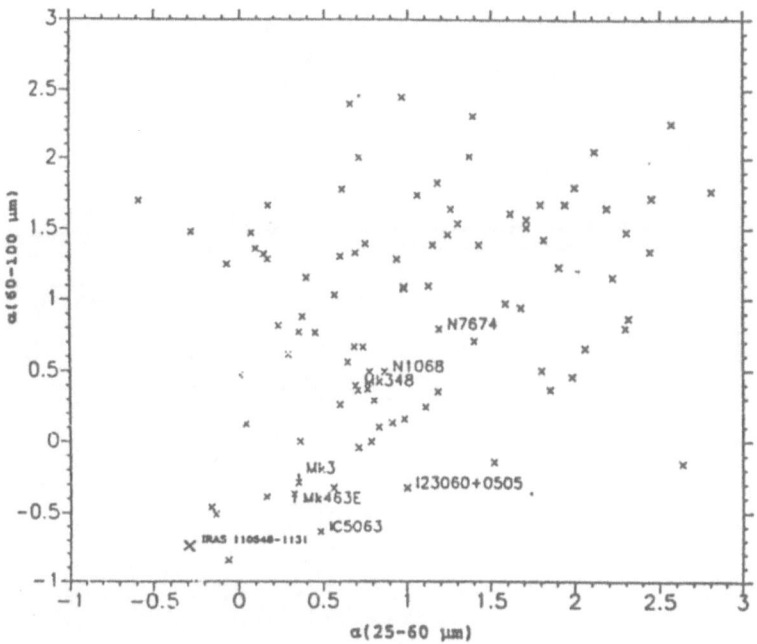

Fig. 8. Plot of $60 - 100\mu$ m versus $25 - 60\mu$ m spectral indices for Seyfert galaxies. A number of the Seyfert 2 galaxies which show scattered broad lines are marked. Note that they tend to cluster to the bottom left, indicating that their dust is on average warmer than the typical Seyfert.

luminous nuclei making the detection of the scattered flux easier (Inglis *et al.*, 1993).

2. **Relatively high radio luminosity.** The radio properties of Seyfert 1 and 2 type galaxies have sometimes been advocated as an aspect-independent demonstration of intrinsic differences between the two classes of AGNs. Figure 9 shows the well known correlation between radio luminosity P_{1415} GHz and [OIII] luminosity for Seyferts, with the known hidden BLR objects identified. The objects with scattered BLRs tend to have high radio luminosities and broader lines.

3. **Concave Wavelength Dependence Curves.** All the objects with scattered BLRs show a steep rise in polarization to longer wavelengths as well as to the UV (Figure 10). Usually the gradient towards the IR is much steeper than towards the UV; in the most extreme case the polarization at $K(2.2\mu)$ rockets to > 10 compared to the optical. If one builds models of the wavelength dependence of polarization and polarized flux then, in many cases, it proves difficult to match this result with simple dilution models (Young, 1994). Introducing appreciable extinction of the scattered flux as well as of the direct

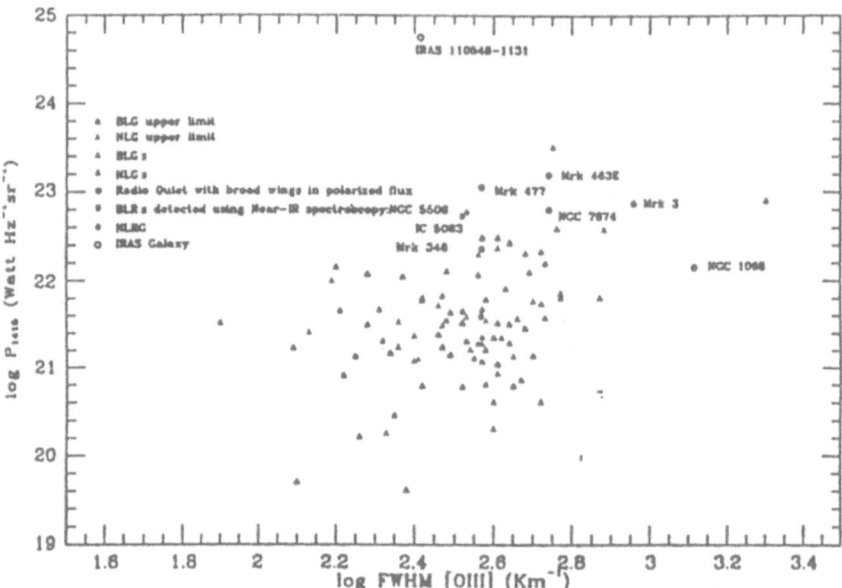

Fig. 9.   Radio power correlation. A plot of radio luminosity P_{1415} versus [OIII] line width. The objects with scattered BLRs tend to have high radio luminosities.

flux from BLR one can do much better, particularly in the IR (Hough *et al.*, 1991). If this interpretation is really correct, then there must surely be Seyfert 2's in which the obscuration of the scattered BLR is so high that we cannot see it at all in the UV or the optical, but will see it in the IR due to the reduced extinction. Some of the results on IRAS selected Seyferts certainly favour this picture, as do the existing polarimetry data on Cen A and other nearby radio galaxies which we describe below. IR spectropolarimetry will allow this idea to be tested.

4. **ENLR and Linear Radio Sources.** All the galaxies with scattered broad lines have both linear radio sources and ENLR. However, not all ENLR galaxies appear to have hidden BLR. Since the number of Seyferts with linear radio sources is not that common (5% of known Seyferts—Kukula, 1993), and ENLR are conspicuous by their absence in sources with compact radio structures, this starts immediately to raise statistical problems for GU. This intimate correlation with the ENLR raises an interesting question. Why do some ENLR objects appear not to have a hidden BLR? One could imagine, for example, that there is insufficient scattering column to give us a detectable reflected component against the dilution background starlight, especially in low luminosity objects. Another possibility is that electrons are the primary

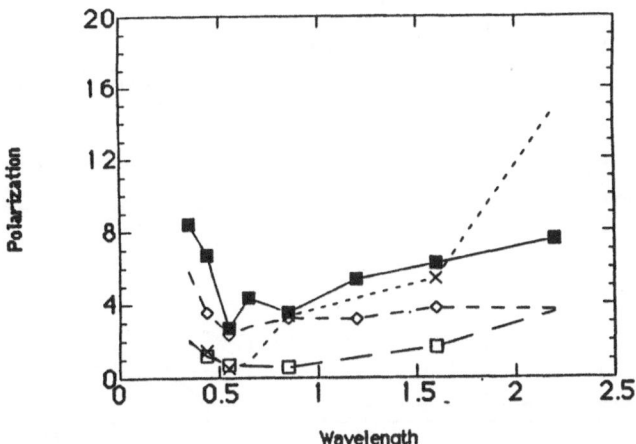

Fig. 10. The wavelength dependence of polarization of Seyferts with scattered BLRs illustrating the steep rise of polarization in the IR as well as towards the UV.

mirrors and the electron temperature in the scattering zone is so high that the lines are just Doppler broadened so that they cannot be distinguished from the continuum. Finally that the scattered light itself might be heavily reddened and dimmed by foreground material. Certainly if one tries to find other physical support for either of the first two possibilities it is hard to find. For example the objects which have the fastest shocks, as indicated by their NLR profiles, might be expected to have the hotter scattering zones, yet they are the ones in which we see the scattered lines (eg NGC 7674). Indeed, on the contrary, the correlation with radio power and [OIII] line width leads one to suspect that the radio ejecta have a role to play in providing a scattering medium.

3.1 THE SCATTERED BLR IN NGC 4151

An important question is whether, as the Unified Model predicts, Seyfert 1's also have tori. As we described above, one general characteristic of the known Seyfert 2's with hidden BLR nuclei is that they all have ENLR. Seyfert 1 galaxies also have ENLR, and following our introductory remarks, by inference there must be an obscuring torus. Moreover, on statistical grounds, the wide opening angles of the ENLR indicate that a galaxy can appear as a Seyfert 1 even if it is viewed at a large angle to its radio axis. At the extreme of this viewing range, when we only just see into the torus, the scattering angles are sufficiently large that one may also detect a scattered BLR component.

NGC 4151 is the nearest and best studied Seyfert type 1. There are several existing pieces of evidence that argue that in fact our viewing angle into its torus

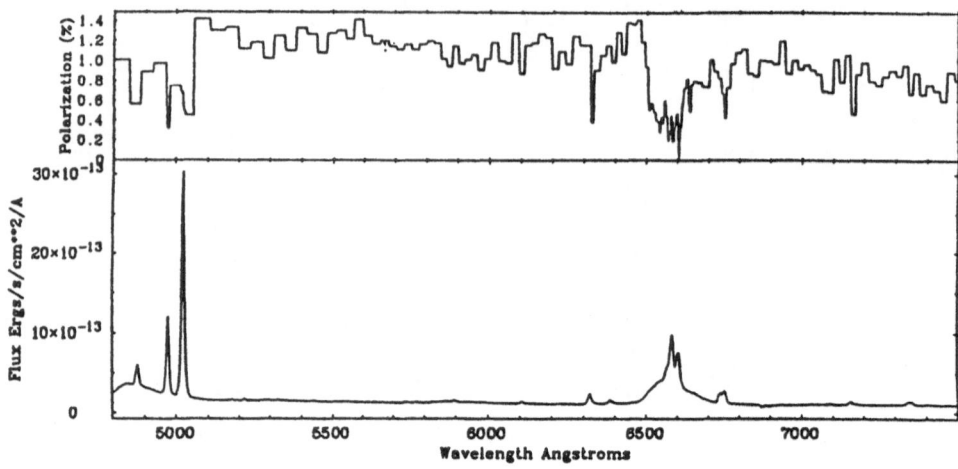

Fig. 11.   Spectropolarimetry of NGC 4151. The polarization has been binned to an accuracy of
0.1%.

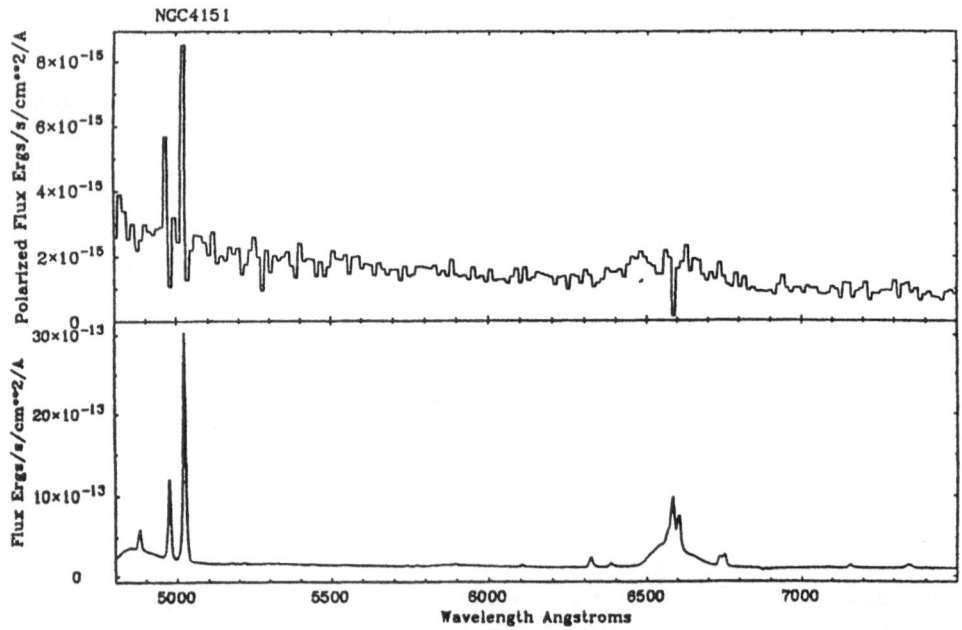

Fig. 12.   The total flux and polarized flux of NGC 4151, showing its *hidden BLR*.

is oblique. Foremost amongst these is the geometry of its ENLR (Penston *et al.*, 1990; Robinson *et al.*, 1994), which takes the form of a narrow linear structure, rather than a cone, and which is displaced from the radio axis by $\sim 15^0$ as seen in projection onto the plane of the sky. Such a morphology arises naturally if the axis of the radiation cone is simultaneously tilted to our line of sight and out of the plane of the disk of NGC 4151, so that it intersects the disk at grazing incidence (Pedlar *et al.*, 1993; Robinson *et al.*, 1994). Figures 11 and 12 show some exiting new spectropolarimetry of its nucleus we recently obtained using the 4.5m William Herschel Telescope. Several things are apparent in Figure 12 which compares the polarized and total flux. First the narrow lines have essentially disappeared, in addition the polarized flux rises more steeply to the blue than the total flux. But the most significant feature is that the broad lines are strongly present in the polarized flux. Here is solid evidence for scattered broad lines in a bonafide Seyfert 1.

4 Polarimetry of Radio Galaxies

As we have seen, the presence of a hidden BLR in Seyfert galaxies is strongly correlated with radio power. It seems natural therefore to try and establish if this trend continues for the narrow line radio galaxies. One would then suppose that the lobe dominated NLRG are objects, such as in NGC1068, seen through an edge on torus. Many intermediate and high redshift radio galaxies showing the alignment effect are known to be highly polarized at a position angle perpendicular to their radio axes (Jannuzzi and Elston, 1991; Tadhunter *et al.*, 1992). Imaging polarimetry of 3C 368 and 3C 405 suggest they have the bipolar polarization structure characteristic of scattering from a point source (Tadhunter *et al.*, 1990; Scarrott *et al.*, 1990).

Radio galaxies at low redshift are also often highly polarized (Brindle *et al.*, 1990a, b, c; Antonucci and Barvainis, 1990; Draper *et al.*, 1993). Imaging polarimetry of a number of these, 3CR 33, 305, 321 and 459, shows them to have also centro-symmetric polarization patterns, but, unlike their high redshift counterparts, the polarized bicones have wide opening angles.

Cyg A is the NLRG best spectropolarimetrically studied. Originally, Tadhunter and Scarrott claimed that it showed a scattered halo, and while this may still be true, on the small scale at least, spectropolarimetry does not reveal scattered broad lines (Goodrich and Miller, 1989; Jackson and Tadhunter, 1993), nor are they visible in direct light in the near IR (Ward *et al.*, 1991).

PKS2152-69 is a particularly important object as it shows a highly polarized blue knot along its radio axis, 10kpc from the nucleus, which is accompanied by an adjacent high excitation emission line region (di Serego Alighieri *et al.*, 1988). It is thought that the knot is seen in the scattered light from a beamed radiation field (Fosbury, 1994). We have made several attempts to get spectropolarimetry across the knot to see if we can detect scattered broad lines, but due to bad weather we have not yet succeeded. Our motivaton for trying so hard to do this is that a direct test of determining the importance of beaming can be constructed if one can measure the angular variations of the polarized continuum in conjunction with those of a scattered BLR. The principle of the method is to use the polarized flux of BLR, which is not beamed, to map the number of scatterers in a given line-of-sight column, and apply this a correction to the flux of the polarized continuum, and thus determine its true angular form. This same idea can, of course, be applied to any of the other nearby radio galaxies with extended polarized haloes. One obvious candidate is Cen A.

4.1 CEN A

In the near to mid IR, steeply rising polarized continua have been detected in the nuclei of three nearby radio galaxies; Cen A, IC5063 and NGC 1052 (Axon *et al.*, 1992; Bailey *et al,*, 1986). Figures 13 and 14 show imaging polarization maps of Cen A at J and K obtained with IRIS on the AAT which nicely illustrate this phenomenon. At J the polarization structure is dominated by dichroic polarization, and the polarization vectors are aligned parallel to the well known polar dust lane which covers the nucleus. In contrast the K band image reveals a radically different picture; the bulge is now virtually unpolarized, but a strong region of polarized radiation is apparent at an angle approximately perpendicular to the radio jet axis.

Originally we suggested that these objects are misaligned Blazars with *tired* spectra, and thus we were seeing synchrotron radiation (Hough *et al.*, 1986). A great deal of interest in determining whether this interpretation is correct has been generated by attempts to model the emission line filaments situated in the Northern radio lobe (Morganti *et al.*, 1991). However, as we noted previously, this steep upturn in the IR polarization is a common characteristic of the IRAS selected Seyferts with scattered BLR. The alternative explanation (Antonucci and Barvainis, 1990) is that the IR polarization is due to scattering, in which case the evidence for a photon beam in Cen A would be somewhat weakened.

One way to distinguish these two hypotheses is to obtain submillimetre polarimetry at such long wavelengths that dust scattering will be negligible, but where one would still see high polarization from a nonthermal source. We recently secured just such observations at 800μ and 1500μ on the JCMT. At both wavelengths, the polarization levels are substantially less than one percent, and thus the polarized nuclear component is due to reflection. As yet no polarized broad lines have been seen, but this is perhaps not surprising because the nucleus and, more importantly, any scattered broad lines are heavily obscured by the foreground dust lane. A key experiment here will be to do nuclear spectropolarimetry in the IR (e.g. Paα) to see if polarized broad wings are present.

4.2 HIDDEN BROAD LINE REGIONS IN NARROW LINE RADIO GALAXIES

If we have yet to find a hidden BLR in Cen A and Cygnus A, we have had success in two other NLRG with quite different radio powers.

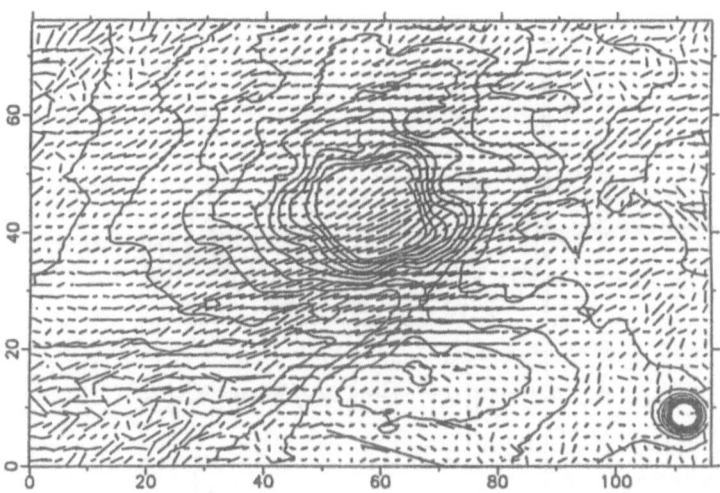

Fig. 13.   Infra-red imaging polarimetry map of Cen A in the J(1.2μ) band. The polarization
vectors are aligned parallel to the dust lane and are dominated by dichroic absorption.

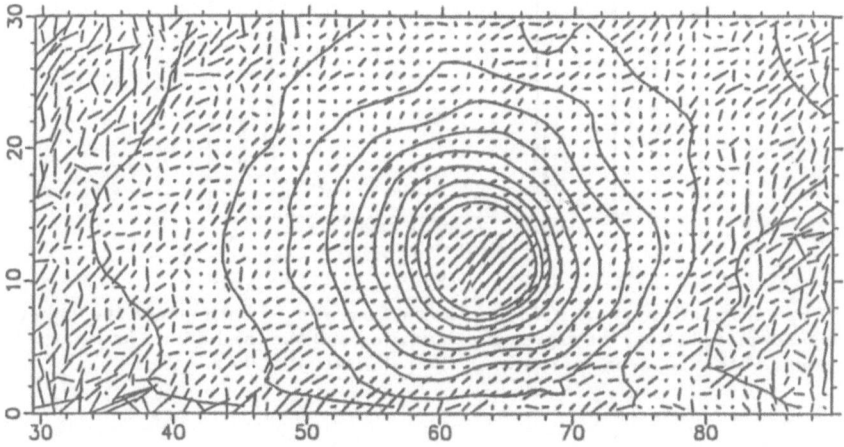

Fig. 14.   Infra-red imaging polarimetry of Cen A. in the K(2.2μ) band. The same region is shown
as in Figure 15. There is a highly polarized region around the nucleus in K, in which the vectors
are aligned approximately perpendicular to the radio jet.

4.2.1   *Evidence for an obscured BLR in the early-type radio galaxy IC5063* .

IC5063 = PKS 2048-572, is usually classified as an SO type galaxy, $z = 0.0110$, with
relatively narrow (350 km s^{-1}) emission lines, similar to those of a Seyfert 2 type
galaxy. As well as its steep IR nuclear component it also has a spectacular ENLR.

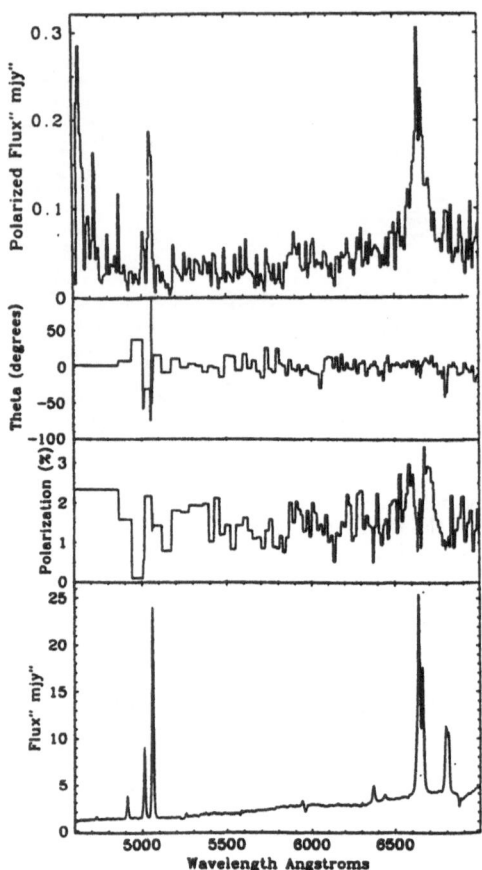

Fig. 15.   Spectropolarimetry of IC5063. Shown from top to bottom are the polarized flux, position angle of polarization, the degree of polarization and the total flux. The polarization data has been rebinned to give a constant error in the polarization of 0.4%

The measured radio luminosity ($\log_P(1.4$ GHz$) = 22.73$ WHz^{-1} sr^{-1} of IC5063, is one order of magnitude larger than typical nearby Seyfert galaxies (Ulvestad and Wilson, 1984). The spectropolarimetry of IC5063 (Inglis *et al.*, 1993) shown in Figure 15 and demonstrates for the first time the presence of a hidden BLR in a low luminosity radio galaxy.

While it is comparatively weak compared to the other examples we have presented, the hidden BLR in IC 5063 is still apparent in the polarized flux of the Hα line and has a FWHM $\approx$ 2500 km s^{-1}.

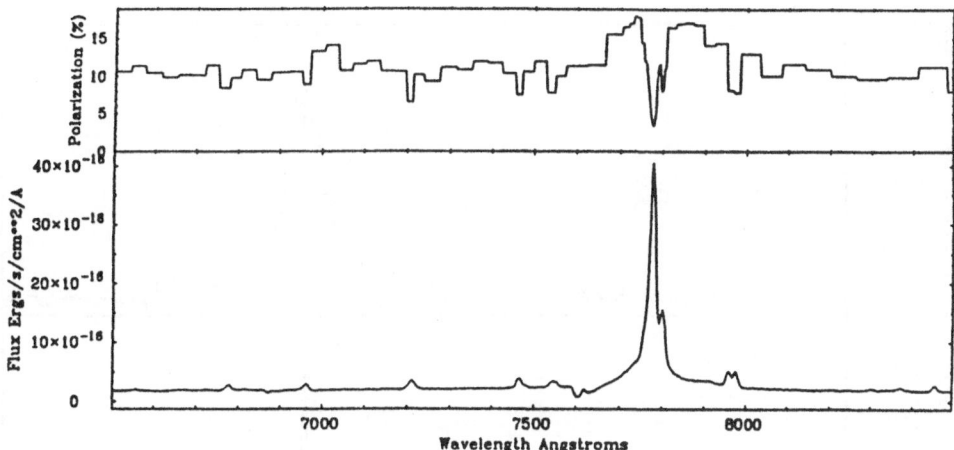

Fig. 16. Spectropolarimetry of 3C 234. The top panel shows the the degree of polarization, and the bottom panel the total flux. The polarization data has been rebinned to give a constant error in the polarization of 0.7%

4.2.2 3C234

The Broad Line radio galaxy 3C 234, is a classical radio triple in PA 68⁰, with a relatively narrow broad line component in the direct flux, and a steep Balmer decrement (Grandi and Philips, 1979). Low resolution spectropolarimetry was originally obtained by Antonucci (1984). His results showed that the continuum was highly polarized ($\sim$ 14%), perpendicular to the radio axis, even in the vicinity of broad Hα. Supposing that the NLR is unpolarized he argued that the BLR is heavily polarized and that 3C 234 is a NLRG counterpart to NC1068 (i.e. the polarization is due to electron scattering) and not dichroism (see also Antonucci and Barvainis, 1990).

We have obtained high quality spectropolarimetry of 3C 234 as shown in Figures 16 and 17. The results are in close agreement with those of Antonucci (1984), but the much higher spectral resolution and signal to noise allow us for the first time to see the details of the rather remarkable scattered BLR in this object. Referring to Figure 17 we see a broad line in polarized flux, stretching out to at least [SII] $\lambda\lambda$6717, 6731 ie half width at zero intensity $\sim$ 9000 km s^{-1}. There is also a hint of a second narrow peak in the line, which would imply the scattering zone is more extended or at least also illuminated by the NLR. There is also a very big difference between the shape and redshift of the BLR in the total and polarized flux, with the blue side of the polarized flux being more heavily peaked and narrower than in the total flux. The large red wing seen in the polarized flux would naturally arise from scattering by moving scatters (e.g. Henney, 1992). Notice also that the BLR in total flux has a much higher equivalent width than in NGC 1068, indicating

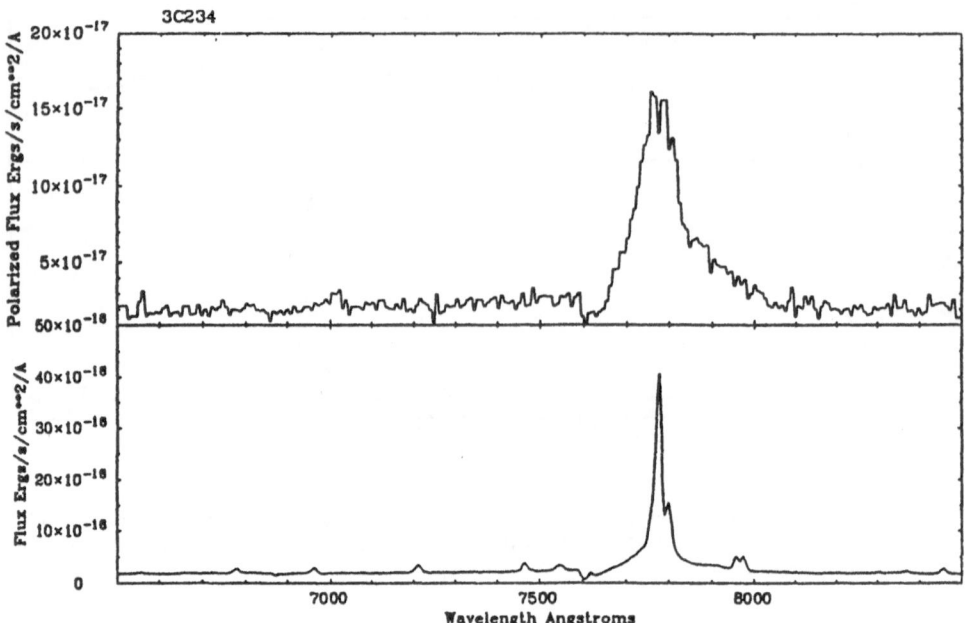

Fig. 17. Comparison between the polarized and total flux profiles of the BLR in 3C 234. The top panel shows the polarized flux and the bottom panel the total flux. The polarized flux data has been binned to a resolution of 3 pixels. Notice the dramatic red asymmetry in the polarized flux BLR line profile. The absorption feature to the blue of Hα is due to the atmosphere.

that the covering factor of the torus is less, either because it is intrinsically less or as in NGC 4151 we have a partial view of it. Tentative evidence to support this last interpretation comes from our IR spectropolarimetry with CGS 4. These show a strong Broad Paα which is more symmetric and wider than both the total and polarized BLR at optical wavelengths, but it appears that Paα is still significantly polarized $\sim$ 2.5%, suggesting that is still a scattered component to the line. However, we think that one must treat this last result with a degree of caution since the existing IR spectropolarimetry is of rather poor quality, and much better data is need to establish what is really happening.

5 Final Comments

One reason for supposing that occulting tori are also present in the radio galaxies and QSO's has been the hope that we might be able to fix the statistical difficulties that the pure relativistic beaming model is known to suffer from. The detetection of hidden broad lines in many more Seyfert 2 galaxies and in 3C 234 and NGC 4151 at least provides evidence that the unified models are partly right. However on a statistical basis, as yet we still see far too few hidden BLR for us to be confident that GU really works. Part of the problem may be that the scattered flux in many of these objects is itself heavily reddened and that much more might be revealed to us when we get a substantial body of IR spectropolarimetry data of the sort we showed for IRAS 23060+0506.

Acknowledgements

We wish to express our thanks to our other collaborators who have made important contributions to the work discussed here, Jeremy Bailey, Clive Tadhunter and Martin Ward. Two of us (SY and MI) were supported throughout the duration of the work presented here by SERC research studentships.

References

Antonucci, R.: 1984, *Astrophys. J.* **278**, 499.
Antonucci, R.: 1993, *Ann. Rev. Astron. Astrophys.* **31**, 473.
Antonucci, R. and Barvainis, R.: 1990, *Astrophys. J. Lett.* **332**, L17.
Antonucci, R. and Miller, J.: 1985, *Astrophys. J.* **297**, 621.
Axon, D. J., Bailey, J. and Hough, J. H.: 1982, *Nature* **299**, 234.
Axon, D. J., Dyson, J. E. and Pedlar, A.: 1994, in *The Nature of Compact Objects in Active Galactic Nuclei*, eds. A. Robinson and R. J. Terlevich, Cambridge Univ. Press, p. 66.
Bailey, J., Axon, D. J., Hough, J. H., Ward, M. J., McLean, I. S. and Heathcote, S.: 1988, *Mon. Not. R. astr. Soc.* **234**, 899.
Bailey, J., Sparks, W. B., Hough, J. H. and Axon, D. J.: 1986, *Nature* **322**, 150.
Bartel, P. D.: 1989, *Astrophys. J.* **336**, 606.
Blanco, P. R., Ward, M. J. and Wright, G. F.: 1990, *Mon. Not. R. astr. Soc.* **242**, 4P.
Brindle, C., Hough, J. H., Bailey, J., Ward, M. J., Axon, D. J., Sparks, W. B. and McLean, I. S.: 1990a, *Mon. Not. R. astr. Soc.* **244**, 604.
Brindle, C., Hough, J. H., Bailey, J., Ward, M. J., Axon, D. J., Sparks, W. B. and McLean, I. S.: 1990b, *Mon. Not. R. astr. Soc.* **244**, 577.
Brindle, C., Hough, J. H., Bailey, J., Ward, M. J., Axon, D. J. and Sparks, W. B.: 1990c, *Mon. Not. R. astr. Soc.* **247**, 327.
Carleton, N., Wilner, S. P., Rudy, R. J., and Tokunaga, A. T.: 1984, *Astrophys. J.* **284**, 523.
Chambers, K. C., Miley G. K. and van Breugel, W.: 1987, *Nature* **329**, 604.
Depoy, D.: 1987, in *Infrared Astronomy with Arrays*, eds. Wynn-Williams, Becklin and Good, University of Hawaii, p. 426.
di Serego Alighieri, S.et al.: 1988, *Nature* **334**, 593.
Draper, P .W., Scarrott, S. M., and Tadhunter, C. N.: 1993, *Mon. Not R. astr. Soc.* **262**, 1029.
Fabian, A. C.: 1989, *Mon. Not R. astr. Soc.* **238**, 41P.
Fosbury, R. A. E.: 1994, in *The Nature of Compact Objects in Active Galactic Nuclei*, eds. A. Robinson and R. J. Terlevich, Cambridge Univ. Press, p. 3.
Goodrich, R.: 1989, *Astrophys. J.* **340**, 190.
Goodrich, R.: 1989, *Astrophys. J.* **342**, 224.
Goodrich, R.: 1990, *Astrophys. J.* **355**, 88.
Goodrich, R. and Miller, J.: 1989, *Astrophys. J. Lett.* **346**, L21.
Goodrich, R. W., Veilleux, S. and Hill, G. J.: 1994, *Astrophys. J.* **422**, 521.
Grandi, S. A. and Philips M. M.: 1979, *Astrophys. J.* **232**, 659.
Henney, W.: 1992, *Scattered Emission lines from Herbig-Haro jets* Ph.D. thesis, University of Manchester.
Hines, D.: 1991, *Astrophys. J. Lett.* **374**, L9.
Hough, J. H., Brindle, C., Axon, D. J., Bailey, J. and Sparks, W. B.: 1986, *Mon. Not. R. astr. Soc* . **224**, 1013.

Hough, J. H., Brindle, C., Wills, B. J., Wills, D. and Bailey, J.: 1991, *Astrophys. J.* **372**, 478.
Inglis, M., Hough, J. H., Axon, D. J., Bailey. J. and Ward, M. J.: 1993, *Mon. Not. R. astr. Soc.* **263**, 895.
Jackson, N. and Tadhunter, C. N.: 1993, *Astron. Astrophys.* **272**, 105.
Januzzi, B. and Elston, R.: 1991, *Astrophys. J. Lett.* **366**, L69.
Kukula, M.: 1993, *The Radio properties of Seyfert Nuclei*, Ph.D. thesis, University of Manchester.
McCarthy, P. J.: 1993, *Ann. Rev. Astron. Astrophys.* **31**, 639.
McCarthy, P. J., van Breugel, W., Spinrad, H. and Djorvski, S.: 1987, *Astrophys. J.* **321**, L24.
Miller, J. and Goodrich, R.: 1990, *Astrophys. J.* **355**, 456.
Morganti, R., Hook, R., Fosbury, R. A. E., Robinson, A. and Tsvetanov, Z.: 1992, *Mon. Not R. astr. Soc.* **256**, 1P.
Morganti, R., Robinson, A., Fosbury, R. A. E. F., di Serego Alighieri, S., Tadhunter, C. N. and Malin, D. F.: 1991, *Mon. Not R. astr. Soc.* **249**, 91.
Nakajima, T., Carleton, N. and Nishida, M.: 1991, *Astrophys. J. Lett.* **375**, L1.
Nakajima, T., Kawara, K., Nishida, M. and Gregory, B.: 1991, *Astrophys. J.* **373**, 452.
Orr, M. J. L. and Browne, I. W. A.: 1982, *Mon. Not R. Astr. Soc.* **200**, 1067.
Pedlar, A., Longley, P., Kukula, M., Muxlow, T. B., Axon, D. J., Baum, S., O'Dea, C and Unger, S. W.: 1993, *Mon. Not R. astr. Soc.* **263**, 471.
Penston, M. V., *et al.*: 1990, *Astron. Astrophys.* , **236**, 53.
Rix, H., Carleton, N., Rieke, G. and Rieke, M.: 1990, *Astrophys. J.* **363**, 480.
Robinson, A.: 1994, *The Nature of Compact Objects in Active Galactic Nuclei*, eds. A. Robinson and R. J. Terlevich, Cambridge Univ. Press, p. 235.
Robinson, A., et al.: 1994, *Astron. Astrophys. in press* .
Ruiz, M., Rieke, G. and Schmidt, G.: 1994, *Astrophys. J. in press.*
Scarrott, S. M., Rolph, C and Tadhunter, C. N.: 1990, *Mon. Not R. astr. Soc.* **243**, 5P.
Scheuer, P. A. G. and Readhead, A. C. S.: 1979, *Nature* **277**, 182.
Tadhunter, C. N.: 1994, *The Nature of Compact Objects in Active Galactic Nuclei*, eds. A. Robinson and R. J. Terlevich, Cambridge Univ. Press, p.13.
Tadhunter, C. N., Scarrott, S. M., Draper, P. and Rolph, C.: 1992, *Mon. Not R. astr. Soc.* **256**, 53P. eds. A. Robinson and R. J. Terlevich, Cambridge Univ. Press, p. 13.
Tadhunter, C. N., Scarrott, S. M., and Rolph, C.: 1990, *Mon. Not R. astr. Soc* . **246**, 163.
Tran, H., Miller, J. S. and Kay, L.: 1992, *Astrophys. J.* **397**, 452.
Ulvestad, J. and Wilson, A. W.: 1984, *Astrophys. J.* **285**, 439.
Unger, S. W., Pedlar, A., Axon, D. J., Ward, M. J., Meurs, E. J. A. and Whittle, D. M.: 1987, *Mon. Not. R. astr. Soc.* **228**, 671.
Ward, M. J. *et al.*: 1994, *in prep.*
Ward, M. J., Blanco, P. R., Wilson, A. S. and Nishida, M.: 1991, *Astrophys. J.* **382**, p115.
Wills, B. J., Wills, D., Evans, N. J., Natta, A., Thompson, K., Breger, M. and Sitko, M.: 1992, *Astrophys. J.* **400**, 96.
Young, S.: 1994, Ph.D. thesis, University of Hertfordshire.
Young, S., Hough, J. H., Inglis, M., Bailey. J., Axon, D. J. and Ward, M. J.: 1993, *Mon. Not R. Astr. Soc.* **261**, L1.

The Starburst Galaxy NGC1808: Another M82?

P. W. Draper, S. M. Scarrott and D. P. Stockdale
Physics Department, University of Durham, Durham, U.K.

Abstract. An optical polarization map of the starburst galaxy NGC1808 shows that on kpc scales there is a coherent galactic magnetic field which follows the spiral arms on the galaxy. The inner 750pc surrounding the central region shows characteristics of a huge reflection nebula illuminated by the various hotspots within the starburst. This is very similar to the archetypal starburst galaxy M82.

1 Introduction

The starburst phase in galaxies—an intense and localised episode of star formation—may play an important role in the development of active galaxies. The more massive starburst galaxies are ultra-luminous in the FIR and show evidence of vast outflows of hot gas and radiation (the superwind) from the star formation regions out into the halos. These superwinds create huge shocks which are seen as both diffuse emission regions and also in the form of a network of tenuous emission-line filaments extending for many kpcs from the centrally located starbursts.

The irregular type II galaxy M82 is the archetypal starburst/superwind galaxy with its large IR/FIR output, central activity and extensive filamentary halo. M82 is unique in terms of its optical polarization properties which show that in continuum radiation, the dusty halo of the galaxy is a giant reflection nebula jointly illuminated by the galactic disc and the central starburst activity (Elvius 1963, Bingham *et al* 1976). Our recent polarization map of M82 in the light of Hα shows the galaxy is a huge bipolar reflection nebula illuminated by a point-like source within the starburst region. This source may simply be the more active part of the starburst region or even an embryonic AGN (Scarrott, Eaton and Axon 1991).

The southern galaxy NGC1808, which is a highly inclined spiral galaxy, shows many of the characteristics of M82 including a central starburst region which consists of several optical hotspots surrounding a visible nucleus which itself has some features typical of Seyfert nuclei. In addition, there are a series of radial dust filaments, kpcs in extent and cospatial with an HI outflow, emanating from the starburst region presumably driven by it. The great similarity between this galaxy and M82 prompted us to look at its polarization properties and we made V waveband measurements on the AAT early in 1993.

2 Results and Discussion

Figure 1 shows a greyscale image of NGC1808 on which are superimposed an intensity contour map of the inner starburst region and a polarization map of the whole

Astrophysics and Space Science **216**: 399–401, 1994.

P. W. Draper et al.

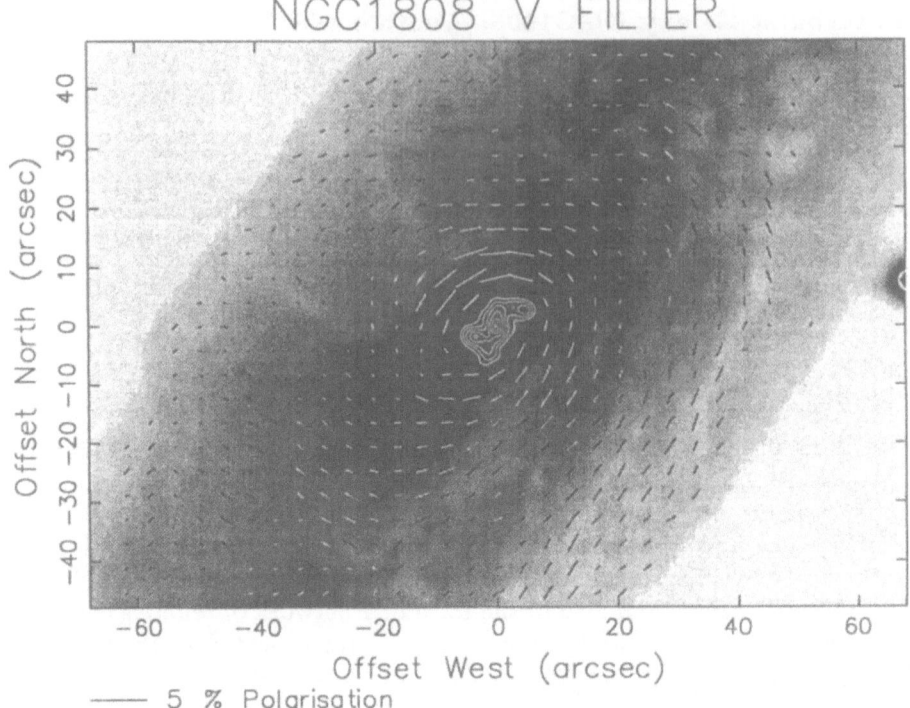

Fig. 1. A large scale greyscale image and polarization map of NGC1808 with an inner intensity contour map illustrating the hotspots in the starburst region. Beyond $\sim 20''$ to the SW of the centre the polarization orientations are parallel to the spiral arms/inter-arms and indicate the presence of a large scale magnetic field within the galaxy. The inner $20''$ shows a generally circular pattern typical of a huge reflection nebula illuminated from within the starburst region. This is reminiscent of M82.

galaxy. The greyscale image and contour map show that the optical morphology of this Sbc pec galaxy consists of a highly knotty central region surrounded by the spiral galaxy seen at a high angle of inclination.

The polarization map shows that, to the SE and at distances $> 20''$ from the centre, the polarization orientations form a rather regular pattern which follows the spiral arms and inter-arm regions of the galaxy. We believe this polarization arises from dichroic extinction by magnetically aligned dust grains in the galactic arms/inter-arms of NGC1808. Such patterns in optical polarization measurements have been seen in other galaxies (e.g. M51 and NGC1068—Scarrott *et al* 1987, 1991) and there is also evidence from 6cm radio observations (Dahlem *et al* 1990) for a large scale magnetic field in NGC1808 itself. Our results confirm the existence of a coherent large scale magnetic field in disc/arms of NGC1808. This spiral pattern of polarization is not repeated in the off-nuclear regions to the northeast because the arms are hidden by the foreground dusty galactic halo.

Within ~ ±15″ of the nucleus the polarization orientations follow a general elliptical arrangement typical of a reflection nebula with an extended illuminating source. A detailed map of the inner 15″ (not shown here) indicates that the brightest knot of the starburst region, which corresponds to the galactic nucleus, is the dominant source of illumination but there are localised contributions from other hotspots. The reflection nebula appearance/extended source/polarization asymmetry between opposite sides of the galaxy, seen in here in NGC1808, are also seen in M82 when viewed in polarized light in a similar waveband.

As far as polarization properties are concerned NGC1808 is another M82.

References

Bingham, R. G., McMullen, D., Pallister, W.S., White, C., Axon, D.J. and Scarrott, S. M.: 1976, *Nature* **259**, 463

Dahlem, M., Aalto, S., Kelin, U., Booth, R., Meobold, U., Wielebinski, R. and Lesch, H.: 1990, *Astron. Astrophys.* **240**, 237

Elvius, A.: 1962, *Lowell Obs. Bull.* **5**, 281

Scarrott, S. M., Rolph, C. D., Tadhunter, C. N. and Wolstencroft, R. D.: 1991, *Mon. Not. Roy. Astr. Soc.* **249**, 16p

Scarrott, S. M, Eaton, N. and Axon, D. J.: 1991, *Mon. Not. Roy. Astr. Soc.* 252, 12p

Scarrott, S. M., Ward-Thompson, D. and Warren-Smith, R. F.: 1987, *Mon. Not. Roy. Astr. Soc.* **224**, 299